QUARTZ AND FELDSPAR

Quartz and Feldspar

Dartmoor: A British Landscape
in Modern Times

MATTHEW KELLY

JONATHAN CAPE
LONDON

1 3 5 7 9 10 8 6 4 2

Jonathan Cape, an imprint of Vintage Publishing,
20 Vauxhall Bridge Road,
London SW1V 2SA

Jonathan Cape is part of the Penguin Random House group of companies
whose addresses can be found at global.penguinrandomhouse.com.

Penguin
Random House
UK

First published by Jonathan Cape in 2015

www.vintage-books.co.uk

A CIP catalogue record for this book is available from the British Library

ISBN 9780224091138

Printed and bound in Great Britain by
Clays Ltd, St Ives plc

Penguin Random House is committed to a sustainable future for our business, our readers and
our planet. This book is made from Forest Stewardship Council® certified paper.

For L.B.

Who's this moving alive over the moor?
From 'Dart' by Alice Oswald

Contents

Acknowledgements

During the five years I have worked on this book, I have persuaded many friends to make the trip to Dartmoor. Je Ahn, Peter Barnes, Joost Beunderman, Simon Blake, John Deans, Ken Edwards, Anabel Rubio Fernandez, Louise Tillin, Maria Smith, Campbell Storey, Julia Szczuka and, especially, Caroline Nye put up with my rapture as well the mud and the rain. I hope they felt rewarded by those exquisite moments when the sky clears, the sun comes out and there is nowhere better to be. Caroline and John took some of the photographs used in the text. Part Three was drafted while I was a Fellow at the Rachel Carson Center for Environment and Society in Munich and I'm grateful to first Peter Coates and then Christof Mauch and Helmuth Trischler (the directors of the Center)—and the German taxpayer—for making it possible. For the *gemütlichkeit* and the great flow of conversation, whether during Alpine hikes or when drinking that sweet Bavarian beer, here's to Melanie Arndt, Wilko Graf von Hardenburg, Claudia Leal, Annka Liepold, Kieko Matteson, Giacomo Parinello, Katie Ritson, Chris Pastore, Frank Zelko and so many others. There was something in the air during that six months and it was a great privilege to be a part of that special RCC alchemy. On my return from Munich, Adrian Smith read Part Three and saved me from some unforced errors and rightly urged that I rein in certain excesses. Caroline Nye cast a sharp eye over other parts of the manuscript, alerting to me to some clunky writing, though like Adrian, she was only partly successful. At a crucial moment in the development of this project, Roy Foster read a full draft of the manuscript and gave exactly the encouragement I needed. His support for my work over many years is a debt that just gets bigger and there is no hope of it ever being repaid. My colleagues in the history department at the University of Southampton remain a wonderful group of friends and scholars. Neil Gregor, Christer Petley and Joan Tumblety are the best of sounding boards, as are historian friends elsewhere like Ian McBride, James McConnel, William Whyte and, well, so many others. And in the year of their retirement from active duty at Southampton, I must single out the extraordinary support shown to me by Jane McDermid and Adrian Smith. I'm also grateful to the Faculty of Humanities at Southampton for maintaining its generous provision of regular research leave. Part Four was got done thanks to this. Clare Alexander, my agent, was just great, and it's hard to exaggerate how wonderful Dan Franklin, my editor, has been. At Cape, I'd like to thank Neil Bradford, Clare Bullock and Ruth Waldram for helping this project along, and Vicki Robinson for a meticulous index. Early in my research, James Paxman, Chief Executive of the Dartmoor Preservation Association, let me root through the Association's archive and I hope this book repays that trust by demonstrating the historical significance of the venerable organisation he represents. On the whole, however, I have ploughed a lone furrow, avoiding the tensions and disputes that cannot but animate the politics of contemporary Dartmoor. This means I have probably made some mistakes, offered interpretations that some might find surprising, and warmed over some best-forgotten controversies. I'm also acutely conscious that whatever a historian says about the present quickly seems dated. As such, I invite all to join the discussion at http://quartzandfeldspar.com/

M.K.
Southampton, March 2015

The National Archive, Kew, granted permission to reproduce images from its files. Here listed are the file references and where they appear in the book: p. 136: RAIL 1017/1/30; p. 176: PCOM 7/109; p. 201: BT 31/27333/183059; p. 228: COU 1/452; p. 230: COU 1/452; p. 235: COU 1/454; p. 304: HLG 127/1200; p. 313: HLG 71/1622; p. 318: COU 1/980; p. 325: WO 32/20944; p. 326: WO 32/20944; & p. 329: COU 1/983.

The lines from 'Dart' by Alice Oswald are reproduced by permission of Faber & Faber Ltd.

Introduction: Okehampton

To stand on the granite mass of Haytor on the eastern edge of Dartmoor National Park is to plant one's feet on rock whose physical characteristics are of a peculiarly distinct antiquity.

Dartmoor is the largest of the five granite bosses formed 280 million years ago that intrude on Devon and Cornwall in south-west England. As the highest and largest upland in the south of England, Dartmoor has enjoyed a particular if not quite unique geological history. During the early Pleistocene, some 278 million years later, a sudden rise in sea level made Dartmoor an island and its edges a coastline shaped by waves. As the waters receded, the Great Ice Age began. Dartmoor's experience was exceptional. It lies to the south of the glaciation that shaped the British landscape north of the line running from the Bristol Channel to the Wash. This makes the granite outcrops known as tors that spot Dartmoor's surface extraordinary survivals. They predate the last deluge and the last ice age. Dartmoor became earth-covered relatively recently, some 15,000 years ago, and human beings began to exploit its natural resources around 7,000 years later.

Dartmoor is often celebrated as unspoilt, but much that is encountered on the upland is evidently the product of human endeavour. It is an anthropic landscape, its tree cover felled by its earliest settlers, its flora grazed by domesticated herbivores – sheep and cattle – and semi-feral ponies, its surfaces patterned by ancient standing stones and abandoned tin mines, damp medieval churches and disused railway lines, drystone walls and clapper bridges, farms and villages, quarries and hotels. Granite is the common factor here. Hard, grey, wind-worn, glistening in the sunshine after rain, it is the foundational material of moorland life. In Aberdeen, another place of granite, the architecture of ages seems ever new. On Dartmoor granite's ubiquity and its resilient materiality seem to merge natural history with traces of prehistoric human activity and a more recent industrial past.

Although it is hard to think of something more enduringly real or certain than granite's tough composite of quartz, feldspar and mica, a more malleable Dartmoor of widely contradictory meaning is embedded in the collective imagination. For 200 years men have been incarcerated here, 'sent to Dartmoor'; for the same 200 years people have voluntarily come here to walk or write or paint or photograph. Dartmoor is a place to escape from or escape to. For more than a century books and maps and pamphlets have let tourists in on Dartmoor's secrets, telling ghost stories, recommending walks, domesticating the wild, commodifying its folklore. Since at least 1883, when the Dartmoor Preservation Association was founded, people have worried about the moor, fearing for its future.

Quartz and Feldspar is neither a natural history nor a geological study; nor is it primarily social history, though lives lived in and around the moorscape form a part of the story it tells. Instead, it offers a history of how this unusually distinct landscape has been encountered, imagined and argued over since the late eighteenth century. It is alert to the way environmental historians and ecologists have problematised 'nature' and 'natural', and it is wary of the mesmeric qualities of the Romantic tradition of English landscape and nature writing; its subject is as much the material Dartmoor of granite, blanket bog, rivers, heathland and woodland, of the Dartmoor that provides 'environmental services' as it is the Dartmoor of text and representation. Concern with specificity of place is central to its treatment of the overlapping responses the moor has evoked, be they antiquarian, archaeological, poetic, mythic, folkloric, religious, commercial, preservationist or environmentalist.

Motivating earlier Dartmoor writers was a form of competitive regionalism that sought to prove Dartmoor the equal of European landscapes celebrated for their seductive charm and sensational beauty. Dartmoor's advocates were troubled by the uniquely equivocal place the object of their affection occupied in the British imagination. As the moor accrued sinister associations during the nineteenth century, be they natural, supernatural or man-made, as well as an enduring reputation for awful weather, it could not enjoy the unambiguously positive connotations of the Lake District or the New Forest. Gothic Dartmoor coexisted with bucolic or sublime Dartmoor, generating in its champions feelings of insecurity about its value that did not dissipate until the late twentieth century. *Quartz and Feldspar* traces the diagnosis and progress of this multiple personality syndrome and the changing prognoses of its physicians.

★ ★ ★

Most people encounter the moor from elsewhere, and one gateway is the small market town of Okehampton. The town lies immediately below the spur of the great granite massif of the northern Dartmoor upland and at the confluence of the East and West Okemont, rivers whose origins are found high in the blanket bog. Okehampton is an ideal place to access the most challenging parts of the moorscape; its southern environs were also the location of the most controversial planning decision in Dartmoor's recent history. A glimpse at Okehampton's modern history provides a gateway to the stories this book seeks to tell.

William Bridges, author of an early guide to Okehampton, provided his readers with an elaborate introduction to the town:

> Pilgrim of beauty, whoever thou art, that seekest to amuse and, may we hope, inform thy wanderings over this picturesque old town and its environs, stand with us for a few moments at one entrance of it, in the angle formed by the south abutment of the eastern bridge. How closely in keeping with its ancient structure is everything that one sees there; or, to write more correctly, how time appears to have blended the works of man into harmony with nature! The broad ivy-grown wall toppling over a stream that brawls along its craggy margins beneath; the antique lattices that peep out from attics where slept the menials of some wealthy burgess of the olden day, and whence too his good old dame's waiting-woman might have been wont to cast sly glances at the passengers on the bridge; the umbrageous trees that screen the river where it gushes on as fresh from its mountain rock; the pendulous old sign that creaks welcome to man and horse over the wide archway of the hostel on your left; the suburban farm-yard, with its tenant donkey gazing wistfully through a wicket that opens towards the grateful stream; the little chantry-tower, with its bell, and time-embrowned pinnacles peering over the chimney-tops that cluster thick about it: – gentle wanderer, if thou hast not an eye to detect, and withal a heart to feel, the romantic beauties of the spot, our researches can

avail thee nothing; but if otherwise, thou mayst go on to read and
learn as follows:—[1]

This carefully calibrated evocation of Okehampton's rustic
charms, first published in 1839, was typical of early Victorian writing
marketed at middle-class tourists enjoined to treat their leisure time
as an opportunity for self-improvement. Countless publications
describing the history and antiquities of any number of British
towns followed the same pattern, outlining their aristocratic, eccle-
siastical and civic history, often with a little genealogy mixed in,
and directing readers to key sites. Bridges sent his readers to see
Okehampton Park, where the ruins of the medieval castle, Roman
fort and road could be contemplated, but his didactic intent was
most clear in his comment on Halstock Chapel, which lay just
outside the town on the edge of the moor. 'For the casual spectator,
the spot can possess no common interest; but one who is acquainted
with the local history of this district, who knows each "bosky
bourne and tangled dell", its beauties are enhanced by a thousand
associations.'[2]

'Associations' was a key word. Acquiring factual knowledge was
not an end in itself, but a means to unleash the imaginative faculty:
fact might be combined with feeling, sense with sensibility. Halstock
Chapel, of which there is now little trace, was then an evocative
ruin, a relic of the destructive power of the Protestant Reformation
and a place where the visitor might 'muse' on the transitory folly
of life. Its location between townscape and moorscape, on the cusp
between cultivated lowland and 'natural' upland, was particularly
affecting. As an earlier topographer wrote, the chapel was 'somewhat remote, as if it were banished, without comfort or company'.[3]
Bridges encouraged his reader to picture the 'fallen pomp and
circumstance' of the Catholic ritual once played out in the chapel's
'sacred precincts', and to imagine the 'seclusion' of the moor
tempting a Benedictine brother 'in his black scapulary and cowl'
further from the town towards quiet places where, 'unseen by all

save heaven perhaps', he performed 'his matin prayers and ablutions'. Bridges then breaks the spell, admonishing the reader that 'fancy must not break in on the calmer pursuits of the antiquarian': sense must override sensibility.[4]

If ironic knowingness and a readerly complicity shaped much early nineteenth-century topographical and antiquarian writing, Bridges' celebration of Okehampton as a town being slowly reclaimed by nature also reflected a contemporary sensibility alarmed by rapid urbanisation and industrialisation. Despite this, taking refuge in the compensations provided by the heady mix of picturesque ideal and Romantic feeling had its dangers, and Bridges warned his readers against the kind of Catholic superstition that 'could invest the works of nature as well as those of art with a charm deep and hallowed'. The first part of *Quartz and Feldspar* locates thinking like this within a wider history of how perceptions of Dartmoor developed and evolved over the course of the eighteenth and nineteenth centuries. The moorscape excited antiquarian speculation, was a stimulus to romantic sensation, kept sceptical archaeologists busy and became a site of preservationist concern and activism. Some early-nineteenth-century Protestants were fascinated by the idea that Dartmoor's numinous qualities had made it a centre of Druidism, whereas others insisted that visitors possessing a theologically sound understanding of nature as a manifestation of the divine could come closer to God through their encounters with the moorscape. Affecting nature could corrupt or inspire in equal measure.

Dartmoor's late-Victorian archaeologists prided themselves on their objectivity and rationalism, self-consciously rejecting the 'fanciful' Druidical speculations of their antiquarian predecessors. This collective shift in subjectivity affected how Dartmoor was encountered long into the twentieth century, and late twentieth-century preservationists often claimed a direct line of descent to the late Victorians. And yet, although the break with the antiquarians brought new methodologies and perspectives, the men and women of the 1880s and 1890s, like their predecessors, were of their time, and their work was saturated with the new racial science. *Quartz and Feldspar* tries to avoid condemnation in favour of context-ualisation, locating historically the successive Dartmoors of eight-eenth-century luminaries like William Borlase and Richard Polwhele, and nineteenth- and twentieth-century successors such as Nicholas Carrington, Eliza Bray, Samuel Rowe, Sabine Baring-Gould, William Crossing and Sylvia Sayer.

Back in Okehampton, Bridges advised readers venturing to Halstock to 'tread lightly . . . for the ground is yet holy', but a little beyond they might 'commune' with 'undecayed Nature in the wild glen beneath'.[5] Many encountering Dartmoor in the early nine-teenth century found its 'undecayed Nature' an affront to the gener-osity of God's great bounty. Entrepreneurs sought to 'improve' the moorscape, convinced that particular virtue attached to any profit extracted from such unpromising materials. Part II of *Quartz and Feldspar* traces the fortunes of this early nineteenth-century opti-mism, delineating the peculiar symbiosis that developed on the moor between improvement and incarceration. Dartmoor Prison began life as a prisoner of war depot during the Napoleonic Wars and particular attention is paid to Sir Thomas Tyrwhitt, its insti-gator, and how this institution shaped perceptions of Dartmoor. By mobilising the resources of the state, Tyrwhitt hoped to trans-form unpropitious nature into modern agriculture.

A generation of agricultural improvers shared Tyrwhitt's opti-mism but their ambition far outweighed their achievements. Another great shift in subjectivity ensured that what seemed self-evidently good to the improvers came to be disdained by later generations. By the end of the nineteenth century, new ways of valuing the moorscape predicated on maintaining its 'natural' state came into conflict with the assumption that increasing the economic value extracted from the moorscape was necessarily a moral good. This paradigm shift had multiple causes, reflecting the localist enthusiasms of new civic cultures, the lobbying of antiquarians,

topographers and archaeologists, new commercial interests generated by tourism, as well as the utility of the upland to the army.
Just as the improvers framed their projects in terms of the national
interest, so the preservationists began to regard unimproved
Dartmoor as a scarce asset of national significance. Their localism
was much more than a narrow parochialism, and it carried within
it the origins of the rooted cosmopolitanism – think globally, act
locally – of contemporary environmentalism.

Transformation, however, was inevitable. If Bridges wrote that
in Okehampton 'time appears to have blended the works of man
into harmony with nature', early twentieth-century guides emphasised improvement rather than rustic charm. Much was changed
in 1871 when the London and South-Western Railway brought
Okehampton within four hours of the capital, allowing its development and promotion as a commercial centre. J. W. Besley's The
'Borough' Guide to Okehampton (1919) was packed with advertise-
ments for local businesses and celebrated the town's growing population, widened streets, good footpaths, effective street lighting,
excellent shopping, reliable water supply, sewage system and
'general look of smartness and life'. Outward-facing modernity
rather than somnambulant rusticity was the new boast, and
Okehampton's location on the edge of the moor was the town's
strongest selling point. Besley outlined seven walks into the
moorscape, each more ambitious than the tentative steps into the
edgeland between town and moor encouraged by Bridges. Amid
'wild and beautiful scenery', Besley promised the rambler the
'marvellously revivifying quality of the air' that 'braces the nerves
and banishes the depression of the spirits felt at lower levels'.
Dartmoor as 'nature' was a 'tonic', allowing essential respite from
oppressive modernity, 'the noise of motors on earth and in air'.[6]
From 1931 the guide was subtitled 'The Capital of the Devonshire
Highlands', and in 1936 its cover illustrations depicting the ruins of
the castle or the town's golf course were replaced with a drawing
of an emancipated woman rambler, fit, healthy and comfortably

dressed, looking out over a highly stylised north Dartmoor land-scape. Writing in the early 1950s, W. G. Hoskins described the view from Yes Tor (2,028 feet), a few miles south of Okehampton, as 'extraordinary', stretching west as far as Bodmin Moor, north over most of Devon, and to the south 'a vast, awe-inspiring, and deso-late prospect into the deepest recesses of the Moor'.[7]

Shifts in orientation were also evident in the foldout map tucked into the guide. The scale of the 'district map' of the 1936 edition was one inch to four miles and Dartmoor was about a quarter of the total space depicted; the scale of the 'rambler's map' of the 1937 edition was one inch to the mile and Dartmoor was about two thirds of the total space depicted. Changes in mapping such as this were another way in which the moorscape was re-valued, suggesting that moorland south of Okehampton had become more esteemed as a site of leisure than the pleasant agricultural land to the north. Without the easy availability of detailed and sophisticated maps, Dartmoor's development as a site of mass leisure and recreation in the interwar period was inconceivable. Okehampton was re-valued too. The picturesque curiosity was now 'a Health Resort'.

Mapping was important to the preservationists too. A landscape feature, natural or anthropic, realised a new status when it was inscribed onto a map, achieving a particular form of validation when included on the state's Ordance Survey maps. Mapping also played a vital role in the development of new ways of treating Dartmoor as a resource in the twentieth century. Part III of *Quartz and Feldspar* chronicles the growing role of the state to prevent or permit in the context of Dartmoor's history as one of the first national parks in England and Wales. As Dartmoor became established as one of the most valued rural landscapes in Britain, so the 'barren waste' condemned by nineteenth-century improvers became subject to intense conflict over land use. Was it legitimate to use the northern plateau as an artillery range? Should local water undertakers be permitted to establish reservoirs in the upland river valleys? Was it acceptable for the Forestry Commission to establish non-native conifer plantations on the sites of old upland farms or to fell native woodland and replace it with stands of fast-growing pine? Dartmoor's status as a national park suggested it was a 'governmentalised' locality, but enthusiasts for the national park ideal were quickly disappointed by the limited powers of the new park authority. If anything, the exploitation of the moorscape increased in the post-war period, though, as *Quartz and Feldspar* shows, the novel statutory framework ensured that new interventions generated protracted and bitter political conflicts.

Nothing was more illustrative of the uncertain status of the national park than the long battle provoked by the decision to route a section of the Okehampton bypass inside its boundaries. Pages could be dedicated to unravelling the lineaments of this controversy, but even a brief excursus along its many byways reveals much about how late twentieth-century Dartmoor was comprehended.

Post-war Okehampton, unkindly described by Hoskins as 'singularly dull', was blighted with high unemployment and poverty but was most notorious as one of the country's worst traffic bottle-necks.[8] The A30 trunk road connecting Cornwall to the M5 (completed in 1977) at Exeter followed a winding route through Tedburn St Mary, Cheriton Bishop, Whiddon Down, Sticklepath, Okehampton, Bridestow and Lifton. Throughout the year heavy goods vehicles choked Okehampton with noise and fumes, while the tailbacks that built up outside the town during the holiday season made trips to the west famously miserable. Tourism was an essential part of the Cornish economy and the complaints could not be casually dismissed as the grumbling of the metropolitan middle classes; moreover, the introduction of regulations restricting the time commercial drivers could sit at the wheel meant delays added significantly to the cost of doing business in the West Country. By the early 1960s few doubted the A30 urgently needed a comprehensive upgrade and that Okehampton should be bypassed; few can have foreseen that when Devon County Council proposed a bypass in its County Development Plan of 1963, legislation enabling its construction would not be passed for another twenty years. At question was whether the bypass should be routed to the north of Okehampton or, as the government preferred, to the south, passing through the national park. When the government finally forced the Okehampton Bypass (Confirmation of Orders) Bill through parliament in November 1985, it ended a process that had included private discussion with interested parties in the 1960s, public consultations in the summer of 1975, the announcement of the government's preferred route in 1976, the publication of a draft order in 1978, an epic local public inquiry held at Okehampton lasting ninety-six days in 1979–80 and a hearing before a parliamentary joint committee in February 1985.

Little controversy had accompanied the staged construction of the dual carriageway from Exeter to Tongue End Cross, a mile or two from Okehampton. Rather than connecting the villages, the new route followed natural contour lines, bypassing the villages and for short stretches taking the traffic away from the national park boundary. Possible northern routes for the bypass, of which many were proposed and none was thoroughly investigated, took the projected dual carriageway through medium-grade agricultural land. The southern route passed through the moorland and park-land immediately south of Okehampton, cutting off Okehampton Castle, stretches of the East and West Okemont, the golf course and woodland from the rest of the national park. The southern route was cheaper, shorter and favoured by the Department of Transport, Devon County Council, local district, town and parish councils (though not always by large majorities), and special interest groups like the Country Landowners' Association, the Road Haulage Association, the farming lobby and other parties with a commercial interest in the decision. On the opposing side were the Countryside Commission, Dartmoor National Park Authority, the Council for the Preservation of Rural England (CPRE) and numerous conservation and amenity organisations, including the venerable Dartmoor Preservation Association. Politically and ideo-logically, the dividing lines could not have been clearer.

A simple principle seemed to be at stake. Was it legitimate to build a dual carriageway through a national park? A government circular of 1976 stated that investment in trunk roads should avoid upgrading existing roads or building new roads for long-distance traffic in the national parks, 'unless it has been demonstrated that there is a compelling need which would not be met by any reason-able alternative means'.[9] The opponents of the southern route thus thought they had moral principle and government policy on their side. The public inquiry of 1979–80 did not express its conclusions in terms of the nebulous quality of reasonableness, but it did find 'the total price that would be paid for avoiding the National Park . . . too high'.[10] In particular, it found the environmental conse-quences of the northern route more severe. This last claim, perplexing to the opposition, rested on the idea that routing the (southern) bypass directly below the upland would mean much of it was invisible from the moor, while noise levels and the visual effect of the road would be minimised by careful landscaping and planting 70,000 trees. If the northern route was chosen the bypass would be visible on a clear day from Dartmoor's highest points, spoiling iconic views of north Devon. This argument relied less on ecological evaluations than scenic judgements and encouraged opponents of the scheme to denigrate the landscapes of the northern route.

Opponents of the southern route insisted on the ecological integrity of the park, held its borders to be sacrosanct, and were impatient of the claim that it would be unacceptable to delay the decision by commencing a full inquiry into the feasibility of the northern route. Proponents of the scheme showed increasing impatience with the segregationist attitudes of the Dartmoor lobby. Robin Maxwell-Hyslop, MP for Tiverton, was not alone in reminding the Commons that the national park boundary was somewhat arbitrary and not infallible testimony to the exceptional quality of the land within, which included 'some areas of great ugliness'.[11] Many proponents, alert to how their opponents had manipulated public sympathies by exaggerating the bucolic quality of the threatened land, drew attention to the railway, the viaduct, the reservoir and the quarry that already disfigured this most industrialised part of the national park.

Visitors to the disputed site could come away underwhelmed. A humorous account of the select committee's visit by David Mckie, the *Guardian* newspaper's political sketch writer, was typical of the metropolitan encounter with Dartmoor. 'Most members of the Joint Select Committee were wearing their Wellington boots; which was just as well, since the path across Bluebell Wood to the bubbling West Okemont river was treacherously swampy. Sometimes the mist came down and blotted out promised panoramas; later the wind blew fiercely, and in time it began to rain.'[12]

McKie was not the first visitor to Dartmoor to be a little

nonplussed, the extraordinary experience promised confounded by mud, mist and rain. George Bishop, also writing in the *Guardian*, did not pretend to uncertainty. Utilising every cliché of the motoring journalist, he derided Dartmoor as 'a scruffy piece of moorland' where nothing grows and which was 'of no use to anyone except the Army, who fire guns there'.[13] *The Times*, by contrast, portrayed what it drolly termed the 'Battle of Soakampton' as pitting patrician against patrician: class politics was never far from post-war disputes about Dartmoor.[14] Angela Rumbold, closing the parliamentary debate for the government, described the threatened land as poorly managed and 'rather unsightly', before making a more substantial point about this part of the moor: 'No one is suggesting that the countryside south of Okehampton is not beautiful Devonshire countryside. Of course it is. It is certainly not a wilderness, and when people write in defence of the last wilderness left in southern England, it merely serves to show that they have never been there.'[15]

This was indisputable, and the bypass controversy is a reminder that Dartmoor's modern history has been troubled by the discursive tension between Dartmoor the wilderness, the last place in southern England where nature was ascendant, and Dartmoor the anthropic landscape of shifting meaning and value. *The Times* indirectly captured something of this aspect of Dartmoor's history in a leading article following the House of Lords debate on the Bill. Impatient with the parliamentary manoeuvrings of the scheme's opponents and glad that local people would finally be satisfied, the newspaper argued: 'It is a simple fact about major building projects that they represent the ethos of the time that they were conceived better than that of the time at which they are completed. The choice between two Okehampton routes neatly mirrors the shifting conflicts between the interests of ecologists and farmers. But constantly to justify changes in planning decisions in terms of shifts in national preoccupations is a recipe for chaos.'[16]

Britain's planning system might induce such fatalism in any

observer, but the broader shift identified by *The Times* is important. The environmental turn in late twentieth-century British politics had a profound effect on how landscape was valued: as one Labour peer put it in 1984, 'things that were done 10 years ago would never be tolerated in the present climate'.[17] That proponents of the southern route sometimes minimised its likely effect by observing that the bypass would not pass over open moorland or common gives some indication of the way different parts of the national park were valued. To the agricultural improvers of the nineteenth century, a common was a valueless waste in urgent need of dynamic human intervention if its potential was to be fulfilled; to the environmentalists of the early twenty-first century, a common is an exceptionally important but fragile ecosystem in urgent need of careful human intervention if its biodiversity is to be restored. The last part of *Quartz and Feldspar* relates the rise of environmentalism to the modern history of the Dartmoor common, tracing a shift in subjectivity of equal importance to the preservationist turn of the late nineteenth century. Particularly important is how the commoners, long resistant to outside interference regarding their use of the common for grazing, have been co-opted by the state as conservationists. In the early twenty-first century grazing is as likely to be explained in terms of 'ecosystem services' as it is meat production.

When work began on the Okehampton bypass in November 1986 a local politician made an awkward appearance balanced 'on a bulldozer and triumphantly waving a tin helmet'; when the bypass was opened on 19 July 1988, Peter Bottomley, minister of roads and traffic, was on hand to snip the ribbon.[18] The councillor protecting her head and the minister with his scissors were peculiar symbols of misplaced triumph. They simplified complexity, obscured division and displayed certainty where a show of resignation or even contrition was more appropriate. Politics does not allow for doubt, even when a dual carriageway is driven through a national park.

Decades later the traffic surges along the A30, most motorists oblivious to the controversy or the towns and villages bypassed, the upgraded trunk road having proved something of a mixed blessing to communities that once benefitted from the passing trade. It is less easy to miss the dark massif of Cosdon Hill, whose great domed summit heaves darkly into view between Whiddon Down and Okehampton. Compelled to get a better look, the driver leans forward, peering up through tinted shatterproof glass. The threat of an accident stills that giddy sense of wonder and he sits back, eyes on the road ahead. Environmentalists, armed with the apocalyptic findings of modern science, recognise that the sense of wonder roused by the natural world is powerfully instrumental, inspiring human beings to take better care of it. And wonder stimulates the desire to know more. Getting to know the modern history of a marginal upland in the south-west peninsula of a small island in the North Atlantic is as good a place as any to start. Rooted cosmopolitanism once again.

I

Antiquarianism and Archaeology

Perambulation

> I was agreeably surprised to find among the copper plates several
> of my old acquaintances, which I lately degraded to no better rank
> than boundaries of parishes, or unmeaning heaps of stones; but for
> the future, I shall esteem them as the sacred relics of the worship
> of our ancestors.
>
> Thomas Pennant to William Borlase, 15 June 1754[1]

Travelling west along the B3212 from the edge of the moor in the
direction of Princetown, it is good to stop at Postbridge. A cyclist
destined for the hostel at Bellever will already have struggled up
onto the moorland road from Exeter. Conquered first was the long
winding tree-canopied climb to Moretonhampstead, the Bridford
and Cockford Woods hugging the road tightly, the town marked
on the way in by sombre almshouses and on the way out by a
mechanic's workshop. Beyond, the ups and the downs are gentler,
maintaining altitude, before the culminating climb to Shapley
Common is signalled by the rattle over a cattle grid, a sharp right
and the shock of a sudden steep hill. In a panting moment the
stone wall marking the boundary between the newtakes – small
enclosures of common land – and Dartmoor's great central waste
is passed. At the hill's first summit, some 400 metres above sea level
and beyond enclosed land, the cyclist emerges from tree, hedge
and wall-lined roads into a massive landscape.

The road swoops on undulating for three miles until reaching
the Warren House Inn, Dartmoor's highest pub. The sudden
descent into the wide depression of the East Dart river valley sees
the bicycle's spendthrift wheels squander a hundred metres in not

many more seconds. Rapidly tracking, the landscape becomes more domestic, the road again tree- and hedge-lined and fenced, the land behind enclosed, dotted with houses, Bed and Breakfasts and a pub or two. But this is not lowland pastoral but upland plantation country. On the Ordnance Survey map the newtakes are only a narrow strip along the road's edge marked in white, contrasting with the green of the plantations and the much more expansive yellow of the moorland beyond. The conifer mass of Soussons Down hunkers to the south, the green-dark shadow of the Bellever plantation rises ahead, and the map tells of the great Fernworthy plantation and reservoir beyond Stannon Tour to the north.

Postbridge, cool and sheltered, is an oasis. It has a car park (motorists are asked to voluntarily deposit a pound in a hefty stone collecting box), a national park noticeboard and shop, and a post office/general store with a defunct petrol pump. It takes its name from the medieval clapper bridge that crosses the Dart and lies parallel to the modern road bridge.

Access to the yellow parts of the map is easy from Postbridge. Paths in most directions can be picked up from here or from Higher Cherrybrook Bridge, the other side of the Bellever plantation. The bridleway towards Fernworthy is tempting. Strongly marked on the map by green em dashes, it follows the River Dart for two miles or so. As the Dart chuckles on the left and cows chew watchfully, you push through dense scratchy gorse. Offended sheep haul themselves to their feet, their great woollen rumps bobbing as they lumber up the valley slopes.

Where the East Dart veers to the left, heading west and then north to its head near Cranmere Pool, the path goes straight on, gradually reaching higher, drier, yellower ground, the landscape becoming more like its cartographic representation. To the left – to the west – the landscape now opens up, the waste stretching into the hazy distance; to the right, the plantation and glimpses of enclosed farmland are coming into view. The track's gradual ascent peaks some sixty metres below Sittaford Tor (538 metres). On this hot and bright day the only sounds are conversation, the *thrunch* of footfalls, the light wind, flowing water, the incessant song of the larks and meadow pipits, and the humming sawing of insect life. Conversation ebbs, thoughts find a groove, and breathing and step synchronise, overlaying that indistinct but generalised moorland ambience. To be here in the winter, especially in the snow, when there is no birdsong or insect hum, is to know how much quieter it can be.

Near this highest point on the path are the Grey Wethers, two intersecting stone circles. A bracketed addition on the map says they have been 'restored' but here, on the ground, no authority has felt the need to erect an explanatory sign. Walking among the stones, they take on the mobility of sculpture and it is difficult to take a photograph that squeezes these stunted stones into a single frame and captures their mystery and precise layout.

The path continues. Wistful backward glances testify to a piqued curiosity.

In the distance Teignhead Farm emerges from the landscape just beyond the North Teign River. Surrounded by open moorland, the stone walls and the mature trees shielding the farm buildings from the wind suggest a defiant exercise in enclosure, setting an example few saw fit to follow. A harder stare reveals that it is abandoned, the OS, confirming that the farm is a 'ruin', once again determining what is seen. Established in 1780, it was an extreme example of late-eighteenth-century attempts to 'improve' Dartmoor and led to the enclosure by drystone walls of 1,551 acres of open moorland. So vast was the undertaking that these walls now provide the walker with an excellent way of orientating herself. The farm was requisitioned by the War Office in 1943 when the military extended its use of the moor and has lain derelict ever since. Its crumbling walls, Eric Hemery predicted in 1983, 'will soon go "back-to-moor"'.[2]

Fernworthy plantation lies on a north-west axis and bulges towards the track, which now takes a sudden right directly into the muffling atmosphere of woods. This is a different kind of environment. Stacked logs, deep tyre tracks in dried mud, well worn stony roads, tags marking the trees: it is carefully managed, one of the Forestry Commission's extensive Dartmoor interests and responsibilities. At a clearing in the plantation there is another stone circle. It is just there, a circle of granite stones, once out on open moorland, now surrounded by conifers, the decision not to plant in this space the only indication of its significance. Again, it is hard to photograph effectively; foregrounding the stones distances the conifers, exaggerating the size of both.

At the reservoir, itself three-quarters encircled by the plantation, there are a number of options. From Gidleigh Park, roads can be picked up that lead into the densely plotted lanes fanning out from the ever-fashionable village of Chagford, but for walkers tied to their cars the best option is the Hurston Ridge cutting directly south back to the road. Quickening the step is promise of the Warren House Inn at its terminus, which overlooks the old commercial rabbit warrens that shape the land falling gently away to the south towards the Soussons Down plantation. From here, fortified, the pleasant twilit amble back down the hill into Postbridge completes the circle.

That evening the day's curiosities can be followed up. It was not the disused farm, the plantations or the old warrens, barely discernible to the unpractised eye, which stayed in the mind, but the stone circles, one out in the waste, the other among conifers. Samuel Rowe's *A Perambulation of the Antient & Royal Forest of Dartmoor* comes to hand. The rare first edition (1848) can be downloaded, printed and bulldog-clipped together. The *Perambulation* opens with a reproduction of a painting of the 'Borders of Dartmoor, Scene on the Taw', a spot in the far north of the moor just south of Belstone, a dramatically located ancient village. In the background of the picture dark and sombre mountains rise, reaching pointy peaks; in the mid-ground, cattle graze on the plain through which the Taw flows north: the contrast between the high moors and the lowlands is very sharp. In the foreground lie large flat boulders and on one is a pair of day trippers, one lounging, one decorously perched.

A page on and the frontspiece identifies the author as the vicar of Crediton, a little town near Exeter, and the publishers as J. B. Rowe of Whimple Street, Plymouth and Hamilton, Adams and Co. of London. Dominating the page is a lithograph titled 'Drewsteignton Cromlech'. In the foreground a simple stone structure composed of three upright stones topped by a large flat stone; in the background a mountainous scene. Turning to the index, references to the Fernworthy Circle and the Grey Wethers are easily found, and the reader is quickly plunged into the middle of an obscure discussion of temples, sacred circles and 'the struggle between ophiolatry and solar worship'. The Grey Wethers, Rowe says, is the largest such circle in Devonshire. So affecting is their mysterious power, he imagined them inspiring the work of a great poet. He cites a passage from Keats's *Hyperion* describing a group of overthrown Titans:

> One here, one there,
> Lay, vast and edgeways, like a dismal cirque,
> Of Druid stones, upon a forlorn moor
> When the chill rain begins at shut of eve
> In dull November.[3]

Rowe's discussion also mentions a *via sacra*, 'a processional road of Druidical worship, according to the Arkite ceremonial', and refers to the 'great Sacred Circle' on Scorhill Down. Of this Rowe writes that it is 'by far the finest example of the rude but venerable shrines of Druidical worship in Devonshire', adding that it 'may successfully dispute the palm' with better-known circles, such as those at Castle Rigg near Keswick and at Rollright in Oxfordshire. The 'rugged angular appearance of the massive stones' that form this 'rude hypaethral temple', he suggests, can be contrasted with the 'squarer . . . more truncated' stones of the Grey Wethers.[4]

Scorhill Down is just a mile or two north of the Fernworthy plantation, a short distance from Gidleigh Park. Once the meandering country lanes that skirt the North Teign have been negotiated, it is easily accessible from a small car park. West and south-west of this point there is no road for, respectively, thirteen and eighteen kilometres. The 'sacred circle' is just a short walk down a grassy granite-strewn trackway.

Samuel Rowe and his Antiquarian Antecedents

When published in 1848, Samuel Rowe's *Perambulation* was the most ambitious book ever written about Dartmoor. Announcing itself as a 'topographical survey' of Dartmoor's 'antiquities and scenery', it also boasted 'notices' relating to natural history, climate and, tellingly, Dartmoor's 'agricultural capabilities'. Shaped into two long sections, the first and shorter of the two consisted of thematic discussions. These provided a physical description of the moor, an outline of the historical roots of its various systems of landholding and rights, an account of its 'aboriginal inhabitants' and their religious practices and, finally, a long section, some two thirds of the total, on Dartmoor's 'monumental relics'. Having read these sixty pages of preparatory material, the conscientious reader would be properly equipped to accompany Rowe as he set off across the moor, describing his perambulations and elaborating on his themes as prompted by the points of interest encountered.

Rowe was not the first to so structure his account of the moor, but his became the paradigmatic example of a genre of Dartmoor writing that made the subject – the landscape itself – manageable by approaching it this way. Walks allowed Dartmoor to be treated as a series of narrative poems that flowed allusively over the moor's surface but required annotation and commentary. The reader might take impressionistic pleasure in the 'poem', but its full meaning could only be grasped when accompanied by skilled topographical, antiquarian and historical commentary.

Rowe explained that the moor should interest the geologist, who could make a study of its tors; the botanist, whose curiosity would be particularly roused by Wistman's Wood; and the antiquary, who could hardly miss the importance of the 'aboriginal circumvallation' of Grimspound.[5] If landscapes like Dartmoor were now the province of a wide range of specialists, Rowe's book announced itself as a pioneering attempt by a clerical amateur to draw this multidisciplinary knowledge into a coherent whole. Rowe's readers were united – or so he implied – by their developed sensibility to the aesthetics of landscape. The 'Devonshire Highlands', he asserted, provided 'scenes of unexpected loveliness and grandeur' which rivalled the 'far-famed scenery of North Wales'.[6]

Rowe recognised that north Wales, like the Lake District, the Scottish Highlands and continental landscapes such as the Alps and the French *campagne*, were long established as landscapes that could satisfy Romantic sensibilities. The development of these tastes has been traced to the great shifts that occurred in the social, political and cultural life of Europe during the eighteenth century. Early modernity, experienced through new forms of production (agricultural enclosure and early industrialisation) and the new forms of urbanised life that flowed from this, stimulated new attitudes towards the landscape. The gradual shift in aesthetic taste saw the popularity of the formal garden fall and enthusiasm for places thought less affected by human intervention rise. Much was paradoxical, not least the readiness of the owners of Britain's great gardens to dedicate great resources to manufacturing a fashionable illusion of naturalness.[7] The new cultural value attached to nature evolved into a taste for wild places in all their apparently untamed variety. Dartmoor, Rowe understood, could be promoted as a desirable commodity.

Rowe had to convince a discerning consumer. Since the late eighteenth century an ability to appreciate landscape had become de rigueur for the cultured members of the new middle classes.

William Gilpin's influential essays of 1792 on the picturesque taught his readers to appreciate the English landscape as they would a painting. Being able to distinguish one picturesque view from another, explaining its strengths and weaknesses, became an important new source of cultural capital. Cultures of competency like this allowed possessors of new sorts of wealth, accumulated through manufacturing, commerce and the professions, to both emulate the disinterested exercise of taste manifested in the gardens of the gentry and the aristocracy and to challenge outright aristocratic dictation in questions of taste. Subjecting the landscaped garden to aesthetic judgement was an assertion of cultural authority over privately owned land, while celebrating 'natural' landscapes, in the context of enclosure, brought into question the degree to which the gentry and aristocracy could be trusted to preserve the nation's landscape heritage.[8] Though the search for the picturesque as a craze was short-lived, it bequeathed the English a set of pre-industrial rustic ideals that long continued to constitute what the urban middle class thought was best about rural England.

When ideas of the picturesque coalesced with Romantic notions of sensibility particular landscapes became celebrated for the feelings and sensations they evoked. New ideas about beauty further brought into question established aesthetic standards. Edmund Burke's essay on the sublime (1757), little noticed when first published but eventually of great influence, described the 'delight' we experience when we have before us 'an idea of pain and danger, without being actually in such circumstances'.[9] Certain kinds of terrain came to be considered sublime, particularly mountainous regions, their apparent resistance to domestication revealing the awesome power of nature. As the cult of nature generated a cult of the sublime, so landscapes were celebrated that were aesthetically quite distinct from those admired in terms of the rustic ideals of the picturesque. If a landscape could be celebrated because it was 'awful', the unproductive terrain of Dartmoor, brought to the attention of the Romantic, became a landscape of desire.

Symptomatic of this landscape thinking was a competitive region-alism, and Rowe was keen to establish Dartmoor as a Romantic landscape equal to the best. Rowe quoted non-Devonian observers who could be relied upon for their objectivity, like the Scottish jour-nalist who admitted that when 'wandering' Dartmoor he 'often forgot' he was not holidaying 'among the blue bonnets of Auld Scotland'. High praise indeed. Or William Howitt, who in his *Rural Life of England* (1838) found in Dartmoor evidence of the sublime: 'If you want stern-ness and loneliness, you may pass into Dartmoor. There are wastes and wilds, crags of granite, views into far off districts, and the sounds of waters hurrying away over their rocky beds, enough to satisfy the largest hungering and thirsting after poetical delight.'[10] Howitt described the 'melancholy music' of a 'deep dark river', and 'gnarled oaks' overlooked by 'glowing ruddy tors standing in the blue air in their sublime silence', but the key words or ideas to hold in the mind are 'want', 'satisfy', 'hungering' and 'thirsting': Howitt was introducing Dartmoor to a readership craving Romantic sensation.

The extract concluded: 'one sole woodlark from the far ascending forest on the right filled the wide solitude with his wild autumnal note. At that moment I reached an eminence, and at once saw the dark crags of Dartmoor high aloft before me.'[11] This is romantic Dartmoor, always out of reach, a site of infinite regression, each eminence leading ineluctably to another. Coming close to the moor flooded the visitor with desire, but ineffable nature, reached for through exultant prose and poetry, imminent but never grasped, could only expose her to the agony of non-consummation.

Rowe's promotional romanticism was wedded to a more prosaic purpose. The moor needed to be preserved against the ravages of industrial modernity: 'But this rocky citadel is no longer secure. Quarries are opened on the heights of Dartmoor – powder-mills are projects in the very heart of its solitudes – cultivation is smiting its corners – steam is marshalling his chariots of iron and coursers of fire, panting to penetrate its fastnesses – and the most interesting vestiges of antiquity are in hourly danger of destruction.'[12] Howitt's

inaccessible Dartmoor was proving penetrable. The conditions exercising his Romantic desires were threatened by quarrying, milling
and the railway. Even new systems of cultivation wrought a metallic
violence on the moor. Not merely subject to the benign interest of
learned specialists, Dartmoor now inspired entrepreneurs keen to
exploit its natural resources. It was thus necessary to record the
evidence being lost to 'multiplied population, increasing commercial
speculation, and economic improvements'. Rowe hoped his book,
by bringing the antiquarian riches of Dartmoor to national attention, might help mobilise action against the 'impending assaults of
the mason's hammer, and the excavator's pick', saving what ought
to be of 'interest in the history of this country and of mankind'.
This, an early expression of Dartmoor preservationism, is the subject
of a later chapter. First, though, we must understand what it was
that Rowe thought should be preserved, and it is here that his
thematic discussions, particularly those concerned with Dartmoor's
'sacred' relics, are most important. They will take us back to the
eighteenth century and another source of interest in wild places.

Rowe's survey of Dartmoor's 'monumental relics' began with the
'circular temple, or sacred circle', the most 'conspicuous' of
Dartmoor's 'Druidical antiquities'. Although admitting that nothing
on Dartmoor approached the 'massive proportions' of Stonehenge,
Rowe thought Dartmoor's relics were more interesting on account
of their greater antiquity. This was evident in their 'rude simplicity',
which he contrasted with the 'artificial trilithons' of Salisbury Plain.
Stonehenge, famous for a century, is thus diminished because it was
a work of skilful derivation rather than unselfconscious expression.
He continued in this vein, drawing attention to Dartmoor's stone
avenues or parallellitha, arguing that though neglected by previous
writers, they had played an important role in Druidical ritual. 'Rock
idols', logan stones and rock basins also attracted his attention. Had
the aboriginal people of Dartmoor worshipped or idolised
Dartmoor's more striking rock formations? What role had logan

stones, remarkable structures in which one stone rested precariously on another, played in those same rituals? Though no logan stones remained intact in 1848, legend had it that a single man, exerting maximum force, could not move the balanced stone, but it would 'rock' or 'vibrate' at the lightest touch. What role was assigned to the basins, sometimes several feet wide and deep, found in the rocks, particularly on prominent tors? Continuing to classify, Rowe listed kistvaens (stone tomb-like structures laid on the surface of the moor), cairns (burial sites formed from large piles of broken stone), barrows or tumuli (earthen burial grounds) and rock pillars or 'rude Stone Obelisks' (long narrow stones placed upright, perhaps of some memorial or ritual purpose). He also identified the remains of stone huts, as evident in the surviving small circles of stones that he thought had once been built up into walls with a roof, and pounds, or circumvallations, large stone enclosures often containing within them evidence of stone structures. Ancient trackways, roads, track lines, boundary banks, bridges, forts and entrenchments also got a look in. Most remarkable of all was the cromlech at Drewsteignton.[13]

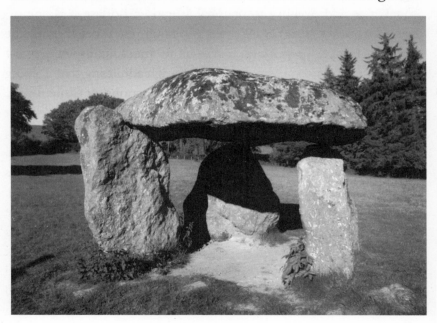

Rowe's impressively learned and erudite text had begun life as a rather plain lecture of 1830 cataloguing Dartmoor's Druidical antiquities delivered to the Plymouth Institution. This worthy institution, typical of early-nineteenth-century middle-class associational culture, was dedicated to 'the promotion of Literature, Science, and the Fine Arts, in the town and neighbourhood'.[14] Rowe drew on the work of contemporaries like Sir R. Colt Hoare (1758–1838), famous for his research on the antiquities of Wiltshire, but he was most influenced by the antiquarian writings of William Borlase and Richard Polwhele, two West Country clergymen. Polwhele (1760–1838), a Cornish clergyman who later became vicar of the east Devon parish of Kenton, was the author of *Historical Views of Devonshire in Five Volumes* (1793) – only the first volume was completed – and *The History of Devonshire* (1797), as well as comparable works on Cornwall and many published sermons and poems. His work on Devon paid particular attention to Dartmoor. Borlase (1686–1772) had been active during the heady days of mid-eighteenth-century antiquarian debate. A correspondent of William Stukeley, who brought Wiltshire's Stonehenge and Avebury to national attention, Borlase was one of the leading antiquarians of his time. Though he did not write directly about Dartmoor, his work on Cornwall was profoundly influential, inspiring later writers who believed the ancient history of Devon and Cornwall – the region known as Dunmonia in the Roman sources – was distinct from the rest of Britain.

As Rowe set out on his perambulations, obsessively noting down Dartmoor's antiquarian and topographical features, he encountered a moor that had been given meaning by these earlier antiquarians. The relics he discovered during his perambulations, often listing them for the first time in print, were used to elaborate a set of contentious theories that had been in circulation for the best part of a century.

Though antiquarian enthusiasm proved infectious in late eighteenth-century Britain, it was not universally admired. Then, as now, the antiquarian was thought obsessively and eccentrically preoccupied with the collection of obscure detail, a harmless pursuit that gave insufficiently busy clergymen in isolated parishes something to do.[15] Eighteenth-century historians, tasked to produce eloquent discourses on the lessons to be learned from the study of Graeco-Roman classical civilisation, were particularly contemptuous. David Hume, for instance, thought it 'rather fortunate' that the barbarian prehistory that provoked the curiosity of the antiquarians was 'buried in silence and oblivion'. What possible benefit could be gleaned from the study of the primitive?[16]

High-handed dismissals like this gradually became less plausible as antiquarianism's challenge to early eighteenth-century notions of what were the proper objects of historical study came to resonate with wider shifts in public opinion. Antiquarians asserted with greater confidence the virtue of national self-knowledge, promoting their work in nationalist terms, a sentiment that intensified during the Napoleonic Wars.[17] Research into prehistory provided increasingly celebratory accounts of an indigenous ancient British culture, apparently 'free of the mannered artificiality and corrupt lifestyle' of late-eighteenth century life.[18] Such enthusiasm was reinforced by the fashionable primitivism of the time, most often associated with the writings of Jean-Jacques Rousseau. With the primitive conceived as innocent and uncorrupted, as distinct from the barbarian, so the notion of civilisation itself came under scrutiny. Most obviously influential was *The Decline and Fall of the Roman Empire*, Edward Gibbon's monumental historical study published in six volumes between 1776 and 1788. Gibbon argued that great civilisations declined when they lost contact with their foundational principles and became decadent and ill disciplined. To be ignorant of a nation's history was to be blind to what made it great and might, given continued vigilance, ensure that it remained so.

When iconic poems celebrating British resistance to Roman tyranny, like William Mason's 'Caractacus' (1759) and William Cowper's 'Boadicea' (1782), were added to the mix, ancient Britain could be embraced as an authentic and heroic component of the national story.[19] The poets made the ancient Britons defenders of liberty, a people who hung on to their own traditions until, crucially, their conversion to Christianity. Then 'spiritually and with commendable docility' they 'surrendered to the teachings of Christ'.[20] In this way, the British never were slaves, for they resisted Roman hegemony through the maintenance of their customary practices. The ancient Britons could be celebrated for their patriotic rectitude rather than scorned for their prehistoric barbarity.

These patriotic interpretations chimed with the emergence of stadial theories of human development, according to which each civilisation was thought to advance through a series of recognisable steps. Prompted by imperial encounters with the non-European colonised 'other', one strain of thought suggested that these peoples were not irredeemably barbarian, primitive or savage, but at an early stage in the development of their civilisations. If the development of all peoples followed a similar pattern, as this nascent comparative anthropology seemed to suggest, then an understanding of primitive peoples in the present could help the antiquarian decode the meanings of the material survivals of Britain's own primitive past.[21]

None of this was uncontentious. Mid-eighteenth-century antiquarian debates were animated by disputes concerning the religious practices of the ancient Britons and the degree to which continuities could be claimed between pre-Christian and Christian practices. Dartmoor's tremendous prehistoric riches – those 'monumental relics' catalogued by Samuel Rowe a century later – made it an important focus for antiquarian discussion. Was this landscape the preserve of a British heritage that testified to the great continuity of a distinctive British civilisation? Might those relics even be seen as proto-Protestant? Or did these relics signify a more alien British past that was barbaric, perhaps Celtic, and possibly akin to Roman Catholicism? Who were the people who had built those stone circles? Where did they come from? What did they believe? Could the free-born Briton claim them as ancestors?

Cornish Contexts and Druidical Dartmoor

William Borlase was born on 2 February 1696 at Pendeen House in the parish of St Just. His family were second-rank Cornish gentry, just the type of people who expected to hold positions of local responsibility or to make their way in the professions. Borlase's father had been educated at Exeter College, Oxford and married the daughter of a respectable Devonshire family. Borlase followed his father to Exeter in 1713, receiving his BA in 1716 and his MA in 1719. In a period when many Oxford undergraduates left without taking a degree this was indicative of his commitment to academic work. Like many a third son, Borlase went into the Church, becoming rector of the Cornish parish Ludgvan. At Exeter he had become a friend of John St Aubyn, scion of one of Cornwall's grandest and wealthiest families, and this led to other good connections in the county. Encouraged in his work by influential friends, Borlase corresponded with two Oxford dons about the possible uses of rock basins and attracted the attention of the poet Alexander Pope, who sought his help in sourcing the rocks and minerals needed to construct a grotto in his garden. In 1751 Borlase was elected a fellow of the Royal Society following the good reception of his treatise on Cornish crystals. Not bad for a relatively humble clergyman. Most immediately pertinent was his friendly correspondence with William Stukeley, Britain's most famous, if not always most respected, antiquarian.[22]

Stukeley had made his reputation with books on Stonehenge (1740) and Avebury (1743), helping establish them as the two most

famous ancient British sites in England. His work is still admired
for the accuracy of his observations and the quality of his draw-
ings. Inspired by the seminal work of the mid-seventeenth-century
writer John Aubrey, Stukeley believed both sites were prehistorical
Druidical temples, and he developed an elaborate and highly contentious theory of Druidical practice and ritual to reinforce his claims.[23]
Fundamental to Stukeley's thinking was the long-established distinction between 'natural' religion, arrived at through reason, and the
religion suggested by the Old Testament prophets and 'revealed'
in the life and teachings of Jesus Christ. Natural and revealed religion were both aspects of the same overall divine plan. Druidical
religion, Stukeley claimed, recognised one god, and the Druidical
veneration of the heavenly bodies, the earth and the four elements
was not polytheism – the worship of many gods – but the worship
of the most extraordinary manifestations of this single deity.

As Alexandra Walsham states, the Protestant Reformation in
Britain had not banished the idea that 'God used nature as a supple-
mentary text of revelation'.[24] Ancient man, Stukeley believed, had
grasped the fundamental 'unity of the Divine Being' and represented it in stone circles, so 'expressive of the nature of the deity
without beginning or end'.[25] Thus, Stonehenge and other ancient
British sites were transformed into evidence of the integrity of
monotheistic religion in prehistoric Britain. Ronald Hutton, drawing
together these various strands, argues that Stukeley believed 'in a
primeval religion which had been shared by all the peoples of the
remote past, simply because it was the natural one for primitive
humanity to embrace'. Natural religion was not spread by mission-
aries 'but arose spontaneously from "the same common reason in
mankind"'.[26]

Borlase was not convinced, and in his *Observations on the
Antiquities Historical and Monumental of the County of Cornwall*,
published in 1754, he implicitly disputed Stukeley's patriotic conviction that the ancient British had remained true to a set of religious
practices continuous with Christianity. *Observations* was a lavish

production, notable for the many high-quality lithographs that illustrated the text, and beyond the purse of a Cornish curate. As with many similar works, Borlase relied on an extensive list of subscribers, headed by in this case Her Royal Highness the Princess Dowager of Wales and containing many Oxbridge fellows and college libraries, as well as numerous clergymen and private citizens.

Early in the *Observations* Borlase made it clear that he regarded 'Druid superstition' as a form of idolatry and a distortion of true religion. The hundred pages Borlase then devoted to explaining the origins and meaning of idolatry formed an apparently necessary preface to his full analysis of its material precipitates in Cornwall. Much of his explanation drew on contemporary theories about the racial or ethnic origins of the earliest inhabitants of Danmonia, part of the wider debate regarding the origins of the first inhabitants of the British Isles. Theological needs were at the core of these disputes, and even Stukeley, faced with the need to accommodate the biblical truth that God had taught Abraham the essentials of the true faith, diluted his belief that the British came to their primeval religion through natural processes. Instead, a means had to be found by which Abrahamic teachings could be communicated to the ancient British. Stukeley embraced the idea, long subject to historical discussion, that Britain was first colonised by the Phoenicians, Mediterranean seafaring traders.[27] Thus, Abraham effectively became the first Druid, and, once in Britain, the Phoenicians maintained the purity of their Abrahamic practices.[28]

Borlase thought this poppycock. It defied logic that the Phoenicians, a mercantile people, were the 'first planters of our isle'. They would only have come to the south-west peninsula if there were people to trade with. Nonetheless, Borlase did believe that traders from this seafaring people had made it from the eastern Mediterranean to Dunmonia. As trade links developed, they settled in Cornwall, bringing with them their native religious practices.[29] Druidism, then, had eastern origins, but rather than celebrating these proto-British Druids for the purity of their practices, Borlase treated the Druids of ancient Dunmonia as typical of their time. Without thinking comparatively, he wrote, 'we shall be apt to think that the Druids stand alone in all the instances of barbarity, magick, and grove-worship laid to their charge'.[30] With the Danmonii cleared of an exceptional tendency towards corrupted religion, Borlase turned to what concerned him most: the causes of idolatry and its surviving material precipitates.

Borlase's reasoning was an elaboration of conventional religious thinking, but unfolding his argument helps make sense of how his thinking shaped perceptions of the Devon and Cornwall moors. Idolatry, Borlase explained, had simple all-too-human origins. It stemmed from the difficulty of living according to the 'strictness of life and manners' demanded by a proper recognition of the 'transcendent purity of God'. Those 'who delighted in violence and wand'ring lusts' were happy when the 'inventive and powerful' convinced them 'new sorts of Deities' existed 'who were to be pleas'd upon easier terms'. As a consequence, the 'natural' religion of the ancients, which had contained within it the fundamentals of 'revealed' or Christian religion, was vulnerable to corruption. Unlike Stukeley but like Isaac Newton, Borlase believed that in ancient times true religion survived only within 'the little nation of the Hebrews': only the Jews had sustained a belief in the existence of the one true God.[31]

Idolatry generated a shift from monotheism (the worship of God) to polytheism (the worship of gods). Borlase maintained that polytheists believed the immortal ghosts of 'good, great, or ingenious men (whether good or wicked)' were capable of intervening in the mortal world and, therefore, should be worshipped. 'Planets, Devils, Brutes and senseless Images' including first, and most importantly, the sun, represented these ancestors and were thus appropriate subjects for worship.[32] Polwhele later claimed that a number of Dartmoor place names like Bellever and Belstone stemmed from this ancient worship of Belus or the sun. Borlase explained how it became 'customary to make gods of every thing which appeared either capable of doing harm, or necessary and beneficial to human life'. Beginning with the elements – air, earth and water – but soon expanded to include plants and herbs, flowers and trees, ancient people soon 'made rude and shapeless stones the representatives of their fanciful Deities'. Critically, as the distinction between the absent deity and its physical representation dissolved, the people came to adore the symbol itself, 'the huge lifeless lump of Stone'. Proliferating rituals and the erection of altars – each new god apparently required a distinct ritual – were accepted because they were adaptations of good and familiar religious practice. Borlase believed that the 'gloomy kind of awe, and religious dread' associated with Druidism, which was manifested in 'Grove, and Nightworship' and artificially heightened the affectivity of Druid ritual, was an attempt to simulate the 'true fear of God', which was incompatible with polytheism.[33]

Further evidence of how 'a false superficial purity' supplanted 'a true purity of heart' was found in the centrality to Druidical ritual of white garments and purified water. Human sacrifice, that most heinous Druidical innovation, had developed from the innocent and holy practice of animal sacrifice thanks to the false belief that the gods required what was 'most precious to the heart of man'. Indeed, 'when Fire became a Diety the children of the Idolater were offered and burned, that the Deity might have them, and be propitiated'. Ritual culminated in 'Luxury and Debauch'.[34] Much eighteenth-century antiquarian commentary took a prurient delight in detailing or alluding to the sensuality and sinfulness that was integral to idolatrous practice, and contemporaries associated this with the Bacchanalia thought to follow the popular rituals of Catholic penance, another false form of purification.[35] Borlase found it particularly disturbing that the dead were worshipped in a manner that echoed their behaviour in life; a sinful life generated sinful religious practices. 'If he was cruel and bloody, he was to be sacrificed unto by human victims; if he was lustful or drunken, prostitution was to attend his festival, and his propitiation was to be a scene of intemperance and debauch; if he had been avaricious, the innocent and weak were to be plundered to make a rich offering to his altar.'[36] Sex, drink and plunder. These were the associations Borlase attached to the ancient monuments found in Devon and Cornwall.

These lengthy generalities were only a preface to Borlase's extended discussion of the particularities of Druidical practice. At ninety pages or so, this far exceeded the combined length of the available classical sources,[37] which was typical of a tradition that had already extrapolated on what could be learned from the classical writings. Reiterating his belief that Druidical practices were common to eastern idolatry, including human sacrifice and the ritualistic use of oak and mistletoe, Borlase nonetheless found much that was peculiar to Druidism in Britain, Ireland and Gaul. It was the basis of an entire social system, in which the elite was formed of three orders. The Druids were a kind of priesthood responsible for spiritual matters; they were supported by the Bards, who provided the community with its collective memory, and the Eubates or Vates, whose business was to foretell future events.[38] Although classical authors suggested a form of Druidism might have been found in Germany and Spain, Borlase concluded – contrary to much contemporary thinking – that it was principally a phenomenon of British origin that had spread to Gaul. French antiquarians, he pointedly noted, were reluctant to admit that 'their forefathers [were] indebted so much to this island'.[39] A provoking claim given the role the Gauls and Druids had played, long before Asterix, in a patriotic French myth of resistance to Roman tyranny.[40]

Borlase agreed with Stukeley that the Druids, as a learned teaching order, ran schools, taught philosophy and applied themselves to research in astronomy, geography, physics and the natural world. Distrusting the written word – a neat explanation for the absence of written records – they committed everything to memory, working in the vernacular rather than Greek. Teaching and study took them to quiet places like caves, woods or cairns, which became their particular domain.⁴¹ There they contemplated the ultimate problem, the immortality of the soul and its migration at death to another human body. Borlase's discussion became an exercise in comparative religion, emphasising how typical such beliefs were in ancient times.⁴² These ideas were taught through 'the ancient Oriental manner of Allegory and Mythology', which protected 'their great and sublime truths' against them becoming 'cheap and contemptible'. This was an important claim, for like other writers on Druidism Borlase believed the Druids deliberately manufactured an air of mystery in order to maintain their authority. The parallels with Protestant critiques of Catholicism need hardly be emphasised. Drawing on Seneca's claim that 'a grove thick set with ancient Oaks . . . immediately makes you think it the habitation of some God', Borlase acidly commented that Druidical ritual occurred in groves because 'without this solemn scene of shade and silence, the mind could not be disposed to embrace so readily all the fabulous relations of their false gods, much less to comply with all the absurd and detestable rites of their idolatrous worship'.⁴³

Around the same time Burke made a similar claim in his *Essay on the Sublime*. 'To make any thing very terrible,' he suggested, 'obscurity seems in general to be necessary.' Obscurity, or insufficient knowledge, generated fear, a state of mind in which 'all motions are suspended', whereas full understanding – enlightenment – tended to diminish fear, allowing individuals to rationalise the threats they faced. Despotism, according to Burke, was sustained by 'the passion of fear', and this relied, at least in part, on keeping the chief obscured 'from the public eye': the chief's power lay in the fact that he was kept unknowable, a claim that chimed with contemporary ideas about 'oriental despotism'. The same applied to the maintenance of many religions, hence the fact that 'almost all heathen temples were dark'. 'For this purpose too,' Burke added, 'the druids performed all their ceremonies in the bosom of the darkest woods, and in the shade of the oldest and most spreading oaks.'⁴⁴ Reading Borlase through Burke, particularly as evident in his preoccupation with darkness and deception, suggests the ways in which Protestant enlightenment thinking structured his argument.

If fear sustained Druidism, then nothing could be more horrifying than human sacrifice. Borlase reckoned that convicts were usually used, though in their absence an innocent might be found or, should the circumstances be thought sufficiently grave, children or even a prince were selected. The unhappy victim – shot with arrows, crucified, impaled, burned or bled to death – was first assured that through sacrifice their soul would be translated to heaven and their remains become a holy relic. When stabbed to death, the direction in which the body fell was thought prophetic. Most 'monstrous', victims were crammed into a huge image, usually a bull, made from wicker, which was set on fire, 'consuming that, and the inclos'd in one holocaust'. Some of the remains, according to Pliny (says Borlase), were eaten; the rest were burned on the altar. 'Intemperance in drinking generally clos'd the sacrificing'; the altar was reconsecrated by strewing oak leaves on it.[45]

Borlase's page or two on human sacrifice was a small part of a much longer discussion of Druidical rites that included 'Superstitious Rounds and Turnings of the Body', 'Holy Fires' and 'Divination, Charms and Incantations'. He noted the great resemblance between Druidical and Persian superstition – another nod towards the oriental – and reminded his readers that Phoenician traders were responsible for bringing all this ghastliness to Britain. When, on page 152, he finally turned the attention of his enthralled – or exhausted? – readers to his discussion of 'Rude Stone-Monuments', he commenced with portentous comment: 'The precariousness of human life, and the uncertainty of worldly affairs, taught people very soon after the Creation to endeavour by some memorials to perpetuate the remembrance of those persons and events, which had been of importance in their time.'[46]

The reference to 'very soon after Creation' is a reminder that Borlase, like his contemporaries, thought according to a time frame determined by scriptural authority. Notoriously, Bishop Ussher had calculated that the earth was created in 4004 BC and the flood occurred in 2448 BC. Mankind, therefore, could not have settled in Britain before 2400 BC. This meant that this history of pre-Roman Britain had to be 'squeezed into about eighty generations'.[47] Only in the later decades of the eighteenth century did new thinking begin to overturn biblical authority and generate notions of geological time.[48]

Borlase's discussion of the monuments built in those pre-Roman years was extraordinarily learned, drawing extensively on biblical and classical texts, modern authorities (notably William Camden and John Toland) and comparisons with Ireland and Scotland. Methodologically, it was a masterclass in the methods of eighteenth-century antiquarianism, demonstrating how classical texts, carefully read and compared with the help of modern interpretations, could be used to explain the former uses of ancient monuments. Although every sort of monument could not be directly accounted for, the degree to which the antiquarian could demonstrate his mastery of the authoritative texts determined the degree to which his speculations were taken seriously. Admitting an analysis was speculative provided cover for the ventilation of ideas with no real basis in the source material and was a way of displaying a kind of antiquarian sophistication.

Borlase observed that ancient stone monuments had a variety of uses. They could be 'of a truly Religious Institution', either as symbols of gratitude, a religious covenant or commitment, or sepulchral, being a monument to a burial; or they could be of civil significance, marking boundaries or functioning as memorials to contracts made or military exploits. Crucially, Borlase argued that a holy site could outlive its original meaning or purpose and come to be used by civil institutions as an appropriately august location for courts, councils or official ceremonies. Where such places were found to be associated with the superstitious beliefs of the present-day peasantry, antiquarians interpreted these beliefs as vestiges of earlier, more formalised belief systems.[49] By drawing on folklore often ignored by earlier topographers and antiquarians fearful of preserving superstitious ideas alongside classical, biblical and more recent scholarly texts, Borlase's work was both at the methodological cutting edge and indicative of a Protestant confidence.[50]

Rowe's concern with logan stones, stone circles, cromlechs and rock basins, particularly basins cut into the top of logan stones, echoed Borlase. Taking his lead from Toland, Borlase argued that the pretence that only a Druid could move the logan stones meant that the 'ruffling or rest' of the purified water deposited in the basin could be used to find the truthful answer to their questions. Toland had claimed that this elicited confessions from a presumed guilty man; Borlase concluded that the sole business of these 'Arts of Magick' was to deceive in order to increase private gain and establish 'an ill-grounded Authority, by deluding the common people'.[51] Stone circles, by contrast, were dedicated to large-scale religious or festive ceremonial, including sacrifice. They provided space for rituals seeking godly intercession or knowledge of the future, and they sometimes doubled as sepulchres, with a cromlech positioned in the centre. The essential variable was size, which reflected either the importance of the ceremony or the status of the officiating Druid.

Determining the purpose of cromlechs did not cause Borlase much difficulty. He rejected Stukeley's idea that they were altars on which burned offerings were made because the Druids would have needed some kind of ladder to reach the top – not very dignified – and, according to his own experiments, the stones would have cracked under the heat. Instead, the 'quoit' or cover stone, because person-sized, suggested a cromlech probably marked the grave of an eminent druid or prince.[52] Rock basins deserved closer attention. Carved out of the upper surfaces of rocks, the presence of lips or channels indicated that they were of human design. Perfunctorily surveying a number of possible uses, Borlase settled on his greatest *idée fixe*, namely that rock basins were used to acquire the purified water central to Druidical ritual.

Nothing, he claimed, was more universal in ancient belief and custom than the idea that the soul was 'defil'd' by the impurity of the body. Ritualised 'sprinklings and washings', known as lustration, purified the body by washing away sins. Borlase maintained that every formal action in ancient society, including religious activity, was preceded by 'either total or partial Washings'. This was not, as some had claimed, 'borrowed' from the Jews, but had 'been the Customs of Mankind before the dispersion, and passed into all Countries with the first Planters of Nations'. Only the purest water would do, and this was sourced in heaven and delivered in the form of snow, rain or, best of all, dew. Thus the 'people who perform'd Sacrifice to the infernal Deities were sprinkled with dew' – 'Celestial Liquors' – collected from the rock basins fashioned for this purpose. Borlase's conclusion relied on a circular argument. The location of the rock basins 'on the tops of Hills, on the Crags, or Karns, on places which have the vestiges of every kind of Druid Superstition' indicated that they were 'subservient to the purposes of Paganism, as taught by the Druids'.[53]

Ronald Hutton observes that Borlase's work had the effect of enchanting the Cornish landscape, 'fillings its hills, heaths and moorlands with impressive vestiges of a vanished culture'.[54] Enchantment did not necessarily have positive connotations, and, Hutton adds, Borlase did much to shatter the positive image of the Druids propagated by Stukeley.[55] Indeed, Borlase met the idea that Druidical practices demonstrated the great continuity of religious life in English history with a more pessimistic and evangelical perspective. Idolatry, enforced by priestcraft, was not only a constant threat but, prior to the Reformation and owing to the weakness of man, had been a constant of life in Britain. Cornwall's and, as we shall see, Dartmoor's ancient monuments, could not be embraced, as Stukeley said of Stonehenge and Avebury, as sacred relics dedicated to true religion. Instead, these stones had witnessed the abuse of power, the manipulation of the naive and ignorant, the slaughter of innocent victims, and much frenzied drunken sex.[56]

Writing of the system of stone circles at Botallak in Cornwall (since destroyed), Borlase did not merely imagine an orgy of horror and degradation, but systematised horror and degradation perpetrated on a purpose-built multi-purpose site, to which the material resources of the community were dedicated:

> Some of these [circles] might be employ'd for the Sacrifice and some to prepare, kill, examine, and burn the Victim, others allotted to Prayer, others to the Feasting of the Priests, others for the station of those devoted to the Victims; Whilst one Druid was preparing the Victim in one Place, another was adoring in another, and describing the limits of his Temple; a Third was going his round at the extremity of another Circle of Stones; and, likely, many Druids were to follow one the other in these mysterious Rounds; Others were busy in the Rights of Augury, that so all the time, and under the inspection of the High-Priests; who, by comparing and observing the indications of the whole, might judge of the will of the Gods with greater certainty: Lastly, that these Circles intersected each other in so remarkable a manner as we find them in this Monument, might be, to intimate that each of the Holy Rites, though exercised in different Circles, and their own proper compartments, were but so many Rings, or Links, of one and the same chain, and that there was a constant dependence, and connection betwixt Sacrifice, Prayer, Holy Feeling, and all the several parts of their Worship.[57]

Richard Polwhele, born in Truro in 1760, claimed descent from Empress Matilda's chamberlain Drogo de Polwhele, on whom a grant of land was bestowed in 1140. While still at Truro School, he published two poems, one of which, 'The Fate of Llewelyn, or, the Druid's Sacrifice, a Legendary Tale' is evidence of his precocious interest in local legend and things Druidical. In 1778, he went up to Christ Church, Oxford, where he kept his terms but did not take a degree. He entered the Church of England in 1782, first taking the living of Lamorran, near Truro, where he was married, and then the curacy at Kenton in west Devon, near Powderham Castle, the seat of Viscount Courtenay. Access to the library at Powderham made the ten years he spent in Kenton his most productive: he became involved in local literary societies, and wrote poetry, anti-Methodist polemics and his Devonshire histories. In 1795 his bishop transferred him to Manaccan, near Helston in Cornwall, possibly in response to a dispute arising from his literary activities. From Manaccan he continued to write, not least reactionary letters to the *Anti-Jacobin*.[58]

Just four years separated the publication of Polwhele's two works on the history of Devonshire, evidence of what one biographer considers his 'fatal fluency of composition'.[59] Despite some overlap, they are significantly different works. The earlier of the two, *Historical Views of Devonshire*, took a mainly chronological approach, explaining the first migrations to Britain and how the mix of cultures this produced evolved into Druidism, largely as outlined by Borlase. His *History of Devonshire* was livelier, bearing a closer resemblance to a modern guidebook. Among other things, several pages were dedicated to describing the air quality of different towns and regions. On the north coast the air was salty, which strangers found 'fretting and acrimonious', though the 'keen' air more generally characteristic of the north and north-west accounted for the 'strength, activity, and longevity' of the natives. Outsiders who found this air 'too sharp', might prefer the 'balmy and salubrious' air of the south coast, which was comparable to that of the south of France (the South Hams continues to attract well heeled holidaymakers). In general, however, the coastal aspect of Devon, in contrast to more inland parts of the country, ensured that the 'atmosphere is never loaded with exhalations from bogs or marshes' because the sea breeze soon dispersed 'floating vapours'.[60]

Polwhele's plan to write the first of these histories had got around. In December 1788 he received a letter from Sir George Yonge, secretary of state for war in Pitt's government and owner of family estates at Colyton in Devon. He provided Polwhele with a steer, urging that two folios would allow for a full discussion of the waves of migration that had shaped the history of Devon. Normans, Danes, Saxons, Romans, Phoenicians, Greeks and the original British (whoever they were) should all be included. The Phoenicians, he asserted, had found the 'western parts peopled and traded with them'. Had Polwhele read Borlase? Yonge was also keen to see the antiquities of the region fully detailed in order to strengthen evidence for the primacy of the west of England in the national history. Yonge's interest in Polwhele's work marked the second time he had offered his patronage to an antiquarian writing about Devon. With Sir Robert Palk and Sir John Chichester, Yonge had taken an interest in the earlier attempt of the late Reverend S. Badcock. Such patronage was a characteristic feature of antiquarianism, not least because county histories often provided detailed lineages of property ownership. Badcock had been reluctant to accede to Yonge's requests, questioning the quality of Borlase's work on ancient Dunmonia. Polwhele would show no such qualms, and he does seem to have shared Yonge's priorities, though he did not take on his putative patron's advice that the work be produced without 'too much haste'.[61]

Following his antiquarian predecessors, Polwhele challenged the conventional view that the 'Aborigines of Britain' were Celtic settlers from neighbouring Gaul, preferring the theory that 'our primitive Colonials' were earlier émigrés from the east. Citing evidence from the ninth-century *Anglo-Saxon Chronicle*, the writings of the twelfth-century annalist Geoffrey of Monmouth and local custom, Polwhele concluded that these sources, 'distinct in themselves', all met at the same point: the earliest people in Britain were the Armenians or southern Scythians, and they had landed on the site of the town of Totnes (then on the coast owing to higher sea levels) some time after the dispersion from Babel. These were the first people of Britain, and south Devon had a good claim to have been colonised 'whilst the rest of the island was yet a desert, and even the opposite continent of Gaul and the greater part of Europe were uninhabited'. This gave the ancient Dunmonii an extraordinary position in the history of European civilisation, placing them 'among the *most ancient Nations* in the world'.[62] Like Borlase, Polwhele gave precedence to the Dunmonii over the Gauls and, by that measure, to the British over the French.

With this laboriously established, Polwhele thereafter casually referred to the Armenian Britons as the first of several pre-Roman-conquest migrations, which included, in chronological order, the Phoenicians, the Greeks and then the Gauls, often known as the Belgae. The Phoenicians, thanks to their similar geographical origins, easily assimilated to the Armenian-originating Dunmonii because their religious practices and language, manners and customs were very similar. It was the tin and lead trade that drew the Greeks to Dunmonia, and Polwhele reckoned the apparent presence of Greek words in Dunmonian was evidence that some of the Greeks must have settled. Offended by the suggestion made in John Whittaker's *Genuine History of the Britons Asserted* (1772) that when the Belgae came to Devon and Cornwall all they found was 'wild forest', Polwhele argued that the Belgae in fact forced the Dunmonii, an 'Asiatic' civilisation, from their ancient seats, causing them to retreat inland or emigrate to Ireland.[63] At first glance, this suggests that high Dartmoor became the refuge of a displaced people, ethnically distinct from the new inhabitants of the lowlands. But if Gaul had been originally colonised by the Dunmonii, then this invasion could be seen as a reverse migration.

More generally, Polwhele did not add much to Borlase's account of the Druids. He outlined their role in early British society, drawing on Tacitus to emphasise their role as lawmakers who conducted their business 'in the open air and on high places'.[64] He highlighted the role sacrifice played in their religion, asserting that 'after the Phoenician colonies had mixed with the primeval Britons, this degenerated priesthood delighted in human blood'.[65] With this sleight of hand, the Armenians, practitioners of an acceptable Druidism, become 'primeval Britons', and Phoenicians the

corrupting foreigners. More important was that Polwhele's identification of Dartmoor as a location for Druidical practice transformed the upland into a once-sanctified landscape. This would
shape how Dartmoor was encountered for decades to come,
bequeathing a set of easily accessible canonical locations, namely
Grimspound, Bowerman's Nose, Crockern Tor and, above all,
Drewsteignton – 'the town of the *Druids, upon the Teign*', according
to Polwhele's etymology. Once these sites were understood, the
visitor could understand the myriad ancient relics she might
encounter elsewhere on the moor.

Grimspound, in the parish of Manaton and overlooked by King
Tor, Hookney Tor and Hameldon Tor, can be easily accessed by a
path linking enclosed land to the road running south from the
B3212 towards Challacombe Down. If scale be the sole measure of
importance, Grimspound is Dartmoor's most significant prehistoric
site.

Having argued that ancient Dartmoor was divided into four
districts or *cantreds*, Polwhele happily identified Grimspound as both
the judicial seat of Durius, an ancient name for the Dart, and 'one
of the principal temples of the Druids'.[66] He attached a slightly
different significance to Crockern Tor. At the time Polwhele was
writing, the lord warden of the stannaries could still summon to
Crockern Tor representatives of the subordinate stannary courts
regularly held at Ashburton, Chagford, Tavistock and Plympton. Up
in that exposed spot the stannary law could be set, tin works and
mills registered, petitions heard and penalties imposed. Known as
the parliament of the moor, Polwhele regarded Crockern Tor as 'a
strange place for holding a meeting' given its exposure 'to all the
severities of the weather'. Lately, he noted, because it was 'too wild
and dreary a place', only symbolic gatherings had been held there
before the participants retired to one of the more comfortably
located courts in the low country. Crockern Tor's status, Polwhele
believed, reflected the 'peculiar sanctity' that had been attached to
the site 'from the earliest antiquity', when it had been the seat of

the judicature for the ancient *cantred* of Tamara. Its role in more recent times reflected an inherited respect for the site, the origins of which were long forgotten.[67] This was a classic example of antiquarian reasoning. Present-day customary practice, otherwise inexplicable, could be explained as a vestige of earlier practices. Polwhele's claims also served a broadly nationalistic purpose, transforming the ancient rocks of the moor into symbols of the continuity between antiquity and the present of British civic life, a powerful lesson in the midst of the French Revolution.

By the early nineteenth century, evidence suggests the moorstone tablets and seats of Crockern Tor had been destroyed.[68]

If, as Borlase had claimed, the Druids believed that their divinities resided in natural rock formations – the rock idols – then, as Polwhele caustically observed, Dartmoor would be 'one wide Druid temple', its 'dark waste' transformed into 'consecrated ground'. A readiness to bow before the 'granitical god' on Hayne Down, which could be taken for granted, would logically generate a similar urge to submit before any number of other tors. This circular argument stemmed from the belief that the natural characteristics of the moor were prone to generate in people 'the effect of enchantment'. If 'Druidical scenery', as Polwhele opined, 'inspires even in the cultivated mind . . . a sort of religious terror' – Burke again – it was possible to imagine the effect it had on 'superstitious minds under the direction of the Druids'.[69] Like Toland and Borlase, but again in contrast to Stukeley, Polwhele's Druids were calculating and manipulative, persuading their followers that the natural 'rude grandeur' of the moors was somehow their creation.

By this reading, Dartmoor's most affecting landscapes were not the open land of the high moors, all rolling hills and craggy tors, but the 'fantastic scenery' of the east Dartmoor river valleys, most particularly the Teign Valley. Drewsteignton, lying just to the north of the valley, was in an area peppered with 'many Druidical vestiges', the most important being the cromlech at Shilstone Farm. It was 'placed on an elevated spot . . . overlooking a sacred way', as marked out by since-destroyed stone pillars and circles running parallel to the valley. Rowe, as we have seen, reproduced these exact ideas half a century later. Polwhele maintained that it was the valley's natural drama – its wooded sides plummet to the river below – which attracted Druids to the area. He elaborated his thesis by speculating about Dean Burn, a river valley some distance south of Drewsteignton and south-west of Buckfastleigh. Though comparatively modest in scale, Polwhele thought it 'united the terrible and the graceful in so striking a manner, that to enter this recess hath the effect of enchantment'. The Druids, he speculated, ever alert to the potential of an affecting natural landscape, surely exploited the sensations provoked by the 'enormous rocks', 'the deep foliage of venerable trees' and the 'roar of torrents'.[70] By so turning from the Teign valley to Dean Burn, Polwhele made a significant move in his argument. Grimspound, Bowerman's Nose, Crockern Tor and Drewsteignton, sites whose significance was apparently irrefutable, were treated as ideal types whose meanings could be applied to innumerable other sites on the moor. Given that Dunmonia was the first part of the island to be colonised, the ancient monuments and the most striking features of Dartmoor now took on a tremendous national significance.

Polwhele had more to say about Dartmoor. Air quality interested him. Moretonhampton, for instance, was known to be 'particularly healthful' because, sheltered by surrounding hills, 'it hath the dry-air of the moor without its keenness'. More generally, Dartmoor was prone to 'remarkably thick and sudden fogs' (true) and high rainfall (all too true). Also interesting was the quality of the soil, which deteriorated the further the moor was penetrated, the tors (some were 'like so many exhausted volcanoes, surrounded with the stones which they have formerly emitted, whilst the more ponderous masses remain on their summits as too heavy to be thrown off') and the different types of granite which formed the moor (he distinguished granite, moorstone, whinstone, and red granite).[71] More intriguing were his comments on Dartmoor's 'mineral evaporations', as exemplified by the 'vast number of luminous appearances' to be seen at night 'dancing in the goyles and hollows of the down' below Haldon Hill. Similar emanations could be seen at night in his home parish of Kenton, when 'globules about a foot in diameter, exceedingly light, resembling fleeces of wool' would rise about five feet from the earth before falling back and then rebounding: the 'exhalation seemed too feeble to ascend far into the atmosphere'.[72] Such exhalations, here scientifically accounted for, fed other imaginations.

Much of this learning provided raw material for Polwhele's *Dartmoor; a poem*. It mixed topographical description, historical knowledge and didactic commentary, peppering elaborate descriptions of the landscape and the weather with esoteric allusions to moorland myth and legend. Aware that what particularised the scene for the knowledgeable reader can only have obscured it for others, Polwhele appended notes at the foot of the page. These

referred readers to his own and other works, or they explained his
poetic intention, telling the reader what they ought to find
happening in a particular passage. This academic apparatus reflected
his awareness that he was engaged in an attempt to make Dartmoor
a suitable subject for poetry. By repackaging his earlier work, the
poetic form allowed the same knowledge to convey different meanings. His erudition gave the landscape meaning, creating a
mythology worthy of poetic sublimation.

Polwhele divided his poem into three parts. The first personified
Dartmoor's natural characteristics, suggesting that it was perpetu-
ally caught in a struggle between fiendish tempers and an angelic
sweetness. The second part blended historical narrative and topographical description with commentary on the moral dignity of
the moor's inhabitants. The third part, much the shortest, was a
dedication to the Crown and a declaration of Polwhele's hopes for
Dartmoor's future.

From the outset, the moor – or the 'Highland' – is characterised
as a body, with a 'monumental head', 'granite ribs' and 'stern
brows'. It 'scowls' on the 'little hills and valleys' that surround it.
The reader is placed in those lowlands and needs to strain her eyes
to see much beyond the moor's 'sullen sweep' with its 'shaggy
sides'. What kind of a hybrid monster is this? Making the effort to
see at 'close of day' is to be rewarded by the moor transformed,
the sunset reddening its summits, drawing from the 'scatter'd heath-
flowers' a 'yellower gleam', from the 'reedy stream' a 'pale radiance',
the 'glimmering specks' of distant crags and cairns turned crimson
and gold. Night comes soon enough, the darkness broken only by
the moon's 'track of lurid light' and soon after the chill signs of a
gathering storm. This strikes a strong Gothic note, suggesting that
Polwhele was au fait with the latest fashionable reading.

Leaves rustle; unaccountable 'charnel' sounds affect the 'Fancy':
the hailstorm breaks, and the imagination gets to work. Riding the
storm, the poet sees one personification of the moor, a 'heavenly
maid / Pavlion'd on the bosom of the air!' Her brow is open, her

hair is bright, her limbs graceful, 'her blue eyes ineffably benign'; she is a symbol of 'Courtesy', 'Candour', 'chaste Desire', 'Patience', 'ardent Faith' and 'Love of human kind'. Her alter ego, the 'Forest-fiend', cannot accept this intrusion. Shaking his 'pyritic robe', he raises his crest and demands, 'Who breaks upon my sacred solitude?' Before an answer can be heard, the fiend commences a long raging monologue on the nature of his being and the nature of the moor itself. His crown is hailstone, his sceptre flame, the moor a place of sulphurous sterility, of 'the stony root' and the 'sapless branch', of 'venom'd plants' and 'unpeopled wastes'. Dartmoor is malevolently active, luring the 'bewilder'd shepherd' into the wilderness, turning the wrathful bison on the hunter, sending birds of prey against the infant. The fiend's rhetoric seems inexhaustible, but soon the sweet harmonies of the maid soothe the scene. 'Slowly the troubled Phantom sank away: / His cloud dissolving, in a thin white fleece.'

The fiend's raging develops an elaborate picture of the moor as a Gothic landscape where nothing good or life-giving thrives. But it takes just a few lines for the pacifying maid to see off the fiend, restoring peace. Almost all writing about Dartmoor from the eighteenth century through to today takes as a theme the rapidity with which violent storms can batter the landscape, turning a landscape of peace and beauty into one of horror and threat.

Polwhele's footnotes are distancing, reminding the reader that his poem is not merely a work of the imagination but a careful construct assembled from a variety of sources. As a display of knowledge, it is an assertion of authority or power over the landscape. The titanic battle, at best distantly seen by lowlanders, but sometimes no more than intimated by a sudden mysterious surging of rivers whose origins lie in the highlands, is something the poet knows. Polwhele's *Dartmoor* might not be great poetry, but it does appear to be driven by a key insight: as he looks on the moor, or imagines looking, the host of associations that enrich his impressions suggest to him that the moor, like any landscape, has no pure or unmediated meaning.

In the second and the longest section of the poem Polwhele displayed this knowledge fully. It begins by meditating on Dartmoor's creation, a work of 'Nature', celebrating the moment of its first spring when 'Primordial Beauty breath'd, and shap'd the whole.' Polwhele gave abstract forces like Nature and, later, Ambition, Guilt, Fear, Caprice and Piety the power to actively intervene in the life of the moor and the ways of its inhabitants. People, though, also have agency, and Polwhele's brief description of Druidism, alluding to its eastern origins, condensed into a few lines of verse the themes of his historical work. Dartmoor's dramatic landscape is again taken to condition a credulous people, making them susceptible to Druidic manipulations of spiritual truth. Consequently, the 'Historic Muse', a kind of class monitor, 'aghast' at the barbaric rituals conducted on the cromlechs or legitimised by the manipulation of logan stones, is happy when under Roman influence the sites of religious practice are transferred to a 'milder scene', the Dartmoor lowlands. Civilised practices have a better chance in a pastoral setting. Coming down from the moor made paganism's days numbered. Piety's cleansing and purification of the pagan fane was Christianisation itself. Predictably, Polwhele took a potshot at Catholicism, saying that piety was 'soon by pomp obscur'd, by pleasure sunk / She waver'd in the priest, nor warm'd the monk'. In spite of this, the ruins of Tavistock Abbey, destroyed during the Reformation, are mournfully described over ten lines, Polwhele's description of the architectural details generating an idealised medieval stage set for knights now reposed in their tombs, fair damsels and 'crosier'd abbots knelt in prayer'.

With the Reformation efficiently arrived at, Polwhele leaves the lowlands behind him and returns to the moors, now transformed into a place of work and speculation. Miners blast 'new-born treasures' from the rock, but from 'mineral commerce . . . contentions sprang' and the 'hallow'd tors', where once the 'death-doom'd' victims of the Druids met their fate, now become the stannary courts. 'Calm review' led to the development of case law, and the

successive statutes issued by the courts developed into a well regulated government for the moor. Expeditious justice, however, was
a danger, and a footnote informed the inattentive reader that the
line 'dungeon-damps more noisome than his mine' refers to the
notorious hang-first-and-try-later practices of early-modern 'Lydford
Law', named after the eponymous west Devon village and castle
where its unfortunate victims were tried and punished.

Having set a galloping pace, Polwhele moves swiftly on to the
depopulation of the moor, linking it to the decline of hunting and
shepherding, thereby distinguishing an idealised medieval past in
which man was at one with his landscape from the commercial
speculation of the early nineteenth-century present. He alluded to
the Childe of Plymstock story, later one of the most repeated
Dartmoor legends,[73] and wrote sentimentally that although the
shepherd possessed neither land nor comfortable home he could
make a living on the commons, 'Blest in his faithful dog'. The bulls
and boar are gone; the shepherds are now a rare sight, leaving
behind only scattered evidence of their previous existence. As a
consequence, and despite the pockets of mining and speculation,
visitors to the moor must confront their own atavistic responses
to the sublime: finding 'no welcome in the human face' the visitor
finds that 'busy Fancy hath supplied its place'. The natural environment, banished from the mind by modern urban life, renews its
tenancy, taking possession of the imagination: 'beings rise at will',
the imagination's 'wanton spawn'. In the absence of a human presence, the natural environment seems peculiarly alive: the 'quick
bogs quiver', 'abysses yawn', and 'the ghost clanks his chain'. Amid
all this seething activity, Oberon, king of the fairies, 'summons up
his gentler fays; / And frolicksome, they frisk in airy ranks, /
Though mischief, as of old, still prompts their pranks.'[74]

Pixie lore is the folklore most characteristic of Dartmoor, and
much Dartmoor lore tells of travellers on the moor being led astray.
The spell of the pixie can only be broken if the quick-witted,
understanding their ways, come upon an adder skin. Polwhele, of

course, sees all this as superstition and is amused that the tin miners, otherwise so knowledgeable and intelligent, should so readily credit these tales. His brief pixie discursive allows Polwhele to remind the reader of the ways the natural moorland landscape exercises the human imagination, undermining its rational faculty, whether this takes the form of a susceptibility to Druidic religion or folkloric superstition. Polwhele believed Dartmoor exposed the fragility of Enlightenment claims to reason and rationality, the moor's natural characteristics confronting man with his natural state, wherein the imagination and reason jostled for ascendancy, the latter never fully suppressing the former. If so, the Gothic, Romantic and primitivist motifs and ideas shaping his poem remind us that Polwhele's narrator remained in a privileged position, maintaining distance from the fanciful notions of the imagination. Dartmoor could evoke sublime passions, but the modern antiquarian, though determined to make poetry of his knowledge, retained a knowingness that protected his late-eighteenth-century sense of self from the unhindered workings of his imagination. The delicious pleasure that could be taken in these imaginings should not obscure Polwhele's decision to footnote them.

Romantic Reverie and Protestant Purpose

Polwhele's antiquarian and topographical work was not always well received. A scathing review of the *History of Devonshire*, probably written by the novelist and critic Tobias Smollett, appeared in September 1799. Smollett took Polwhele to task for the quality of his classical scholarship, his haste in execution, his inconsistencies, his readiness to see natural phenomena like the rock basins as man-made, his aggrandising of the Dunmonii within the history of

European civilisation and, above all, his theories on the colonisation of the south-west, 'so frivolous an hypothesis so pertinaciously supported'.[75] An anonymous biographer, writing a few years later, was more temperate in expression but equally unconvinced 'that Danmonium was originally colonised by adventurers from the east': 'The notion was worked up with no small degree of ingenuity, and with that glow of colouring which a poetical imagination can lay on in such a manner as to convince those of its truth, who are more apt to pronounce according to their feelings than their judgement.'[76]

In the 1790s Druidical theories were becoming unfashionable, something borne out by the reception of Polwhele's work in metropolitan learned societies.[77] But in the world of popular Dartmoor topography, antiquarianism and travel writing, the Druids were just coming into their own. In a travelogue of 1800 Richard Warner, another Anglican clergyman, associated Druidism with Catholicism and wrote that Dartmoor was an 'apt spot for the exhibition of priestcraft and the celebration of Druidical rites in times of yore': Dartmoor's natural characteristics once again.[78] Three years later the Lambeth topographers John Britton and Edward Wedlake Brayley published the latest of what would grow to be their highly influential nine-volume series, the Beauties of England and Wales. This fourth volume covered Devon and Cornwall and included a substantial section on the 'Dartmoor Mountains'. For its frontispiece, they took a Gothic engraving of the Drewsteignton Cromlech, and their discussion further enhanced the cromlech's emblematic status, heavily investing their account of the high moors with Druidical meanings derived from Polwhele. By offering in counterpoint the pastoral settings of east Dartmoor villages and the landed estates with their extensive parklands, the high moors were distinguished from the lowlands.[79]

If Polwhele's poem was a pioneering attempt to make the moor a proper subject for verse, when Nicholas Carrington came to write Dartmoor: a Descriptive Poem in the 1820s it was still doubtful whether that had been achieved. An unexpected success, the poem was published in a second edition in 1826 with an introduction by the author's nephew H. E. Carrington and extensive explanatory notes by W. Burt. The introduction began by observing that 'Dartmoor is generally imagined to be a region wholly unfit for the purposes of poetry,' before explaining, on the basis of its 'romantic solitude', why it should not be so thought.[80] And yet, just a few years before, as a note in the second edition reminded the reader, the Royal Society of Literature had offered a fifty-guinea prize 'for the best poetical effusion on Dartmoor'.[81] Mrs Felicia Hemans took the prize for her poem *Dartmoor*. Carrington went on to attract an even higher indication of esteem, as *The Times* explained in October 1834: 'A hitherto unrecorded instance of the patronage of which George IV occasionally afforded to genius has just come forth in the memoir of the late N. T. Carrington, the author of the poem of *Dartmoor*. When it first appeared, in 1826, the monarch ordered his opinion of the poem to be transmitted to the author in the shape of a present of 50 guineas.'[82]

Hemans, 'of celebrity in the Literary world', was best known for 'Casabianca' (1826), which opens with the famous line 'The boy stood on the burning deck' and celebrates its protagonist's refusal to abandon his post until ordered to do so. Carrington, by contrast, was a provincial schoolmaster. Keenly observing that his poem had *not* been entered into the Royal Society competition, Carrington (or his editors) generously described Hemans' *Dartmoor* as 'truly beautiful'. Perhaps it was, though it is easy to see why Carrington rather than Hemans has attracted a following among Dartmoor enthusiasts.

Hemans' poem is relatively undistinguished, being described together with her *Wallace* by one sympathetic contemporary 'as the exercises, rather than the effusions, of a mind as distrustful of its own power, as it was filled almost to overflowing'.[83] It had a very weak sense of place; its description and evocation of the moor was highly generalised, with many of the lines given over to a characteristic moralising that was only tenuously provoked by Dartmoor's particularities. Had Hemans even ventured onto the moor? Carrington's *Dartmoor* strikes a more affecting balance between this tendency to pontificate with a precise evocation of place: where Hemans' descriptions of the moor are generalised, Carrington's are tied to a journey by foot across the moor that can be fairly precisely mapped; it was perhaps the earliest and certainly one of the most influential literary Dartmoor perambulations. What these two poems have in common and how they contrast with Polwhele's poem is more significant than any comparison of their respective strengths as poetry. Both began by quoting a few lines of Wordsworth – it was scarcely possible to write landscape poetry without feeling his influence or submitting to his authority – and both treated Dartmoor as a Druidical landscape. But if Polwhele's *Dartmoor* had left the Druids indelibly associated with the moor, or so it seemed in the 1820s, Carrington and Hemans encountered the moor through new ways of seeing and experience.

New was the emphasis both poets placed on two distinct but interrelated themes. First was what early-nineteenth-century social and economic thinkers described as 'improvement', namely the impact human intervention, deploying new scientific methods and new thinking, could have on the economic productiveness of apparently unexploited landscapes and the culture and civilisation of its inhabitants – very often, 'improvers' set about changing how livelihoods were already extracted from the land. Hemans' poem is prefaced with lines from Thomas Campbell's long poem of 1799 'Pleasures of Hope': 'Come bright Improvement, on the ear of Time, / And rule the spacious world from clime to clime!'

Both poets – Hemans more obliquely than Carrington – foresaw a radically different Dartmoor, a Dartmoor conquered by industrial civilisation. Hemans located this within a strongly post-Napoleonic context, and her optimism stemmed from her sense that she was writing in a period of peace following a long Europe-wide conflict; Carrington, living closer to the sharp end of an emergent industrial modernity, was less emphatic, though he too paid due obeisance to the optimistic utilitarianism of his time. Both, however, recognised that Dartmoor's particular experience of improvement had a troubling aspect. Captured French and American soldiers, from the Napoleonic Wars and War of 1812 respectively, had been held at a depot at Princetown. In pondering their plight, Hemans and Carrington both contrasted Dartmoor with idyllic ideas of the French countryside, in Hemans' case with 'the vine-clad hills' of 'lovely southern climes' accompanied by the 'festal melody of Loire or Seine'.[84]

In the 1826 edition of Carrington's poem, H. E. Carrington, the poet's nephew, wrote the introduction. Like Rowe – and it is perhaps from here that Rowe picked up this approach – H. E. Carrington argued that Dartmoor should excite the interest of a range of specialists:

> The poet will picture to himself the remains of sublunary grandeur – will muse amid fallen columns and shattered arches, and sigh over the by-gone renown of fabrics which have passed away from the face of the earth, even as the flower that withereth; the antiquary will image ruined castles lifting high their tottering turrets, and crumbling abbeys with their wind-swept aisles and mouldering cloisters; or he will recognise the relics of a remoter age in the semblance of moss-grown cromlechs and druidical monuments; – the moralist, in contemplating the rude scene, will be reminded of the awful wrecks of human ambition; – and the misanthrope will exult in the solitude of spots where he may indulge his gloomy imaginings undisturbed.[85]

Something for everyone then, but the clichéd language seems knowing, the final reference to the misanthrope indicative of the thread of irony that runs through the whole. In 1812 the cartoonist Thomas Rowlandson had famously satirised Dr Syntax's search for the picturesque; fourteen years later, H. E. Carrington demonstrated his knowledge of the moor and, more pointedly, his familiarity with the intellectual trends of his time. Predicting how the outsider might encounter the moor was territorial, a way of asserting control over knowledge of the moor. This anticipatory resentment suggested Carrington feared that those attracted to the moor by his uncle's poem might then make it *their* subject, challenging the authority of his local, intimate knowledge. With the prospect of publication, the self-appointed gatekeeper was reaching the prickly realisation that no gate actually existed.

H. E. Carrington, as befits one faced with this dilemma, could not help but be an evangelist too. He believed that the moor could be experienced 'in its utmost magnificence . . . when the wintry gale is fiercely howling around'.[86] Echoes of Polwhele's 'Forest-fiend' and Burke's sublime are heard in his claim that 'he who dares, at that fearful hour, to confront the angry spirits of the storm in this their hereditary and undisputed strong-hold, will be amply repaid for his perils by the feelings of grandeur and sublimity which will then steal into his mind.'[87] Further Burkean echoes resound in passages referring to 'alpine grandeur', 'the struggle between barrenness and fertility', and the 'inexpressible charm of freshness and untamed beauty connected with this species of landscape which it would be in vain to seek among the gentler haunts of cultiva-tion'. H. E. Carrington observed helpfully that this 'struggle' was 'a very interesting study for the lover of the picturesque' and then proceeded to demonstrate his own descriptive expertise: 'The heath-flower is blended with the honeysuckle – the fern with the fox-glove – and the dwarf oak, with its hardy boughs and stunted foliage, droops over the purple violet or the meek blossoms of the wild strawberry.'[88]

He continued in this mode, unintentionally leaving evidence of the moorscape's ecology, elaborating his description of the moor's contrasts with the idea that 'Nature, with her frolic breath' had delighted in blowing autumn's 'ripened seeds . . . into the most inaccessible and difficult situations'. Aware that visitors came to

Dartmoor seeking Romantic sensation, he manipulated these ideas, observing that were his readers to make their 'rugged pilgrimage' undaunted by a 'few obstacles', they would worthily pass 'into the sanctuaries of Nature'. A successful pilgrim would find himself 'in the midst of as wild a combination of natural objects as was ever pictured by the most romantic imagination'. Though this slightly arch observation identifying Dartmoor as real satirised those who had nurtured Romantic sensibilities in the drawing room, it was located within a descriptive passage tracing the walk from Shaugh Bridge to Sheepstor that developed into a seductive prose poem in which Romantic effects were reached for. Replete with closely observed and precise descriptions of flora and fauna, it culminated in his arrival at Sheepstor church. Described as 'one of those quaint specimens of ancient architecture' – the church dates back to 1450 – 'met with in secluded situations unmarred by the hand of modern improvement', H. E. Carrington described how the setting sun of a 'calm autumnal evening' touches the 'massy buttresses and crumbling carved work' with a 'mellow light', and a 'feeling of religious placidity then pervades every thing round the old building'.[89] This contemplative mode, in which religious faith is restored by a sensibility for man's proper place in the natural world, recalled Wordsworth's 'Tintern Abbey' (1798), the ur-Romantic poem of place. As such, H. E. Carrington offered a Dartmoor at odds with Polwhele's 'wide Druid temple', finding instead a place of ancient Christian life, a landscape that could be encountered through its granite churches. Granite, the near-imperishable stuff of the rock idol, was also the stuff of the church tower, an older form of improvement.

It might seem strange that the route H.E. Carrington described in such detail in his introduction was in large part that delineated in his uncle's poem, but this simply reminds us that early visitors to the moor, particularly those travelling by foot, were strongly delimited by their starting point and the distance it was possible to cover in a day. Polwhele gave a Dartmoor accessible from Kenton, so the Carringtons wrote a Dartmoor from the perspective of a pedestrian setting out from Saltram, the village just east of the mouth of the Plym and over the water from the then modest town of Plymouth.

Carrington senior's description of 'Devonia's dreary Alps' touched on some familiar places, themes and ideas: there were the 'warrior-dead' of the cairns, the lark's resounding 'voice of joy' and a passage on the way the imagination – 'Fancy' – transformed natural rock formations into man-made ruins; there was Crockern Tor, identified as 'the senate of the Moor', now revered by legislators 'nursed / In the lap of modern luxury'; there were the 'Rough traces' thought to be evidence left by the 'fierce Danmonii', claimed by Carrington as his 'brave fore-fathers'; there was 'the solitary wreck' of Wistman's Wood, the 'Fierce, frequent, sudden' nature of the moorland storm, and references to the Drewsteignton Cromlech and logan stone.[90] Thoughts of Bowerman's Nose at Manaton prompted a slightly awkward passage on the Druids that suggested Carrington had thoroughly absorbed the teaching of Borlase and Polwhele:

> The frantic seer
> Here built his sacred circle; for he loved
> To worship on the mountain's breast sublime—
> The earth his altar, and the bending heaven
> His canopy magnificent. The rocks
> That crest the grown-crowned hill he scoop'd to hold
> The Lustral Waters; and to wondering crowds
> And ignorant, with guileful hand he rock'd
> The yielding Logan. Practised to deceive,
> Himself deceived, he sway'd the fear-struck throng
> By craftiest stratagems; and (falsely deem'd
> The minister of Heaven) with bloodiest rites
> He awed the prostrate isle, and held the mind
> From age to age in Superstition's spells.[91]

In addition to commenting on agricultural and industrial improvement and the plight of the French POWs, Carrington peppered his poem with glancing allusions to Dartmoor legends and demonstrated his expertise as a naturalist and topographer. Much of the first half of the poem and its emotive close were dominated by his subjective experience of a day on the moor; much of the second half was structured by journeys taken across the moor not by the pedestrian but by its rivers. Carrington began by taking the reader to the 'urn of Cranmere' or Cranmere Pool, a fairly inaccessible site in the northern part of the moor that has an almost mystical significance owing to its reputation as the source of the Dart and its proximity to the sources of the Okement, Taw, Tavy and Teign. Striving for a more objective tone, though still punctuating his verse with exultant outbursts (Dartmoor now becomes 'Thou land of streams!'), Carrington separately described the routes of seven Dartmoor rivers: the Teign, the Dart, the Tavy and the Lyd, originating in the north moor, the Plym and the Erme in the south, and the Yealm. In so doing, he placed himself within a literary tradition whose greatest exemplar was Edmund Spenser, another way to increase the value of the moorscape through art.

Throughout the poem Carrington commented upon the significant effects the natural characteristics of the moor could induce, including, as already encountered in his nephew's introduction, purified religious sentiment. Associated with this religious theme was the recurring idea that Dartmoor was a place of freedom and liberty. Beginning his poem with a hymn to the seasonal beauties of Devon, he comments on the 'new-born' impulses that come with spring, as signalled by a bird's 'touching lay of liberty' joyous for those who 'can stray / At will, unshackled by the galling chain / That Fate has forged for Labour's countless sons'. Carrington had only fleeting opportunities 'to taste / How sweet a thing is liberty' before facing 'The hated bonds again.'[92] Freedom was also manifested in the poem as time to read and study (self-improvement), or, as he put it, time to 'Invoke the muse, commune with ages past, / And feast on all the luxury of books'.[93] Carrington's allusiveness was evidence of his erudition where things Dartmoor was concerned, testimony to a little freedom well spent.

Given that the political charge carried by Carrington's *Dartmoor* stemmed from the contrast repeatedly drawn between liberty/ nature and bondage/work, much of the power of his exultantly described perambulation came from the narrative arc outlining a holiday – a single day free from work – spent on the moor. Carrington's decision to spend his holiday hiking up the Plym valley from Saltram, taking in 'Bickleigh's vale romantic', the 'windswept ridge' at Walkhampton, the disused POW depot at Princetown, Tor Royal's 'guardian fence' and 'piny grove' (evidence of improvement), Crockern Tor (reached by noon – it must have been an early start), before arriving at Wistman's Wood, reflected the yearning he had felt since he was a child for this 'wild and wond'rous region' (the most quoted phrase from the poem) 'dimly seen, and priz'd / For being distant – and untrod'. As a child, he had hung upon the speech of 'the half-savage peasant . . . To hear of rock-crown'd heights on which cloud / For ever rests; and wilds stupendous, swept / By mightiest storms' (and so on for a further fifteen lines). As an adult, signs of the moor – like a fresh breeze – were more ambiguous, intimating both the place of freedom hovering distantly to the north and the manacles of work and family that kept him from it.[94]

As such, Carrington's *Dartmoor* meditated on a yearning that the author had carried from childhood into adulthood, imposing on Dartmoor meanings that reflected the plight of a hard-pressed, well educated, lower-middle-class schoolmaster and literary enthusiast. The link posited between liberty and freedom of movement was unambiguous, and the reader was reminded of this throughout the poem. Phrases and words that crop up consecutively over its course include: 'free to rove', 'to tread', 'the rural walk', 'I stray', 'I wander', 'I linger', 'I wandered', 'trod', 'to tread', 'Let me stray', 'the wanderer', 'To saunter', 'I rove', 'I leave', 'With a pensive step', 'I ascend', 'pursues the varied way!', 'I haste', 'as I musing stray', 'I pass', 'I tread', 'vigorous step', and 'stray'.[95] There is also a hint that his movement across the moor was restricted by private ownership: addressing Holne Chase in the Dart river valley, he lamented, 'O that my feet / Were free, Holne Chase, to linger in thy depths'.[96]

Whereas his nephew might ironise encounters with the moor, Carrington wrote with unambiguous sincerity of how

> Here might the man, disgusted with the world,
> Retire, and commune with himself, with God,
> And Nature; – here, unsought, unvex'd, unknown,
> The stern contemner of his race might dwell
> With mountain independence ever; – lord
> Of the lone desert. On his musings free,
> Nor speech impertinent, nor harsh command,
> Nor frivolous talk, consuming half the hours
> Of hollow artificial life; nor hum
> Of populous cities, nor the deafening din
> And shout of savage war, would e'er intrude:—
> But he would live with liberty, and climb
> With vigorous step the heathy ridge, or win
> The vale, and list with rapture to the sounds
> Delightful, that can bless e'en steep and stream
> Of the scorn'd forest.[97]

If the desire to 'commune with nature' now seems clichéd, it is perhaps the form of expression rather than the desire itself that is outmoded. Moreover, if in Britain the landscape traditionally associated with liberty was the greenwood, the demesne of Robin Hood,[98] this Wordsworthian emphasis on the liberties of the mountains was indicative of the ways a European landscape – the Alps – was now associated with a quality – liberty – thought quintessentially English. Indeed, Tim Fulford argues that James Thomson, the early eighteenth-century Scottish poet famous for his 'loco-descriptive verse'

The Seasons (1730), writing against a background of political corruption, came to see in wild landscapes the 'source of a native British
freedom'. Revising his work in the light of large-scale enclosures,
Thomson turned away from Britain and found in Lapland and Russia
landscapes that could nurture this same independence and liberty.
Intensely disillusioned with eighteenth-century progress, Thomson
came to regard Britishness, in its essentials, as confined to the uncultivated rural margins.[99] Coleridge, in the first decades of the nineteenth century, turned to similar landscapes for not dissimilar
reasons. 'Commonland, unmarked by fences that enclosed landlords'
fields,' explains Fulford, 'became a poetic refuge, retirement to which
allowed Coleridge enough freedom from political pressure and social
division to speak for a landscape.'[100] What Thomson found in the
wild margins, Coleridge on the commons and Wordsworth in the
Lake District, Carrington found amid the unenclosed 'Dartmoor
mountains'. Such was the symbolic importance Carrington attached
to the moor, he could peremptorily distinguish between those who
might rightfully sup deep of its restorative powers, and the uninitiated, replete with all their negative qualities of modernity, who
should stay away. Writing of one hidden idyllic spot, a place where
the walker might rest, he instructed:

> Let no heedless step
> Intrude profanely, – let the worldling rest
> In his own noisy world; – far off, – the vale
> Is not for him: but he that loves to pay
> His silent adorations where, supreme
> In beauty, Nature sits, may spend the hour
> Of holiest rapture here.[101]

This can be read literally. Carrington thought the trappings of
man-made modernity – its noise and rush – degenerating, diminishing the person's capacity to recognise in nature God's creation.
To commune with nature took one closer to God.

In a seminal work Raymond Williams traced the subtle shifts in attitudes towards the landscape evident in the evolving 'structures of feeling' manifested in prose and poetry particularly through the enduring contrast between the country and the city.[102] Carrington's work can be understood in a similar way, for he gave expression to a series of familiar urban anxieties, not least the notion that a more authentic mode of existence or being required the repudiation of urban life. It need hardly be noted that this powerfully enduring idea has been vigorously contested by those who take the diametrically opposing view. Nonetheless, modern-day encounters with Dartmoor, at least by outsiders, remain structured by oppositions that would have been familiar to Carrington: rural and urban, tradition and modernity, silence and noise, leisure and work, freedom and confinement: Carrington even mentioned 'cities and their suburbs foul'. A little historical imagination might be needed if we are to accept that Carrington could feel this as intensely as we might, despite the enormous difference in urban population density between then and now. Change, though, there has been, and what must be taken from Williams and other critics is the need to treat Carrington's work with contextual sensitivity, thinking about those aspects of his poem that were distinctly of the early nineteenth century.

Dissecting Carrington's romanticism in terms of oppositions should not be at the expense of an appreciation of the emotional and psychological aspects of his poem. Integral to his literary rebellion against various kinds of urban and familial confinement was the relationship between physical sensation, loneliness and a joyful mood. The poem suggests this was induced by the combination of nostalgia for his childhood imaginings, the temporary satisfaction of adult yearnings and the exhilaration that comes with physical exertion. So powerfully linked was the moor to Carrington's state of mind that he claimed an ability to overcome bleak moments by summoning up memories of his moorland walks. It is tempting to medicalise this relationship to the moor, treating it as a form of dependency, reading this introverted landscape poetry as though spoken from the psychiatrist's couch.

Eliza Bray, writing from a very different class perspective, brought few of these anxieties to her buoyant 1838 work *Traditions, Legends, Superstitions, and Sketches of Devonshire on the Borders of the Tamar and the Tavy, illustrative of its Manners, Customs, History, Antiquities, Scenery and Natural History*. Bray had set herself a wide-ranging task and, as nineteenth-century standards demanded, she spared neither the forests their wood, the printers their ink or her readers their time: her three volumes amassing over 1,000 pages. Thankfully, the book was equipped with an extensive index and – again, according to conventions of the day – each chapter was headed, in small print, with a point-by-point description of its content. These encyclopedic ambitions should not belie how fluent and engaging Bray could be, not least thanks to her prejudices, and how unintentionally hilarious. Like Warner's book, Bray's was composed of letters, in this case sixty-two addressed to Robert Southey, the poet laureate. Oddly, this correspondence is little noticed by Southey's biographers, though the extent and frank tone of his sixty-one replies – copies are held by the British Library – suggest Bray was among his more significant correspondents in the last decade of his life. Indeed, on his last journey in 1836–7, he spent Christmas with the Brays, commenting wistfully in a letter to Katherine Southey that it was the only house in which he had eaten off pewter since he was a child. The weather kept him from the moor, and he thought the snow did not show Tavistock at its best, though he could imagine that under better circumstances it might resemble Cumberland.[103] Bray continued a correspondence with Southey's widow until her death fourteen years later.[104]

Bray's encounter with Dartmoor was strongly shaped by her Anglicanism. This is not surprising given that she came to national attention when her novel *The Protestant* (1828), published at the height of the campaign emanating from Ireland for the repeal of the remaining eighteenth-century laws restricting Catholic life in Britain and Ireland, was accused of pandering to anti-Catholic sentiment. In her autobiography, published a year after her death in 1884, she recalled being accused of 'labouring to excite the Protestants to persecute, and, if possible, to burn the Roman Catholics'. An exaggeration perhaps, though there was no question Bray was fiercely anti-Catholic. Urged to send a copy of *The Protestant* to the anonymous author of a hostile essay on emancipation in the *Quarterly Review*, she found herself acquainted with Southey, who approved of her work.[105]

Although Southey's letters to Bray contained only sparse comment on Dartmoor, he was impressed by her and her husband's antiquarian research. Bray's book, Southey was sure, would 'bring many visitors to Dartmoor'.[106] Given Mr Bray's interest in antiquarian and topographical research, it will come as no surprise that he was another Anglican clergyman (he was vicar of Tavistock); Bray's letters to Southey contained large chunks of prose lifted, apparently verbatim, from his notebooks. In her choice of husband, Bray had form. Her first was Charles Stothard (1786–1821), historical draughtsman to the Society of Antiquaries. In 1820 the Stothards had travelled to Brittany, visiting Carnac, 'that stupendous monument to Celtic idolatry'. Their impressions and theories were published in a series of letters to the *Quarterly Review*.[107] As this suggests, Bray, a cousin of Christina Rossetti, was a minor *littérateur*, at ease among people of letters and well connected.

If Southey offered relatively little comment on Bray's Dartmoor writings, he was more forthcoming on the political questions of the day. A late-flowering Tory, he opposed electoral reform and Catholic emancipation. At the height of the parliamentary reform crisis in June 1832 he was prematurely glad Wellington could not form a ministry to pass the bill and he praised Robert Peel for standing firm against reform, believing this would aid the recovery of a reputation lost by his support for Catholic emancipation in 1829. Hoping for a Tory majority at the forthcoming general election, Southey insisted that the 'revolutionary press must be curbed'.[108] Bray, it seems, must have responded with a melodramatic account of Tavistock's radicals, for in one mordant reply, written on Bastille Day, Southey suggested that they were 'an edifying example of what the Whigs may expect at their hands'.[109]

Bray's letters were by turns flattering, opinionated and more than a little self-indulgent. That she delighted in her distinguished correspondent and assumed she wrote for a wider like-minded audience is obvious, while it seems Southey's approval encouraged her to express her thoughts stridently. Southey's indulgent treatment permitted – or was taken to permit – some comically sycophantic imaginings on her part. On one occasion she imagined herself, her husband and Southey being shown Wistman's Wood, 'the last remnant of a Druid grove', by a local farmer.[110]

Wistman's Wood comprises fourteen hectares of dwarf English oaks on a west-facing steep slope. It is described by Ian Mercer as 'in a dense clitter of very large boulders but with substantial amounts of humus between the boulders and on some of their level surfaces'. It is one of three such English oak woods on the moor, the other two being Black-a-Tor Beare on the West Okemont and Piles Copse on the Erme. Even in this exclusive company, Wistman's can be singled out for its unique natural characteristics. Early nineteenth-century visitors reported that the oaks were the height of a man, suggesting they have since doubled in height. By contrast, the average height today of the oaks at Black-a-Tor Beare and Piles Copse is, respectively, 8.5 metres and 7.5 metres. Evidence suggests that the surface area of Wistman's Wood expanded in the early twentieth century, its present growth only restrained by the appetite grazing sheep have for oak saplings that fall outside its now fenced boundaries. It is salutary to be reminded that a natural site like Wistman's, apparently timeless and with a form that seems so essentially itself, has changed significantly and holds the potential for expansion if local conditions allowed. All three woods boast extraordinary moss, liverwort, lichen, fern and epiphytic flora growth, though Wistman's is the richest, its bent and twisted oaks mean the beard lichen that trickles down from its branches hangs at head height.[111]

Wistman's Wood has been one of Dartmoor's most iconic sites for hundreds of years. Long before the moor attracted sustained interest, it was known to the pioneering topographer Tristram Risdon as one of three 'remarkable' sites on the moor, and it has since then been of enduring fascination to the botanist (especially the lichenologist), the antiquarian (Polwhele noted its singularity)[112] and the folklorist. Much of Wistman's appeal stems from what it suggests Dartmoor might have been like before the great deforestation of the late Neolithic and early Bronze Ages turned Dartmoor into an agricultural landscape.[113] Recently Wistman's attracted the attention of Candida Lycett Green, who, on behalf of *The Oldie*, that bastion of quirky decrepitude, went in search of England's 'unwrecked' landscapes. She dedicated two pages of her ensuing book to the wood.[114]

In Bray's fantasy Wistman's would bring to Southey's mind the Roman poet Lucan's 'Pharsalia', and he would quote the passage describing how Caesar broke the spell cast by a Druidic grove over Roman soldiery. Bray would then chip in with the trivial fact that Nicholas Rowe, Lucan's first English translator (1720), was from Lamerton, a village near Tavistock. Mr Bray, who meanwhile had been musing on William Mason's 'Caractacus' (1759), the poem that popularised ideas about the Druids, would then recite the verses descriptive of a Druidic grove. Bray, appealing, as she put it, to the generous instincts of the author of *Madoc* (Southey's 1805 anti-Catholic ancient Welsh epic) and always keen to advertise her husband's talents, would then recite an early poem he had written about the wood. The farmer thinks this is all a bit daft, though he too makes a contribution to the excitable party, saying how the local people tell that the moor's antiquarian relics are the works of the giants, the moor's former inhabitants.[115]

Classical erudition, contemporary literary culture, a little folk-lore, cultivated company, the sympathetic ear of a great man, the discovery of common tastes and knowledge. Could Bray imagine a more potent fantasy? Southey's presence would validate the moor as a landscape of significance, rescue the Brays from provincial obscurity and erase any notion that their interests were merely the eccentric preoccupations of their matrimonial salon. In an earlier passage she had commented that the 'common observer' saw in the moor only a barren landscape, and evidently she believed their party comprised the sort of people capable of appreciating the moor as a system of signs, which, once interpreted, allowed the study of 'the history and manners of ages long ago passed away'. Striking a Burkean note, she said the lowlands, so pleasantly pastoral, could 'delight and soften the mind', but it took the kind of sensibility capable of 'deep and impressive reflections' to appreciate the high moors.[116]

Bray's regard for Dartmoor as a sublime landscape was informed by her knowledge of the picturesque. Though she readily and repeatedly identified beautiful parts of the moor, she was less interested in guiding her readers towards suitable viewpoints than in describing the different visual effects, and the feelings these evoked, wrought on the moor by variations in light and weather. A 'very fine or rather sunny day' would not produce the 'strong opposition of light and shadow' required by 'mountain-scenery and rugged rocks' to display the 'bold character of their outline' and 'the picturesque combinations of their craggy tops'. Sunshine made the moor 'monotonous', a gathering storm – a '"painter's day"' – made it 'sublime'. Pencil, she argued, was a better medium than ink for representing the moor, presumably owing to the subtle shading lead allowed.[117] Though this advice reflected the popularity of painting and drawing as middle-class pastimes, it also suggested the continuing authority of the picturesque, here merged with Romantic notions of the sublime. In the pages that followed, Bray's sustained description of Dartmoor's 'river-scenery' showed off her ability to delineate its sounds and colours, and the moods created by these 'picturesque combinations'. But by seeking to guarantee that visitors experienced the moor's sublime effects, she imposed on the Romantic experience, often associated with subjective responses by an individual to an external stimulus, the objective outlook encouraged by the picturesque. If the conditions were right and the visitor was not appropriately affected, she implied, this must reflect their own underdeveloped sensibilities rather than any aesthetic inadequacy on the part of the moor itself.

Bray's performance, a derivative display of prosy virtuosity and sensibility, served several purposes at once. It strengthened her bid to be counted among the literati; it was her submission on behalf of Dartmoor to the counsels of the Romantic; and, like all writing, it constituted a claim to power and authority over its subject. At the same time her didacticism manifested certain anxieties. If the moor did not live up to the reputation she was busily creating for it, and Dartmoor's bid was rejected by the gatekeepers of the sublime, so in effect would be her bid to join their ranks.

Bray wrote with greater confidence when making her Protestant readings of Dartmoor. Whereas Borlase and Polwhele generally implied that the idolatry, ritual and behaviour of the Druids resembled Catholicism, Bray operated in a more explicit register. Writing of the Druidic use of excommunication, she described it as 'a religious sentence which has scarcely a parallel in history, if we except that of excommunication as it was once enforced by the tyrannic church of Rome'.[118] Like Stukeley, she accepted the distinction between natural and revealed religion, believing that in groves like Wistman's Wood the Druids learned of the unity of the godhead, the immortality of the soul and the regime of reward and punishment faced by the deceased. This, however, was judged 'too excellent' – too complex – for the people to grasp, and the possessors of this extraordinary knowledge decided instead to teach 'a grosser doctrine, one more obvious to the senses'. The paradox was that if select individuals were capable of coming to a true knowledge of God through their exposure to the magnificence of his creation, their attempts to communicate these truths through appropriate religious rituals had the reverse effect. Ritual came between the people and their capacity to know God; ritual created an illusion of, or a substitute for, real understanding.

As ritual became codified, the cast of people – Druids – who evolved to conduct these rituals became the exclusive possessors of this knowledge, which became the basis of their power over their congregants. Consequently, although regal authority existed in theory, Bray believed that 'all real power was soon usurped by the priests', noting that in both ancient and modern times this 'encroachment' by the priesthood was typical of those who followed 'false or corrupted religion', eliding Druid with priest. For those readers who had not picked up on the implications of her observations, she reminded them that pure worship relied on the submission of the priesthood to the civil government (as the Henrician Reformation had insisted), whereas idolatry, whether 'in ancient times, among the heathen Celtae' or 'in modern, under the popes . . . constantly produced a tendency to quite the opposite spirit'.

Indeed, the priesthood, in its perpetual struggle for power, 'under the Roman pontiffs, as well as under the druids,' aimed to usurp legitimate monarchical power 'with the most arbitrary rule'.[119] Druidism and Catholicism were both forms of despotism.

Crucially, Bray took her argument a stage further, developing Polwhele's belief that the natural characteristics of Dartmoor tended to nurture idolatry. That landscape could have such a profound effect chimes with other ideas in general circulation. For instance, the supposedly vigorous landscapes of northern Europe were thought to have given rise to peoples who were self-reliant, dynamic and Protestant. By contrast, the soft, effeminate landscapes of the European south, with climates effortlessly productive of abundant fruits and vegetables, had produced people who were submissive, indolent and Catholic. Dartmoor, obviously enough, was not a soft, southern landscape, so why did Bray think it was conducive to a Catholic-like idolatry? Drawing on her own experiences and her sophisticated understanding of the sublime and the picturesque, Bray related how the moor, when 'chequered and broken with light and shade', could provoke in the viewer fervid imagining that could convert 'many a weather-beaten tor into the towers and ruined walls of a feudal castle'; more than this, 'even human forms, gigantic in their dimensions, sometimes seemed to start wildly up as the lords and denizens of this rugged wilderness'.[120] The 'impassioned eloquence' of the Druidical Bards could be attributed to imaginations, 'unfettered by rules' and 'impressed from infancy' by Dartmoor's 'wild grandeur', which had followed the 'impulse of nature'.[121]

This markedly Protestant text reiterated the now familiar idea that the natural characteristics of Dartmoor were prone to generate corrupt religion; it was also infused with the vocabulary of early nineteenth-century Anglican reaction, used to highlight Druidism and Catholicism's ghastly similarities. Any apparent mismatch between the jauntiness of Bray's writing and her serious themes can be resolved when it is remembered that she wrote for a knowing middle-class audience that likely shared her assumptions, particularly regarding the credulity of the lower classes. Indeed, her book is best remembered, when it is remembered at all, as an important collection of early nineteenth-century Dartmoor folklore, rich in charming if absurd superstition.[122] The way Bray and Southey promoted the poetry of Mary Colling, a Tavistock domestic – one of Raymond Williams' 'labourer poets'[123] – throws into sharp relief how Bray's encounter with ancient Dartmoor aligned her with Southey's Tory Anglican paternalism.

Were outside observers to remark that Dartmoor's monuments were not impressive when compared to Stonehenge, Avebury or Carnac, the three most celebrated prehistoric sites of the time, Bray had a riposte to hand. Those monuments were striking only in the context of the flat land on which they stood. If Stonehenge was transferred to Dartmoor and its backdrop was no longer Salisbury Plain but the tors, it would dwindle 'into perfect insignificance' like 'a pyramid at the foot of Snowdon'.[124] The Druids had 'moved in the region of the vast and the sublime', she concluded, which explained everything. Bray couldn't hide her pride in the possibility that Dartmoor was Britain's 'most celebrated station of Druidism'.[125]

The Archaeological Turn and a Dartmoor of Race

In *A Perambulation* Samuel Rowe refers to Carrington 21 times, to Polwhele 33 times, to Borlase 12 times and to the Brays 18 times. 'Druid' or 'Druids' are mentioned 38 times; 'Druidism' occurs 19 times, and, most significantly, 'Druidical' 39 times. 'Druidical' is used as an adjectival modifier numerous times and in the following table those phenomena named as druidical are listed from left to right in the order they first appear in Rowe's text:[126]

discipline	antiquities	temples	circles	antiquity
honours	monuments	altars	examples	worship
shrines	ritual	religion	system	structure
vestiges	consecrated grave	assize	government	delusion
rites	associations	influence	grove	monument
tolmen	court of judicature	ceremonies		

As terms like 'system' and 'structure' or, more precisely, 'government', 'assize' and 'court' indicate, Rowe did not merely saturate his text with observations about Druidical antiquities but provided the means to construct a comprehensive picture of Druidical life on ancient Dartmoor. Just as Mr Casaubon's *The Key to All Mythologies* was futile because it was not cut with the learning of German scholarship, so Rowe quickly came to be regarded as the ultimate expression of a strain of Dartmoor scholarship that was quickly becoming outmoded. *Trewman's Exeter Flying Post* might have posted an appreciative notice of *A Perambulation*, following this a year later with extracts from a letter written by a London gentleman describing his visit to see the Dartmoor antiquities adumbrated by Rowe, but more prophetic of the book's fate was the long review published in the *Literary Examiner* in September 1849.[127]

Considered alongside a new translation of Erasmus' account of his pilgrimage to the religious shrines of Walsingham and Canterbury, the reviewer wrote in the arch and all-knowing tone that characterises his profession. Observing that 'jaunting or tour-making' to country locations might be 'a pleasant occupation', the reviewer belittled man's determination to invest pleasure with purpose. If 'religious knowledge or edification' had once motivated people to take to the road, this had 'transmuted into pilgrimages in search of knowledge', be it philosophical, geological, botanical or archaeological. Cocking a snook at contemporary pieties, the reviewer praised Erasmus for understanding that it was the journey itself that was of interest. Rowe, by contrast, inflicted 'upon us a long unbroken bead-roll of antiquarian monuments . . . while the adventures of the allied antiquarian perambulations are carefully suppressed', despite his assiduous evaluation of every 'wayside inn, every market town or village hostelry, in the vicinity of Dartmoor', suggesting fun was to be had along the way.

Equally open to mockery were 'the expectations' the Dartmoor pilgrim 'looked to gratify' when seeking out the 'natural grandeurs' and 'Druidical discoveries' that stocked their minds and 'exalted' their thoughts. And it was the susceptibility of visitors to Rowe's Druidical theories that drew the full force of the reviewer's metropolitan contempt: 'our contemporary pilgrims to the Druidical Dartmoor relics appear to be endowed with quite as indiscriminating a swallow, in the way of credulity, as the earlier pilgrims to the shrines of St Mary by the Sea or St Thomas of Canterbury'. How was it possible, the rhetorical question ran, to know 'whether the faint indications of enclosures or structures' were 'the remains of aboriginal villages and Druid temples, or the work of more recent shepherd-boys for temporary shelter or pastime'? It was not simply that greater caution was needed but that the dominant paradigm shaping thinking about Dartmoor and its antiquities had to be abandoned. The 'crude nonsense of earlier antiquaries about the ark-worship or serpent-worship of the Druids' – as found in Borlase, Polwhele and Bray – were 'only fit to pair off with the monastic legends of Walsingham and Canterbury, and ought to have been recorded with more of the spirit of Erasmus'. More Protestantism.

The reviewer closed with a Chaucerian thought. 'The bold flirtations of the Wife of Bath, and the more coy platonic coquetting of the Prioress, to say nothing of the tipsy jollity of Miller and Cook, are probably not without occasional counterparts in the scientific perambulations of itinerary *virtuosi.*' Were the antiquarians and their perambulating fellow travellers to attract their own Chaucer or Erasmus, 'the light' thrown upon their 'eccentricities', 'solemn foppery' and 'folly . . . will be worth, ten times over, all the specimens and monographs with which museums and libraries are be-lumbered'. It was 'ever thus with our race', wrote this sententious observer: 'what we find exceeds in value what we have been seeking for'. 'Our itinerant sages go in quest of geological and antiquarian fables, and bring home pleasant actual stories, lasting contributions to the life and entertainment of mankind.' By offering 'home', 'pleasant' and 'actual' in counterpoint to 'sage', 'quest' and 'fable' – archaic notions made current by fashionable medievalism – Rowe's reviewer wagged a finger at mid nineteenth-century escapists and their spurious claims to scholarship.

That bad science might produce good literature was a nice enough conceit but it hardly solved the difficulties raised by the antiquarian method. A more constructive answer was found in the constraining empiricism that characterised the emerging discipline of archaeology. If the antiquarian was expected to deploy his imagination to the extent permitted by the classical texts, the archaeologists adopted a grammar of caution and scepticism, tending to ostentatiously distinguish their finds from their interpretations. Those of a bolder turn of mind, however, found explanations in comparative anthropology, drawing on examples from the past or the present to explain their finds. Much of this was made possible by colonial encounters with the less developed 'other' and the ensuing debates quickly becoming ensnared in racial theory and competing ideas about whether the origins of the human species were monogenetic or polygenetic.[128] To reduce the question to its bare essentials, mid-late Victorians asked whether the primitive non-European other was inherently racially inferior to the (northern) European or merely at an earlier stage in its evolution. As Dartmoor

writers drew away from antiquarian methods and embraced archaeology, they still clung to idea that the major shifts in Dartmoor
ways of living were explicable not through evolutionary developments, perhaps accelerated through contacts with other peoples,
but through large-scale migrations which saw one racial group
forcefully displace another.

In the 1860s these new approaches to Dartmoor's antiquities
were occasionally ventilated in the pages of the *Quarterly Review*
and the *Journal of the Plymouth Institution*, but it was through the
membership of the Devonshire Association, founded in 1862 and
dedicated to 'the advancement of Science, Literature, & Art', that
archaeology began to influence popular thinking about Dartmoor.
This was particularly so after 1876, when the association formed a
dedicated Dartmoor committee and then in the 1880s, when it
became preoccupied by the politics of preservation. A spate of new
books written by a new generation of Dartmoor enthusiasts would
follow. The association aimed 'to give a stronger impulse, and more
systematic direction, to scientific enquiry, and to promote intercourse of those who cultivate science, literature, and art, in different
parts of Devonshire'. Membership was on the basis of nomination
and a ten-shilling annual subscription. During the association's
annual meetings the papers to be published in the group's
Transactions were read.

Sir John Bowring (1792–1872), justice of the peace and deputy lord
lieutenant of Devon, but rather better known for ordering the
bombardment of Canton in 1856, delivered the first presidential
address at Exeter in August 1862. He struck a strongly empirical
note, quoting Francis Bacon, 'our great instructor': '"Let observation
be fertile; let authority be barren."' Bowring reminded his audience
that this was not incompatible with religion, for 'genuine' Christianity
was not 'a stationary religion', but had been 'strengthened and
invigorated in all its essential characters by the advance of civilisation'. Applying the inductive method would bring man closer to
harmonising all knowledge and developing a fuller understanding

of God's revelation. 'To shun examination – to deprecate enquiry – is to shew an inner conviction that our opinions are built upon sandy foundations.'[129] And so a new generation of gentlemanly amateurs zealously aligned themselves with the latest incarnation of Enlightenment ideals.

John Kelly's 'Celtic Remains on Dartmoor', a modest but seminal essay, was delivered to the members of the Devonshire Association at Tavistock in August 1866. It concerned the antiquities on the nearby western borders of Dartmoor, the same district the Brays had known so well. Throughout the essay Kelly cautiously distanced himself from Rowe, tentatively suggesting that the hut circles, rather than linked to Druidical worship, might be better understood owing to their close proximity to 'ancient Tin Stream Works'.[130] More forcefully, he described Rowe's identification of the sacred circles on Stall Moor, at Cholwich Town and near Trowlsworthy Tor as 'places set apart for the performance of Druidical religious ceremonies' as 'an error which seems to be commonly prevalent'.[131] Explorations of the tumuli and stone circles at Castle Howard in Northumberland, he explained, had unearthed human remains and beads, urns and flints. Supposing archaeological explorations on Dartmoor unearthed similar relics, Kelly conjectured, this would suggest that the circles, rather than sites of ritual, were simply sepulchral. And though it was clear these west Dartmoor monuments were of great antiquity, until 'relics' were found it was not possible to determine what '"period"' they dated from. If these ancient people were tin miners then they were probably 'acquainted with copper', which meant they belonged to the '"bronze period"'.[132]

Kelly's use of scare quotes – 'period', 'bronze period' – reflected the way this periodising of the prehistoric past still seemed new-fangled. Daniel Wilson, influenced by new Danish thinking, had coined the term 'prehistoric' in his pioneering *The Archaeology and Prehistoric Annals of Scotland* (1851), which described the three-age system of stone, bronze and iron as a 'new historic chronometry'. This important shift in thinking was reinforced a few years later when in 1858 a new fissure was opened during quarrying operations at Brixham near Kent's Hole in Devon. In this cave human artefacts were found alongside the remains of extinct animals, thereby resolving the tension between biblical time and earlier animal finds clearly older than the 6,000 years biblical time allowed.[133] Slowly but surely, advances in geological science and archaeology, indirectly buttressed by German 'higher criticism' of the Bible, were generating acceptance that mankind had been around a lot longer than was previously believed. Much of this was implicit to Kelly's short paper. He ended by proposing that archaeological work be commenced, paid for by voluntary contributions. Twenty or thirty people 'devoted to scientific pursuits' who were willing to contribute five shillings each could get things started.[134] Kelly's cautious expression, his focus on precise description and measurement rather than interpretation, his advocacy of the archaeological method, particularly the need to dig for relics, and the emphasis on civic duty intimated a new departure.

'Celtic' was only mentioned in Kelly's lecture title; the antiquities themselves were not so described, suggesting not only his reluctance to engage in any discussion of origins but also, again, an uncertainty about appropriate language. This was also evident in essays and lectures given by C. Spence Bate and G. Wareing Ormerod in the early 1870s. Both men were stalwarts of the association and took a particular interest in Dartmoor. If the Kelly piece heralded the new archaeological approach, their more substantial essays – Wareing Ormerod's more polemical than Spence Bate's – suggested what this might achieve. In 'On the Prehistoric Antiquities of Dartmoor', read at Bideford in August 1871, Spence Bate described in detail the remains of stone huts, stone rows, sacred circles and cromlechs. Images some way between diagrams and pictures illustrated his published text. Hut circles were shown with the addition of a 'turf roof . . . kept up by a series of rafters or poles';[135] 'parallelitha', cromlechs and menhirs were atmospherically portrayed, but the finest images were of the stone circles at Fernworthy and

on the Erme. Spence Bate drew selectively on Rowe and was not
overtly critical, only discreetly distancing himself from the master.
He mentioned 'what we call Druidical remains (for want of a better
name)';[136] he cited Rowe's view that Grimspound 'presents a more
complete specimen of an ancient British settlement than will
perhaps be found in any other part of the island' but did not repeat
Rowe's Druidical interpretation of the site;[137] he acknowledged the
idea that stone avenues on Dartmoor constituted a *via sacra* but
like Kelly thought the available evidence indicated that these and
the stone circles were no 'more than burial places for the honoured
dead';[138] and he thought Rowe had 'poetically' interpreted the
Drewsteignton Cromlech and though quoting his Druidical ideas
ultimately thought it sepulchral.[139] Where Rowe saw in Dartmoor
antiquities the material precipitates of elaborately ritualised lives,
Spence Bate saw the vestiges of ancient domesticity: as well as huts,
he identified cattle pounds, walled villages and burial sites. The
village at Kestor Rock, for instance, appeared to have been 'inhabited by a people who enjoyed peace', the numerous track lines
crossing the hillside 'evidence of a people whose thoughts were
given to the cultivation of the soil'.[140] Dartmoor, in his hands, looks
less like the 'wide Druid temple' of Polwhele and more like an
abandoned agricultural landscape.

A year later Spence Bate addressed the association again, this
time at Exeter, surveying various Dartmoor tumuli and reporting
on several archaeological finds. He explained that experience had
taught him to expect to find in Dartmoor tumuli an earthen vase
made of baked clay containing calcined bones and a weapon of
bronze or stone.[141] The cairn opened near Trowlsworthy Tor, for
instance, had yielded a stone tool and fragments of an urn. More
significant were the excavations at Hammeldon Down, the site of
Grimspound. There a barrow was also opened up and under five
large flat stones were discovered 'a mass of comminuted bones'
and, most excitingly, a piece of amber work and the bronze blade
of a dagger.[142] Spence Bate maintained these finds told of the 'high

degree of advancement, and also of the intercourse of the people of Dartmoor with those of distant nations'.[143] Contemporary experts he cited thought the amber ornament, inlaid with gold pins, might have been the pommel of a dagger or a sword, while the pattern of the Hammeldon tumulus resembled similar sites recently excavated in Scandinavia.[144]

This expert opinion was music to Spence Bate's ears. His first major essay on Dartmoor, published in the *Transactions* in 1871, had focused on the etymology of Dartmoor place and river names. Drawing on many earlier authorities, including Polwhele, he had identified names from the 'more ancient forms of the Celtic tongue', certain Saxon usages and much that was Cornish. Much of this naming reflected Dartmoor's ancient and early-medieval tin-mining industry rather than ancient religious practices. Steering his readers away from the theory that Dartmoor's many 'balls' and 'bels' (Ballabrook, Coryndon Ball, Bellevor Tor, Belstone, etc.) stemmed from either the Norwegian word *baal* (a funeral pyre), or Baal (a heathen god), he suggested the root might be Balerium, the land of tin, as suggested by Polwhele's *Vocabulary of Cornish Words*.[145] The 'ethnologists', authors of 'traditionary history', had taught Spence Bate to think that Dartmoor place names were determined by 'the earliest inhabitants of the district', but instead of finding names of Phoenician origin – as the antiquarians had led him to expect – he found in the centre of Dartmoor a profusion of names 'from the old Scandinavian race'.[146] Streams, hills, rocks and home-steads on the Dart and Teign had names suggesting that at a 'very early date a horde of Scandinavian adventurers forced their way up' those rivers and occupied the tin stream works at the river heads.[147] How early, Spence Bate could not be certain, though he believed the copper spearheads found near Fingal Bridge suggested 'the earliest stage of the bronze period'.[148] Moreover, he claimed, the names of a recently compiled list of the Vikings – 'the Old Sea Robbers' – thought to have raided the northern and western coasts of Britain bore a close resemblance to many Dartmoor names:

Hamill gives us Hamildon Tor. Grim, Grims Grave, Grims Pound, and Grims Lake. Buthar gives Buttern Tor and Butterton Hill. Brodor gives Bruton, on the Dart. Thorni gives Thornworthy. Hiarn gives Henbury (pronounced Hiarnbury). Thor gives Thurlstone. Sölvar gives Silver River, that flows into the Yealm, and Sivard's Cross . . . Hogni gives Hogam de Cosdon. Bakki gives Bicky River . . . Sweyne gives us Sweyncombe . . . Hengist gives Hingston Down; and Horsa gives Horse Hill.[149]

Thus the ancient tin workings on the West Webber under Warren Tor and below Hameldon Down suggested Grimspound was an early Scandinavian settlement and had nothing to do with eastern Mediterranean settlers. Further place-name evidence allowed further conjecture. The survival of Celtic names for nearby sites, most particularly Trowlesworthy Tor, the evidence of extensive tin workings in Blackabrook stream and the name of the River Cad, deriving from a term for battlefield, suggested 'Grims and his warriors invaded from the more central parts of the moor, and forfeited his life for his temerity.'[150]

Ormerod, in his 'Rude Stone Remains Situate on the Easterly Side of Dartmoor', originally read to the meeting of the Royal Archaeological Institute of Great Britain held at Exeter on 30 July 1873, struck a more strident tone. Acknowledging that Polwhele and Rowe were key texts and had to be engaged with – he pronounced Polwhele the more accurate of the two – he treated their findings with great scepticism, noting that both ascribed Dartmoor's stone monuments to either ancient British or Druidical origins. Citing his own geological work, he insisted that the rock basins, logan stones and rock idols were naturally formed; at the same time he recognised this was the proper limit of geological knowledge and could not prove if they had been worshipped or used in religious ritual.[151] He had little time for Rowe's claim that Dartmoor's clapper bridges, such as at Postbridge and over the North Teign, were 'Primitive Cyclopean' and doubted they were any older than the farms they led to.[152] More radical still, he claimed the stone circles, rather than prehistoric, were most likely erected in consequence of Roman influence, an idea not taken seriously since the sixteenth and seventeenth centuries. Ormerod reminded his audience that John Ferguson, prolific author and sometime president of the Archaeological Society of Glasgow, took the view that such circles were generally erected by 'partially civilised races' after they had come into contact with the Romans, with most of them 'belonging to the first ten centuries of the Christian era'.[153] Ormerod also followed Ferguson in arguing that the parallellitha at Merrivale were representative of armies drawn up in battle formation, providing symbolic protection for the village and a guard for various burial sites in the immediate vicinity. None of this had anything to do with ancient religious ritual.[154] Finally, he offered a long disquisition on Grimspound, surveying the competing theories and, following Mr Shortt, he concluded that it was not a temple, a village nor a fortress, but a 'fold for cattle'.[155]

Spence Bate took issue with this in another essay. According to his conjectures, the wall enclosing Grimspound was double-layered, comprising inner and outer skins, which, once filled with earth, would make for a wall ten to twelve feet thick at the base and eight to ten feet high. This made it much stronger than those of the ancient drift pounds at Erme Head and Dennabridge. It was 'so formidably' built because it was 'erected by a people who were strangers to the place, and therefore built to enable them to defend it against attack'.[156] Taking the opportunity his discussion allowed to extend his etymo-logical argument, Spence Bate suggested that a series of Dartmoor Saxon names (such as Sheep Tor, Vixen Tor, Leather Tor, Fox Tor, Fur Tor, Hen Tor) were corruptions of Celtic and Scandinavian names. This accumulation of evidence – the names, the strength of Grimspound, the bronze and amber finds – suggested that 'in the early bronze period the Scandinavian visited Dartmoor in search of tin'.[157] This is not a view taken seriously by modern authorities, one suggesting that Spence Bate's Grimspound essay is 'fanciful'.[158]

R. N. Worth's 'Were There Druids in Devon?', read at Totnes in July 1880, was a more focused attempt to clear away a great deal of antiquarian detritus, liberating the study of Dartmoor antiquities from the tendency to treat 'the supposed Druidic cultus in Devon as an established fact'.[159] If Polwhele, drawing on Stukeley and Borlase, had originated the problem, Worth believed later writers had taken the thinking to much greater extremes. Antiquarian theories about the use of rock basins, logan stones, rock idols, cromlechs and stone circles were unsparingly mocked. 'If a pile of granite rocks afforded a comfortable seat, canopied by the projecting slab,' he wrote, 'in their eyes it was a throne of the Arch druid, or at the very least a seat of judgement.'[160] Worth set out to demonstrate that evidence gleaned from classical texts, which he quoted at length, the linguistic evidence, folkloric '"survivals"' and archaeology all failed to justify Polwhelean views. Not only did classical authors fail to link Druidism to any particular region of Britain, except north Wales, they did not present a well defined picture of Druidism itself.[161] Caesar's account, he sniped, could not be taken any more seriously than a description written today of 'the religion of a semi-barbarous people . . . whose creed and ritual had been gathered on hearsay by an invading foe'.[162] Caesar had interpreted the Druid according to his own cultural expectations, making him more like 'a compound of the Greek philosopher and the Roman judge than a barbaric reflex of a Pagan priest'.[163] Nonetheless, Worth's answer to his question was not a resounding no. He accepted that there probably was in Gaul and Britain 'an order of men known by some such name as Druid or Druith', 'a barbaric priesthood' who had wielded 'a powerful influence'. The difficulty was that the term itself – Druid – now signified meanings that were fundamentally erroneous. Local folklore, so extensive, was devoid of any particular focus on the oak; mistletoe, though a general feature of British folklore, was so rare in Devon that its claimed centrality to Druidical ritual seemed implausible; the Cornish language, once the language of the whole south-west, did not contain Druidical traces; and Polwhele's claim that Drewsteignton was "'the Druid's town on the Teign" was in utter defiance of the claims of its ancient possessor, Drogo, whose name was prefixed by way of distinction to that particular "Teign Town"'.[164]

Worth's argument relied on his belief that the ancient inhabitants of Britain were too divided and too primitive to have generated a culture as sophisticated as that suggested by Polwhelean Druidism. Following the research of Dr Joseph Thurnam, Worth accepted Caesar's claim that Britain was populated by two races, categorised according to late nineteenth-century racial theory as 'dolicho-cephalic' (long-headed) and 'brachycephalic' (round-headed).[165] The long-headed people were Britain's first settlers and probably from the Iberian peninsula, their modern equivalents being the Basques. Pre-Roman Britain saw this primitive Stone Age people displaced into Wales, Cornwall and the north by the more advanced round-headed men of Gaul, 'the Kelts or Belgae', who belonged 'to the age of bronze'.[166] They in their turn would eventually be displaced by the advancing Saxons. The Romans, this suggested to Worth, encountered 'a collection of savage tribes, resembling in many particulars the North American Indians ere they were brought under European influence, and . . . other existing African tribes of a far lower type'.[167] The Polwhelean Druid needed to be replaced by a 'being akin to the medicine or mystery man of the North American Indians, or the sorcerer or rainmaker of Africa',[168] for he was 'as great an anachronism in pre-Roman Devon . . . as a loco-motive'.[169] Though we might share what was surely the audience's appreciative chuckle, we must not miss the significance of Worth's highly racialised comparative anthropology. His essay, written as the European imperial expansion into Africa was entering its most dynamic phase, was shot through with late nineteenth-century racial ideas. In Worth's hands, prehistoric Dartmoor was no longer a landscape shaped by a struggle between the truths of natural religion and idolatrous corruption but the locus of a Darwinian struggle for racial survival.

The importance of his and Ormerod's papers could hardly be gainsaid. Bypassing debates about the origins of the earliest inhabitants of Dartmoor, they reoriented the landscape's history in broad directions. First, they recognised that ancient people had derived a livelihood from Dartmoor's landscape and that it was principally a place of work rather than a peculiarly dramatic place of worship. By foregrounding tin mining, Spence Bate made Dartmoor a place that could best be understood through its industrial archaeology rather than through the study of ancient religion. Many of its ancient stone structures, rather than the vestiges of temples, were the remains of more humble structures, whose probable uses could be determined by nearby evidence of tin streaming. Second, the emphasis he placed on Scandinavian influences not only oriented Dartmoor's earliest history away from the eastern Mediterranean but also suggested that some of its ancient relics were more modern than had been previously supposed. An appreciation of the apparently Scandinavian dimension of Dartmoor's history allowed for an understanding of its 'ancient' monuments that could be more finely differentiated over time. Dartmoor was becoming a more complex landscape that could be understood in terms of 'Ages' ('Stone', 'Bronze' and 'Iron') and historical periods – 'ancient', 'Roman', 'medieval' and 'modern'. The Borlase–Polwhele–Rowe tradition had homogenised Dartmoor's prehistoric past as biblical time demanded; the citizens of the Devonshire Association were beginning to understand the moor's 'rough traces' in new and exciting ways.

One of the most enjoyable works of non-fiction ever written about Dartmoor is *An Exploration of Dartmoor and Its Antiquities* by John Lloyd Warden Page, first published in 1889. Page was a professional writer and knew what he was doing. His style was chatty and engaging, marrying swift-moving but evocative descriptions and anecdotes to summaries of expert opinion. Quotations from Dartmoor verse, often Carrington, and maps and etchings, developed from his own sketches, frequently leavened his text. Claiming no personal expertise, he generously cited his authorities, readily deferring to their expertise but freely giving his own opinions, seemingly little concerned whether his reader agreed or not. Most indicative of his journalistic art, however, was his preparedness to reproduce outmoded but sensational ideas, allowing his readers to revel in the delicious pleasures of Polwhele without endorsing his thinking.

This was a commercially produced book and became a minor bestseller, but still subscribers were needed, the late nineteenth-century alternative to a publisher's advance. Page's list ran to seven pages of small print and though by no means as grand as that compiled a century earlier by Polwhele, it nonetheless told its own story. There were a smattering of Oxbridge fellows, many people from Devon and Cornwall and a good number of public libraries, the Washington Library of the Supreme Council of the Thirty-third Degree of A. and A. S. Rite being the most curious.[170] More significant were some of the names: the Reverend Sabine Baring-Gould of Lew Trenchard, Devon; Robert Burnard of 3 Hillborough, Plymouth; William Crossing of Splatten, South Brent; G. Wareing Ormerod, FGS of Woodway, Teignmouth; W. Pengelly, FRS, FGS of Lamorna, Torquay; and Frederick Rowe of Belfast House, Exeter. Ormerod we've already encountered; the others would become significant players in the Devonshire Association and the Dartmoor Preservation Association (established 1883). Much more will be heard in these pages about William Crossing, the doyen of a later generation of Dartmoor enthusiasts. These names imply that Page, an outsider who came to Dartmoor clutching his copy of Rowe, eventually came under the influence of these men, adopting their new ways of thinking. That, at least, is what was suggested by the contrast between his intemperate early ideas about Dartmoor, as expressed in *The Western Antiquary* in 1887, and the more considered views of the *Exploration*.

The Western Antiquary; or, Devon and Cornwall Note-Book was first published in 1881, the year after Worth delivered his iconoclastic lecture. Less austere than the *Transactions of the Devonshire Association*, *The Western Antiquary* was to become an important outlet for writing about Dartmoor, first publishing William Crossing's early works. Its mission statement was written by William Borlase's great-grandson W. C. Borlase, Liberal MP for East Cornwall 1880–7, excavator of numerous barrows in Cornwall, and debtor eventually ruined when exposed by his Portuguese mistress. The magazine was intended as 'a medium . . . for the pleasant and fruitful interchange of knowledge and thought between those who are interested in the "History, Literature, and Legendary Lore" of the old Dunmonian promontory'. As might be expected, given his choice of language, Borlase advanced a romanticised view of the uniqueness of the south-west and its importance to the national story. Still, he struck a fashionably Baconian note, breezily observing that the magazine's inductive reasoning differed from the methods of Dr Borlase 'a century and more ago!'[171] Readers were encouraged to write in with their questions in the hope that other readers would provide answers for later issues, functioning just as the more famous *Notes and Queries* (itself no stranger to Dartmoor) did and continues to do.

It was just such an inquiry that triggered Page's first intervention. A reader encountering John Rhys' *Early Britain: Celtic Britain* (1882) was taken aback by the professor's claim that there was 'not a scrap of evidence, linguistic or otherwise, of the presence of Phoenicians in Britain at any time'. If not Phoenician sun worship, the correspondent asked, where did the names of Belstone and Bellever come from?[172] Page swaggered in. The earliest inhabitants of the south-west, he wrote, were 'wild semi-nude Celtic tribes, invaders, it is generally conceded, from the land of the Belgae'. He continued:

As is well known, the priests of this savage people were the Druids. There are those who maintain that this mysterious priesthood had no connection with Dartmoor. But that the Druids did erect their hypaethral altars in this weird wild tract, and worship their gods beneath the shadows of the tors, seems to me most probable. Who can gaze upon the circles of Scorhill Down and Sittaford Tor, and not view them as the work of that wonderful hierarchy who swayed the fierce Danmonii as no priesthood since has influenced a civilised people?

Page found it inconceivable that all the rock basins were natural and rejected 'the band of scoffers' who claimed they were all 'the result of natural causes'. He then repeated Polwhele and Rowe's explanation of the names Bellever and Bel Tors, saying they *were* named in honour of the Phoenician sun god.[173] All the names of Dartmoor's tors, he asserted the following month, could be classified under three headings: Celtic, Phoenician/Druidic-Phoenician or Teutonic/Saxon. The individual names were 'derivable from local sources, from colouring, from resemblance to certain forms, or from the vernacular'.[174] As the journal's readiness to print Page's outbursts might suggest, Rowe hovered over the journal as its ultimate authority. It was particularly revealing that Crossing once responded to a notice that he had contradicted Rowe with the irritated reply that he did so wittingly.

And yet, come the publication of his *Exploration* just two years later, Page had changed his mind. Though maintaining that the Phoenicians did 'visit the West in search of tin', he admitted there was no evidence they '"prospected"' on Dartmoor.[175] The sacred stone circles were now 'so-called', Page adopting the sepulchral line; the 'Bel' theories were not 'indisputable'; Crockern Tor was the site of the stannary court and 'the seat of judicature of the *somewhat hypothetical* cantred of Tamare'; he paraphrased Worth's Drewsteignton argument, though happily provided for the delectation of his readers a long quotation from Polwhele and a summary of what had been written about the cromlech; Spence Bate's Scandinavian take on Grimspound was fully aired, though Page concluded that it was most likely 'an aboriginal village'; and an extract from Carrington invested his description of Bowerman's

Nose with a little poesy, and here Page confessed to preferring the theory offered by 'some author' that its name had Celtic origins: the idea that it was named after a resident from the time of the Norman Conquest was too mundane for a man of his tastes.[176]

Page also revisited the Dartmoor names question in an appendix. Restating the classification offered to the readers of the *Western Antiquary*, he stuck to his guns on the Phoenicians but admitted that 'we do *not* know' if there was any Druid influence, hedging his bets a little by claiming that 'whether he had a habitat upon Dartmoor at all is very much *vexata quæstio*'.[177] Lifting almost verbatim the ironic opening to Worth's 1880 essay, Page commented that it was 'with something akin to regret' that writers now declined to 'admit that the Druids had any connection with Dartmoor at all'.[178] Just as he had found the Scorhill and Grey Wethers stone circles affecting, so now, in more measured tones, Page reflected on the enigmatic parallel lines on Shuffle Down and Mis Tor Common. Worth might have got it about right about the Druids, but 'having dreamt of them so long' it was hard to sacrifice enchantment for 'the prosaic explanations' of men 'whose souls' were 'above mistletoe, criminals, wicker cages, and *id genus omne*'. And yet Page was surely right to observe that while Dartmoor's stone antiquities were principally sepulchral, 'even a neolithic race must have had some religion'.[179]

Perhaps it was Page's capacity to be all things to all people that ensured his book received a very favourable response from the *Western Antiquary*. Though showing due deference to Rowe, it pronounced *Exploration* 'the most comprehensive and valuable' book ever written about 'our loved and lovely moorland'. It would 'do more to popularize the district of which it treats than a more antiquarian and scholarly volume'.[180] Could the same be said of the 'revised and enlarged edition' of Rowe, published with an introduction by his nephew in 1896? It was heftier, weighing in at some 500 pages, 200 more than the original. The new material was dedicated to chapters on Dartmoor's geology, soil, mining, the prison, wildlife, botany, churches and literature. None of this had much to do with Rowe's perambulation, though it does suggest a serious attempt to fulfil Rowe's original multidisciplinary ambitions. More significant was the editor's short preface. 'I have made many alterations and some additions,' wrote J. Brooking Rowe, 'but, although entirely disagreeing with them, I did not feel at liberty to eliminate altogether the Druidical theories of the author, with which the first edition of the book was saturated.'[181] The theories propagated by Rowe, his posthumous editor and nephew explained, had become outmoded, superseded by new scholarship. Excluding the image of the Drewsteignton Cromlech from the title page of the revised edition was one way of signifying the new mood, but more subtle still was the new frontspiece.

As in Rowe's original edition, the first plate is a scene on the River Taw. In the 1848 plate Stepperton Tor is steep, reaching a point; in the 1896 image it has a flatter, more rounded shape. The rocky foreground in the original image has been replaced with heathland and there are no livestock or day trippers to dilute the misty loneliness of the scene. The 1848 image is clearer and precisely detailed, with the tor distant yet sharply delineated, the highest peak of an alpine scene. The 1896 image offered a different perspective and a more impressionistic aesthetic: the tor is less imposing and distant, the surrounding land is flatter. Perhaps this reflected the new realism, the result of the archaeological turn. After all, Page had closed his *Exploration* observing that in Britain there were peaks that 'dwarf almost into insignificance the heights of our Devonshire upland'. Romantic Dartmoor, however, was not dead and gone. The moor still 'asserts itself', Page wrote, 'and men who have ranged the wide world over, who have climbed Alps and Apennines, Pyrenees and Carpathians, will yet return home to these heathery granite-crested hills of the West, with an ardour that has lost nothing by comparison'.[182]

And yet something surely *had* changed. Perhaps it was simply the arrival of the 'useful but unromantic iron horse' which led

Page to contrast the 'unknown' Dartmoor of Carrington's day with the 'accessible' Dartmoor of his own.[183] Karl Baedeker, guiding his readers through England, Wales and Scotland, wrote about a Dartmoor made accessible by developments in public transport. One especially convenient way was to take the train from Exeter to Bovey Tracey, from where coach rides could be taken onto Dartmoor, evidence of how interest in the moor was generating a little entrepreneurial activity. Baedeker explained that each Wednesday, Friday and Saturday the train left at 11 a.m., arriving in Bovey Tracey in time for the 12.30 coach. On Wednesdays this jaunted out to Haytor Rocks and on past Hound Tor, Bowerman's Nose, Manaton and Becky Falls; on Fridays it looped north up through Moretonhampstead and across Dunsford Bridge; and on Saturday it ran south through Ashburton, Holne Chase and Buckland. The excursionist, having tasted something of the eastern edges of the moor, would make the Exeter train from Bovey Tracey at 7.24 p.m., arriving back in the city at 9.26.[184] Good carriage roads could take the more intrepid into the centre of the moor, where there were places to stay. From there, the 'active pedestrian', if they allowed seven hours, might even walk the sixteen miles to Okehampton. That way they could 'explore much of the most characteristic scenery of Dartmoor'. Though the air was 'bracing' and in the summer the climate 'often pleasant and invigorating', the visitor should be careful, for rain was common, and the bogs and mists posed a threat to the unwary.[185] If unaccompanied by a guide, walkers should carry a good map and pocket compass and stick closely 'to the beaten tracks'. Sensible advice. The moor's charms were plentiful, though Baedeker offered a brief and deflating comment on Dartmoor's antiquities: 'The moor also offers much to interest the antiquarian, as it abounds in menhirs, stone circles, and other relics of the ancient Britons, though many supposed ancient monuments are now regarded as cattle-pens and deserted mining-shafts of no great age.'[186]

This last skilfully succinct comment demonstrated not only how effectively its author could accommodate the latest thinking, but also the extent to which Dartmoor had been disenchanted. Sabine Baring-Gould (1834–1924), squire and parson of Lew Trenchard, prolific writer, folklorist, hymn writer and excavator of Grimspound, wrote briskly at the turn of the twentieth century of Mrs Bray's 'tall twaddle', saying all the 'rubbish' she wrote was based on 'supposition'.[187] Rowe was given the benefit of the doubt for he was writing 'when Druids and Phoenicians cut great figures'; Page got pretty short shrift and E. Spencer's *Dartmoor* (1898) was condemned as 'wildly erroneous as to the antiquities'; William Crossing's work, about which much more will be said, was 'free from the arrant nonsense of pseudo-antiquarians of fifty years ago', and B. F. Cresswell's *Dartmoor and its Surroundings* (1898) was 'commendably free from false theorising on antiquarian topics'.[188]

Baring-Gould's self-identification as the embodiment of common sense was evident in his comment on Bowerman's Nose. If it had 'ever been venerated as an idol, the worshippers would assuredly have done something towards clearing this clitter away, so as to give themselves a means of easy access to their idol, and some turf on which to kneel in adoration'.[189] Such a high-handed approach to the errors of past interpreters did not prevent Baring-Gould from giving a great deal of space to the complexities of Devon's supposed racial history. His text, deploying a little comparative anthropology, some Lamarckian thinking and much craniology, distinguished long-headed Iberians from round-headed Celts. The long-heads originated in the east, and he found it instructive to note that the present-day Chinese 'held that a soul, or emanation from the dead, enters into and dwells in the memorial set up, apart from the tomb, to his honour'. The 'distinguishing motive in the life of long-headed Neolithic man,' Baring-Gould explained, was 'his respect for the dead,' which he demonstrated by expending 'vast labour' building 'stupendous tombs . . . out of unwrought stones'. These were the dolmen-builders, men 'who erected the family or tribal ossuaries that remain in such numbers wherever

he has planted his foot'.[190] And though later surpassed by the Celt
or Gael, the dolmen-builder adapted ('cranial modifications') and
his 'dusky skin', 'dark eyes and hair and somewhat squat build'
remained in the Western Isles, the west of Ireland, Wales and
Cornwall, and could still be found in Brittany, south-west France
and Portugal.[191]

The Gaelic invaders, though subjugating the Ivernians, did not
have it all their own way. The long-heads impressed their short-
headed masters with their 'deeply rooted' ancestor worship and
as 'the races became fused' they continued to build dolmens and
megalithic monuments. Stonehenge and Avebury, products of the
Bronze Age, were built by these 'composite people'.[192] Baring-
Gould freely recycled Worth's famous argument, adding a twist
of his own making. The Druids, he argued, 'were the schamans,
or medicine-men, of the earlier Ivernian race, who maintained
their repute among the conquering Celts, and their representatives
at the present day are the white witches who practise on the
credulity of our villages'.[193] Perhaps this last comment was to be
expected of an Anglican clergymen concerned for the spiritual
welfare of his flock, but still it seems possible to discern in his
analysis some Borlase–Polwhelean–Rowe 'survivals': the manipulative Druid, the insistence that the souls of the dead entered into
the stone monuments that commemorated them, the notion that
the contemporary practices of eastern religion provided the key
to understanding the beliefs of the Celtic-Ivernians. Significantly,
the Britons, the next wave of settlers, who eventually drove their
predecessors into what two millennia later would be called the
Celtic fringe, did not interbreed. It was these people, the original
Britons, who occupied what would become Roman Britain and
would eventually convert to Christianity.

Baring-Gould was less overtly polemical than Bray, but still it
was the Celts who built dolmens and practised a dubious religion
in pre-Christian Dartmoor just as, according to many a God-fearing
Englishman, they continued to in the present. In the middle of the

eighteenth century Borlase had disputed whether the antiquities of Dartmoor could be claimed by patriot Britons as part of their national heritage; by the first decade of the twentieth century one of the most influential authorities on Dartmoor had identified those same antiquities as the products of the racially impure Celtic 'other'. Dartmoor's antiquities, in the hands of Baring-Gould, had ceased to be a part of *British* history. Can it be any coincidence that around the same time Crossing was writing a book on Dartmoor's stone crosses, focusing his attention on more recent Christian history?

If the new thinking nonetheless left something of Dartmoor's prehistoric mystery intact, Baring-Gould's encounters with the moor provided other clues to how Dartmoor retained its mystique. For a member of the educated middle class recoiling from modern urban life and in search of the vestiges of an authentic England, Dartmoor folklore provided as much of the extraordinary and apparently ancient they could possibly need. Writing of the Lake District, Baedeker had observed that the 'visitor will be struck by the absence of legend in the Lake Country'.[194] The same could not be said of Dartmoor. As Baring-Gould and his friends embarked on the preservation of Dartmoor's ancient artefacts, so too did they set off into the villages and the hills to record the songs and tales of the moor men and women. Folklore helped fill the gap left by the demise of the Druidical theories that had captured the imaginations of Dartmoor enthusiasts for the previous hundred years.

Becoming Adult

In *Orphan Dinah* (1920), one of Eden Phillpots' epic sequence of Dartmoor novels, Lawrence Maynard, its most troubled protagonist, takes the novel's eponymous heroine on a walk to Haytor. They inspect the quarries, Lawrence wistfully telling Dinah that the stone was said to have been used to build London Bridge and the British Museum, but the real purpose of their journey was to see, as the title of Chapter XIV suggests, 'the face on the rock'. Phillpots says although this formation on Hay Tor was made by the 'chisels of Nature', its 'countenance' was 'malignant': it displayed all 'the brutish semi-human doubt and uncertainty of a higher ape'; 'it burlesqued hugely those beetle-browed, prognathous palæoliths of old time'; its expression 'trembled on the verge of consciousness'. Had it once possessed, Phillpots asked, 'an awe and sublimity we cannot grant it today'? The answer was likely yes and Phillpots explained how we could know.

As children, Lawrence and his sister had been moved by the rock idol, instinctively praying to it for help – they wanted their mother to have a baby boy – and then blaming it for their misfortune – the baby became ill and died. The children's response, Phillpots suggested, tells us much about how early man, 'set forth on the mighty march to conscious intelligence', encountered the rock: 'But it [the sight of the rock] had challenged a boy and girl, who were still many thousands of years nearer to prehistoric ancestors than their parents. For children still move through the morning of days, and through their minds the skin-clad dreamers and stone men are again reflected and survive.'[195]

There's a little evolutionary thought here, for the extraordinary progress of civilisation over millennia was now compressed into the span of a single childhood. As an adult, Lawrence could rationalise his childhood encounters with the natural features of the moor, his maturation mirroring that of humanity. With their different agendas, Borlase, Polwhele, Bray, Rowe, the men of the Devonshire Association and Baring-Gould had all approached the moorscape with similarly demystifying agendas. In the middle of the eighteenth century the questions concerned natural and revealed religion and the extent to which Dartmoor's antiquities could be embraced as part of national tradition. At the end of the eighteenth century the questions concerned the workings of nature on the imagination: Polwhele, a cleric writing within a literary culture saturated with romanticism, uneasily explored how the sublime acted on the primitive, irrational mind. In the 1830s it was the anti-Catholic rhetoric of Anglican reaction that gave Bray's writings its grammar. Rowe wrote in a similar atmosphere, albeit in a more muted key, innocently generating an epic work that for all its scholarship was already outmoded – Baring-Gould later recalled how it had been his Bible during his youth. The men of the Devonshire Association, indirectly grappling with the challenges raised by Charles Darwin and the developments of mid-Victorian science, struggled to find new ways of anchoring their thought about Dartmoor. Before R. N. Worth's revolutionary polemic – revolutionary within this small world – comparative anthropology and etymology had only partially displaced Rowe and his predecessors. Worth offered what quickly became a new orthodoxy, and his words were borrowed, attributed or not, by many future writers. Baring-Gould, insufferably pleased with himself, proved susceptible to racial science, showing himself to be no less a man of his time than those predecessors he was so quick to condemn. Only Carrington invited the moor to work its effect on him, though this, of course, was characteristically Romantic. He found what he was looking for.

In 1927 T. D. Kendrick published *The Druids. A Study of Keltic Prehistory*. He wrote not as a member of a provincial association but from the august environs of the British Museum's Department of British and Medieval Antiquities. Benignly examining the 'inextinguishable affection' with which the 'popular imagination' thought about the Druids, Kendrick admitted that the efforts of the 'learned' had seen 'the priesthood' lose 'little, if any, prestige'; it remained 'securely enthroned in the fancy of the people'. Kendrick dismissed any notion that the resilience of the tradition could be attributed to folk memory, insisting that the historical record showed it to be a modern invention. What he could see was why the ideas appealed to 'common sense', for they satisfactorily provided a 'solution to a definite puzzle'. Stukeley, in particular, had proved a brilliant communicator. His 'natural ebullience' and the strength of his conviction ensured that the 'propagation of his doctrine', 'a splendid hotch-potch of invention and surmise', was achieved 'more suddenly and completely' than the teachings of any other archaeologist Kendrick could think of.[196]

It would be hard to dispute any of this, and Kendrick's notion that Druidism was an 'invented tradition' chimes nicely with much contemporary historical thinking. Stukeley, however, remains useful in other ways. His attempt to integrate British antiquities into a Protestant British history threw into sharp relief the continuing strength of a post-Reformation desire to 'evacuate the numinous from the landscape'.[197] The writings of Borlase, Polwhele, Bray and Baring-Gould showed them to be zealous Anglicans, determined that the West Country would not be a focus for what they perceived to be the idolatry or superstition of the ever-menacing Counter-Reformation. Dartmoor's antiquities, near-imperishable when left alone, were made to tell a story not about what had been lost but what had been gained.

But had they been left alone? When Baring-Gould and his friends took it upon themselves to restore the stone circles and other monuments, placing upended stones upright, they held mindless vandals, farmers and road-builders responsible for the destruction they found. Doubtless there was some truth in this, particularly when a recently disappeared standing stone was rediscovered at the entrance to a field with a gate attached to it. It's likely, however, that their acts of restoration also reversed deliberate acts of desecration and iconoclasm, possibly prompted by Druidical theories.[198] If so, rubbishing Rowe and becoming adult made Dartmoor's antiquities safe for preservation.

2

Improvement and
Incarceration

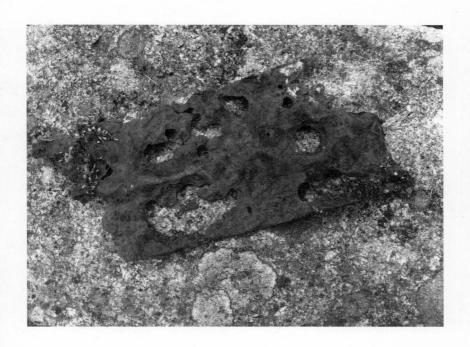

Perambulation

It has rained overnight and the air, heavy with moisture, is cool on the face and wet in the mouth. I'm heading south. To the left of the irregular but well worn path out of Princetown towards South Hessary Tor is a drystone wall. Beyond the wall is the long narrow strip of the Tor Royal plantation; on this side are the browns, yellows and greens of open moorland. The path is mostly wet sandy soil, worn by foot and hoof. It is punctuated with grassy knolls and dips and hollows muddy from the previous night. There's something satisfyingly frictional as my boots, mud encrusted from before, twist slightly with each step in the wet sand: they produce a masculine, workmanlike sound. As the path ascends, my glances back at Princetown form a photo essay exploring its gradual dissolution in the fine morning mist. Ahead, the rectangular bulk of South Hessary Tor comes solidly into view. I remember an abandoned midnight walk through the pitch-dark night with my younger brother. We momentarily mistook two granite gateposts for people conducting arcane rituals and lost our nerve.

A lightness of mood carries me swiftly down the first depression of the moor's great undulating decline towards Plymouth. Above hang layers of cloud white and grey and mauve, but more distant skies show blue, and the blowy Atlantic wind suggests a day of mild but changeable weather. South Hessary now marks the northern horizon, and a crossroads gives me three choices. I could turn left and opt to explore the disused Whiteworks tin mine, and then either take the easy loop back into Princetown, a brisk jaunt to Hexworthy or the great hike to Ivybridge. Striking out south gives many options, and the path towards pretty Sheepstor church and the pub at Meavy is tempting, but I choose to go sharply right along the bridleway. It traces a line through old tin-mining country towards the conifer plantations that abut the northern banks of Burrator Reservoir. The OS map helps the untrained eye pick out the remnants of this ancient Dartmoor industry, orientation helped by two restored crosses along the way. Suddenly, over a ridge from the direction of Ditsworthy Warren, a Royal Marines Commando unit comes into view. Thoughts flit. If they were after me, what chance would I have? Have they spotted me? They're spread out in some kind of formation. What would make good cover?

Burrator Reservoir lies in the valley towards the south-west and is surrounded by plantations finely differentiated on the map by the preservation of old names: Raddick, Stanlake, Norsworthy, Crofts, Peekhill, Beechcroft, Roughtor, Middleworth. Furze, as gorse is hereabouts known, grows by the path and is yellow with blossom. As the foliage changes on the gradual descent into the river valley, so the senses note subtle variations in air quality and climate. It's milder down here, and a little elementary botany says the abundant moss and lichen on the drystone walls is proof of exceptional air quality.

At Leather Tor Bridge, which crosses the Meavy, the scudding clouds allow through the pale sun of late winter.

Deep into the plantation, walking through Norsworthy and into Crofts, there are proliferating signs that this is a worked landscape. Stacked logs carry warnings that they are not to be climbed; the leats, man-made channels dug to divert river water, have simple mechanisms to help control the water's flow.

Visitors who come here will have ventured a little further than most from the pleasant walk around the reservoir. Sometimes they come for their own private reasons. At the stone cross below Leather Tor a woman and a child have left simple memorials to a husband and a dad. Their gesture in this quiet place renews familiar terms, giving meaning to shop-bought sentiment.

On through the Peekhill plantation and below is the grim little town of Dousland. The narrow track leading to the Yelverton road brings into view the elegant tower of distant Walkhampton church. It's a characteristic Dartmoor contrast. In one direction winding lanes dotted with pretty villages and granite farmhouses – the lovely Huckworthy Bridge is not far off – and in the other houses built for workers by the local authority, arranged in rows, using materials that don't stand up well to the weather. A signpost points away from Dousland and towards Welltown.

A little way up the road I pick up the dismantled railway line, whose meandering way leads back to Princetown. It is on a high bank at this point, the product of much heavy shovelling. Before long the contour is found, and for a few miles the line roughly marks the boundary between the enclosed lowlands of Sampford Spiney and the open moorland of Walkhampton Common. A stretch is lined with hawthorn trees bent into shape by the wind. To me nothing in Dartmoor is more evocative than these sinewy trees, their skeletal branches and bony fingers sharply defined against the sky.

The varied nature of the western edges of the moor can be seen here. Walkhampton Common, to the east, has evidence of Bronze Age settlement and ancient tin mining. Lying low to the west is the wooded valley of the River Walkham. Single-track lanes knot together a string of small Dartmoor farms, some now abandoned, each earning a living from their enclosed fields and rights of access to the common. Ahead I can see the great southern peaks of the north moors: Cox, Middle Staple, Great Staple, Roos and Great Mis Tors.

Quarrying has left a deep mark on this district; in part, this was why the railway was built. I come around Ingra Tor, an exposed spot where a sharp sou'westerly blows hard, and encounter mounds of extractive waste and an abandoned quarry. A halt was established here as late as the 1930s. Venturing into the pit of the quarry, its flat sides rise up on three sides, claustrophobic rather than dramatic.

A fine granite bridge crosses the railway line. It has outlived its usefulness. Rough tracks lead north and south. This extraordinarily sturdy but now redundant structure, like the spoil heaps that line the track, will last a great deal longer.

More dramatic are the remains of the quarries at Swelltor, the ruined buildings making the site desolate. It is said that in the 1840s, 600 men worked these quarries. Sleepers half buried in the earth, the best-preserved section of the old railway gives material substance to the 'Dismantled Railway' marked on the OS.

The dozen corbels that have lain by the side of the track for a century are evidence of the value once attached to Dartmoor granite. Hundreds were cut from the moor and dressed in 1903 and

used to build London Bridge. Now they're in Arizona, whence the
bridge was moved, stone by stone, in the late 1960s.

As the railway line rounds King's Tor, I can see to the north the
deep cut into the hillside below Great Mis and Great Staple Tors.
This is the quarry at Merrivale. Further on and across the common
– the rain lashed sharp for five minutes – the Foggintor quarries
are spread out below North Hessary Tor and the television mast.
Nelson's Column in Trafalgar Square was cut from this stone;
Dartmoor Prison, in its earliest incarnation, was also built of
Foggintor stone; a mission hall once stood here, and it is said thirty
families lived in cots on the site.

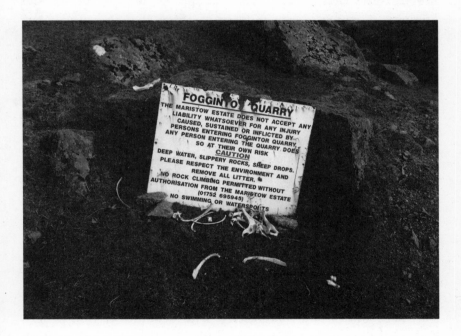

Warnings decorated with sheep bones should be heeded. DEEP WATER,
SLIPPERY ROCKS, SHEEP DROPS. From this threatening spot the line
follows the contour for a tranquil mile or so to its terminus marked
by an old railway building behind Princetown's modern fire station
and close to its small brewery. The last train drew out of here in 1956.

From the prison to the dismantled railway to the quarries, this route cuts through the moorscape most radically altered during the nineteenth century. Here the Industrial Revolution came to Dartmoor, natural resources attracting intensive commercial speculation and exploitation. Here too new penal regimes brought the modern state, turning the upland into a site of incarceration. Dartmoor became known; it became of national interest.

The debris at Swelltor is not the remains of a quarry but the remains of Swelltor itself, 'decapitated and disembowelled by the quarrymen'.[1] The rubble of Foggin Tor and the gash in the hillside that was once Ingra Tor tell the same story. It is hard not to think this fascinating landscape spoiled, yet to see the quarries as purely destructive or to think that the prison has distorted the popular sense of the moor is to suppose that conservationist sensibilities have always shaped how Dartmoor is thought about. In the late eighteenth and for much of the nineteenth century writers predicted a Dartmoor shaped by industrial modernity. Even the poets got in on the act. The Dartmoor works of H. E. Carrington, Mrs Hemans and Joseph Cottle, all Romantic poems of place published in the 1820s, imagined a radically different Dartmoor. Hemans wrote enthusiastically of agricultural improvement, Carrington imagined an industrious soundscape created by quarrying, and Cottle reckoned the traveller coming upon the prison at Princetown would be deeply affected by the contrast between the moorland and the 'long, formal houses' and their 'few adjacent enclosures, *looking green from cultivation*'. Cottle hoped 'the spade and the plough' would transform Dartmoor until it was 'intersected with beautiful *English* hedgerows, and all the other accompaniments of prosperous cultivation'. Although Hemans and Carrington were less credulous – Carrington celebrated the common as a place of liberty – in all three poems there was tension between a Dartmoor that was irredeemable 'waste' and a Dartmoor of opportunity, attractive to the entrepreneur and inviting transformation. The peculiar way in which improvement and incarceration became entangled is one of the most important stories to be told about modern Dartmoor.

Prisoners of War: Dartmoor becomes a National Question

Explaining why Britain needed to build prisoner of war depots in the first decade of the nineteenth century is easy. During the French Revolutionary Wars (1792–1802) prisoners were housed in existing military prisons at Millbay, Stapleton, Forton and Portchester; in 1797 pressure on capacity saw a further prison built at Norman Cross in Peterborough. Much greater capacity was required during the Napoleonic Wars (1803–15) and the British–American War of 1812. The unprecedentedly large number of POWs these wars generated reflected the novel attitudes underpinning how they were fought. Napoleon's revolutionary notion of the 'nation in arms' dissolved the distinction between combatant and non-combatant. In 1803 he interned all British male civilians aged between eighteen and sixty resident in France, reasoning that they were eligible for service in the militia and so potentially active enemies of France. Napoleon also abandoned the regular prisoner exchanges that had been a routine aspect of war, believing imprisoning captured combatants for the duration weakened the enemy's capacity to fight back.[2]

Britain responded in kind, increasing the number of prison ships moored at Plymouth, Portsmouth and Chatham from twenty in 1806 to fifty-one by 1814. Perhaps a third of French prisoners in British custody were held in these notoriously cramped and filthy 'hulks'.[3] New land prisons were also established, for ships converted to prison use were expensive, had high operational costs and a relatively short life. Service on the hulks also kept officers, seamen and marines from active duty. It was more cost-effective to deploy militia as prison guards because they had to be posted, paid and fed whether they were on active duty or not. In 1809 a naval report concluded that more land prisons like that being built at Princetown on Dartmoor would solve the problems faced by the Admiralty's overstretched Transport Board.[4] Other prisons were built at Valleyfield and Perth in Scotland.

It is less easy to explain why a POW depot was established on a piece of common land 2,000 feet above sea level, several miles from a significant town, with poor transport and communication links and in an environment prone to heavy mists, rain and snow. Isolating the prisoners made sense. Holding enemy combatants in hulks on naval bases close to explosives and weaponry carried obvious risks,[5] but isolation alone does not explain why the government selected this inauspicious site. Further clues can be found in press coverage announcing the decision. In October 1805, when the Admiralty issued masons and builders an invitation to tender, some newspapers outlined the scale of the project and noted the local availability of building materials. Moor stone for the buildings and the boundary walls could 'be broken from the scattered rocks on the spot'.[6] Other newspapers gave the project greater meaning. Reports explained that the 'Forest of Dartmoor', on the order of the Prince of Wales, was 'rapidly improving'. As Duke of Cornwall, the Prince of Wales had title to a large proportion of the Dartmoor upland, itself part of the 200 square miles of mainly Devon and Cornwall that since the fourteenth century had constituted the Duchy of Cornwall. Several thousand acres had been 'grubbed up for planting' and soon on Dartmoor's 'bleak and comfortless bogs and mountains' would 'arise neat, habitable dwellings, fit for farmers and cottagers'. Thanks to the prince's intervention, 'barren heath' would be 'converted into productive land'.[7] *Trewman's Exeter Flying Post* linked the improvement of Dartmoor to the establishment of the depot. Soon 'acres of barren heath' would be 'converted into as many of oats, barley, and wheat, for the benefit of society'. His Royal Highness, it was said, had contemplated doing this for some years, and was 'now determined to have it carried into execution without delay'.[8]

These reports gave the project prestige and legitimacy, enhancing the prince's fragile reputation for duty and patriotism, aligning his contribution with the broader moral and political dictums of improvement. Thomas Tyrwhitt, the prince's private secretary, probably orchestrated the press coverage, ensuring that the announcement was pitched in the most desirable way. Indeed, after the Transport Board had advised the Admiralty that Dartmoor was a suitable location, press reports identified Tyrwhitt as having originated the idea. It was he who proposed transporting the prisoners from Plymouth up the Tamar – 'that beautiful river', according to one newspaper[9] – and landing them at Lopwell Quay just a few miles from the depot's proposed location at Princetown.[10]

Tyrwhitt remains an obscure figure, and the same basic facts about his life are often repeated. Born into a distinguished Lincolnshire family in 1762, he was befriended by the prince while at Christ Church, Oxford and consequently enjoyed a career that saw him become deeply involved in Devonshire public affairs. Initially auditor (1786) and secretary to the Duchy of Cornwall and keeper of the privy seal (1796–1803), he was later lord warden of the stannaries (1803–12), vice admiral of Devon and Cornwall (1805) and gentleman usher of the black rod (1812–32); he was member of parliament for Okehampton (1796–1802), Portarlington (1802–6) and Plymouth (1806–12).[11] His activities stemmed from his decision in 1785, aged just twenty-three, to lease from the duchy 2,300 acres of Dartmoor common. The area was ripe for investment following the opening of the turnpike between Tavistock and Moretonhampstead in 1772. Bisecting the moor, the new road profoundly affected its future development.[12] Tyrwhitt linked his new interests to the new road by developing the track that led to

Rundlestone, thereby triangulating Yelverton, Tavistock and Princetown, his new town. By the opening years of the new century he had built a house (Tor Royal), an inn (the Prince's Arms, later the Plume of Feathers), a mill (at nearby Bachelors Hall) and some labourers' cottages.[13] Trees were planted to shield his newtakes and efforts were made to cultivate flax and other crops. Tyrwhitt's newtakes and plantations can easily be seen in the depression below the moorland tracks connecting Hexworthy to South Hessary Tor. William Crossing later insisted, contrary to what was sometimes said, that Tyrwhitt's enclosures deprived the commoners of good pasturage. And it is true that the newtakes gradually established along the Tavistock–Moretonhampstead road – which Tyrwhitt helped inspire – cut off the southern from the northern parts of the moor, effectively cutting the common into two.[14]

Tyrwhitt's fellow improvers were not much worried about commoners' rights *qua* rights. Taking instruction from the leading improvers of the day, they believed these rights had inhibited the development of British agriculture and had kept the commoners in a state of moral degradation and backwardness. As a parliamentary select committee of 1795 argued: 'The idea of having Lands in Common, it has been justly remarked, is to be derived from that barbarous State of Society, when Men were Strangers to any higher Occupation than those of Hunters or Shepherds, or had only just tasted the Advantages to be reaped from the Cultivation of the Earth.'[15]

Arthur Young, the most influential agricultural improver of the day, saw the persistence of 'waste' as a national disgrace. Improving empty land at home, he believed, would help address concern at the high rate of emigration to America. Why colonise America, he asked, when there was so much opportunity at home?[16] Young hoped for the 'complete transformation of the rural landscape . . . central to which was the disappearance of common land, the reclamation of the unproductive "wastes", the expansion of arable, the consolidation of farms, and the creation of new farm buildings and larger fields'.[17] Central to this vision was faith in new scientific methods. More productive land would generate higher yields and increased profits; properly invested, profits would bring a primitive peasantry towards civilised and enlightened ways of living. For men like Tyrwhitt, willing to invest over the long term in land thought unproductive, enlightened self-interest was the name of the game: private profit, when generated by developing unproductive land, would bring socio-economic progress and cultural uplift for all. That, at least, was the theory.

Much of this thinking was exemplified by Charles Vancouver's *General View of the Agriculture of the County of Devon* (1808), a semi-official investigation produced for the Board of Agriculture. This immensely learned survey text combined elements of a perambulation with densely argued evaluations of contemporary agricultural practices. Teeming with carefully calibrated recommendations for the future development of Devon's various soil types and micro-climates, a chapter simply titled 'Wastes' dedicated its first twenty or so pages to 'Moors and Commons'. Classifying the commons as waste was of tremendous polemical significance. '"Waste",' Sarah Wilmot explains, 'not only carried connotations of a barren and unprofitable area but was also imbued with undesirable moral qualities such as "barbarity".'[18] Vancouver's survey divided Devon's 'wastes' into five classes of district, as defined by their climate and soil type, and dedicated a separate section to Dartmoor, 'this stupendous eminence'. The detail of his five types can be passed over, though Vancouver was left in no doubt 'of the propriety of enclosing and cultivating these old moors and waste lands.'[19] He was particularly encouraged that thirty acres of Great Torrington Common in north-west Devon had been 'wrested from the inter-commoners at large' and granted to a company for the purpose of establishing a 'woollen manufactory'. This was but a small proportion of the 260-acre common, 'subject to the claims of every pot-walloping inhabitant'. Enclosing Great Torrington, Vancouver estimated, could yield £582 per annum, sharply contrasting with the £60 it currently generated in obligations to the landlord. Vancouver thought much the same could be said of other commons, moors and downs. Hatherleigh, for instance, open to 'unlimited intercommonage', was worth 7s. 6d. an acre, but Vancouver reckoned that once enclosed and developed that figure would rise to 35s. an acre.[20] Improvers, almost by definition, wished to see the dissolution of common rights.

The most powerful political tool at the disposal of the improving landlord was enclosure by parliamentary act, and Vancouver was in no doubt that landlords should use parliament to suppress rights of common.[21] Many did. Between 1700 and 1844 approximately six and a half million acres or about one fifth of the surface of England was enclosed following the successful passage of more than 4,000 private bills of enclosure. Enclosure also proceeded on a similar scale on the basis of private agreements between landowners, tenants and leaseholders, as had occurred around the low-lying Dartmoor town of Moretonhampstead in the late medieval period.[22] Enclosure established clear rights of ownership, ending long-standing customary practices that sometimes obscured which stretch of waste belonged to which parish.[23] Thus, private and parliamentary enclosure saw huge swathes of English common divided into individual farms, creating thousands of new property holders, whether as owners, leaseholders or tenants. This dynamic process generated complex procedures and negotiations overseen by commissioners appointed to represent and reconcile the property rights of the landlord, the tenants and, to a very limited degree, the customary rights of the commoners. Considered by radical historians as a grotesque process that saw the common folk dispossessed of ancient rights to the land, revisionist historians writing in the 1950s and 1960s thought the transformation was less revolutionary, less iniquitous and less rigidly predicated on the rights of private property than was once thought. According to this view, parliamentary enclosure accelerated an ongoing process based on a rough and ready fairness which generated much work for landless labourers – someone had to dig the new drainage ditches, erect the new fences, raise the new stone walls and plant the new hedges.[24]

In an eloquent riposte E. P. Thompson argued that landless labourers, with modest rights of common but no formal rights to land under the enclosure acts, were reduced to relying on the goodwill or self-interest of the landlords if they were to continue to have somewhere to graze a cow or two. Moreover, Thompson argued, reducing access to commons left the landless labourer more dependent on the parish during times of hardship, thereby offsetting some of the economic gains afforded by enclosure.[25] And though the communities of pre-enclosure should not be idealised, customary forms of cooperation and access were profoundly undermined by the transformation of the commons into privately held lands. Moreover, the moralising languages of improvement invested self-interest with a high moral purpose, valorising some forms of rural life and condemning others as barbarian, primitive and unpatriotic.

At first glance Devon did not provide the improver with a little capital at his disposal many opportunities. In contrast to the large unenclosed tracts of land in central England, much of Devon's best land had been long since enclosed. At the time of Domesday some 750,000 acres of Devonshire country had not been cleared for agricultural use. In the centuries that followed, Devon developed very rapidly, with over 200,000 acres enclosed between 1205 and 1348. At least another 200,000 acres were enclosed between 1550 and 1800. Remarkably, as W. G. Hoskins argued, the basic structure of the Devonshire landscape was established by the Elizabethan age, with much subsequent enclosure brought about through the expansion of already existing farms into the relatively low-lying moor and heathland scattered throughout the county.[26] To eighteenth-century improvers, the relative antiquity of the Devonshire landscape was of dubious value. Devon's farms and fields were too small to be efficient; its hedges occupied too much space; its roads were twisty and narrow; and its farming practices outmoded. William Marshall,

Arthur Young's great rival, went so far as to identify a specifically 'Danmonian husbandry', a strangely alien and antique agricultural culture that he believed – drawing on antiquarian thinking – could be traced back to the arrival of the peninsula's original settlers from France.[27] Such sentiments can even afflict the most sophisticated of early twenty-first-century observers. 'To an easterner like myself,' writes Francis Pryor, 'the landscapes of the south-west seem distinctly "foreign".'[28]

Devon, then, was a victim of its precocity. If the Midlands counties could be imagined as a blank canvas onto which the most advanced agricultural methods could be painted, Devon was afflicted by a primitive agricultural culture reinforced by custom and entrenched by proprietorship. Hoskins, writing at the height of World War II, took pleasure in those eighteenth-century Devonshire landowners and farmers who enjoyed a 'rough comfort', ignored all talk of improvement, lived for little more than hunting and the bottle, and were satisfied by knowledge of their family's long if not necessarily distinguished lineage.[29] Indeed, Vancouver had grumbled of Devon's low-lying wastes that the problem was less their potential than 'the want of means or inclination in their present owners or occupiers to improve'.[30] In this context Tyrwhitt could picture Dartmoor as a place of opportunity. He was a friend of the landowner, none other than the Prince of Wales, and the commoners were insufficiently organised to pose any serious opposition to him taking a lease. Because primitive and backward, Dartmoor could function as a laboratory of seemingly limitless possibility, prospecting a fantastical imagined future in which the moor became Devon's most modern and progressive agricultural landscape.

Tyrwhitt's ambitions were looking pretty flimsy by the first years of the nineteenth century. A medal awarded by the Bath and West Agricultural Society for cultivating flax on the Tor Royal estate could not disguise the fact that his twenty-year experiment was failing.[31] Perhaps his thinking was flawed, but the greatest immediate difficulty he and all Dartmoor improvers faced was the huge labour cost of bringing moorland into cultivation. How could this be met when the land could not be expected to generate any income for years to come? How could such a risky venture attract investment? The opportunity to establish the POW depot at Princetown promised improvement by cheap prisoner labour. As such, the Admiralty's decision in 1806 to establish the depot at Princetown precipitated the most important act of enclosure in Dartmoor's modern history. And not only did the depot sustain Tyrwhitt's ambitions. It proved more significant than the antiquarians, the topographers and the poets to the production and diffusion of knowledge about the moorscape, setting in train events that would shape discussion about the moor for decades to come.

More immediately, Dartmoor offered a more exciting news menu, no longer limited to the dramatic weather reports that occasionally cropped up in the local press.[32] Some events, like the death in custody of General Jago, 'a black Frenchman' who had received in action 'upwards of twenty wounds', had an exotic edge.[33] More matter-of-fact reporting could be sensational too, such as coverage of the coroner's report concluding that Pierre Ayers had committed an act of wilful murder when he stabbed a fellow French prisoner.[34] Other stories, often involving escapes, violence within the prison or acts of rebellion, were more exciting still. In October 1809 *Trewman's* reported: 'Several of the French prisoners at Dartmoor lately got beyond the boundaries of the prison walls, into the military pass, an alarm was immediately given, when an officer's servant of the Royal Lancashire Militia unfortunately came without his usual accoutrements, and being mistaken for one of the Frenchmen, was stabbed by his comrade in so desperate a manner, that he expired two days later.'[35] Or there was the murder and suicide of April 1812:

A most determined act of revenge, accompanied by self-destruction, took place last week at Dartmoor prison. One of the French prisoners having had his feelings greatly irritated during a quarrel with another prisoner, swore that he would be revenged, and accordingly, at night, when the object of his diabolical resentment lay asleep, he stole like an assassin towards him, and having plunged a dagger, or some sharp instrument into his body, retreated to his own cot, where he put an end to his existence by cutting his throat from ear to ear. Hopes are entertained of the recovery of the wounded prisoner; but the other was found stretched in his own gore and quite dead.[36]

Sensation was only one form of knowledge generated by the depot. It directly exposed thousands of people, whether guards or prisoners, to the climate and topography of the moor; it made the moor subject to Whitehall scrutiny, parliamentary debate and press comment; and, following a notorious 'massacre' of American prisoners in 1815, it became the focus of transatlantic controversy, becoming established in American republican memory as the site of an arbitrary act of monarchical violence. These encounters generated bureaucratic, polemical and literary responses, each offering alternative ways of knowing Dartmoor. Was the view of the moor from behind the granite walls and wire of the prison any more constricted or subjective than that of the agent who ran the prison, the government employee at the Admiralty struggling to get a clear picture of what was going on, the point-scoring politician on the green or red benches or the journalist providing copy? No particular point of view is singularly authoritative: the establishment of the depot made early nineteenth-century Dartmoor a landscape of widely contested meaning and significance.

In December 1808 the Morning Post reported that the prison was ready to receive its first prisoners, which were expected in mid-January.[37] This proved impossible, and in early 1809 impatient letters landed on the desk of Isaac Cotgrave, the first of the depot's two agents. What state were the floors of the prison buildings? Had the drains, pavements and guttering been finished? How were the new plantations progressing? How many prisoners could they receive?[38]

Under great pressure to expedite the work, Cotgrave was instructed on 14 April to supply weekly progress reports. Two days later he received a peremptory note to the effect that 2,500 prisoners would arrive by 1 May. It took another eleven days for the Transport Board to accept that the prison was not ready and the agent was granted more time. On 17 May a similarly peremptory note informed him that the necessary guard had been raised to escort the prisoners to the depot, leaving Plymouth imminently.[39]

The guard were mainly volunteer militia units, typically posted to Dartmoor on five-month cycles. The first to be barracked at Princetown were Lord Fitzwilliam's West York Militia. *The York Herald*, taking an interest in the prospects of local men, ominously noted that the prison was situated six miles from either town or village – an observation that might have irked Tyrwhitt. Two hundred West Yorkists marched to Princetown in advance to prepare the barracks and familiarise themselves with the layout of the depot; 300 more followed two days later, escorting the first 500 inmates.[40] In August another thousand prisoners were marched to Dartmoor, this time escorted by the men of the West Essex Militia. These prisoners had been removed from the prison ships at Hamoaze to make room for the arrival of 700 men captured at Zealand.[41] Towards the end of October, the West York Militia was relieved by the Royal Lancashire Militia.[42] This coming and going of militia groups, each facing the march from and then to Plymouth, was an essential part of the depot's history, bringing thousands of men from all parts of the country to Tyrwhitt's struggling town. At no other point in Dartmoor's history were so many strangers from so many different parts of the country exposed for a significant period of time to the moor's climate and topography. When the suitability of the moor as a location for a prison became subject to national debate, or indeed when the prospect of improving the moor was discussed after the peace, in pubs throughout the country might have been heard the opinions of men from Kent, Nottinghamshire, Cornwall, Devon, Cheshire, Gloucestershire, Hereford, Perthshire, Lancashire, Norfolk, Somerset and else-where.[43]

Anxiety about the nature of the moorscape can be read in the official records that were a by-product of the need to monitor and evaluate the day-to-day administration of the depot. They did not comment directly on the moorscape but concerned the main-tenance of humane standards within the depot, suggesting much about the particular challenges posed by the environment. Were the rights of the prisoners being met? Were the buildings being suitably maintained? Were the contractors meeting their obliga-tions? On an almost daily basis, the agent – the prison governor – was asked to explain his actions, accounting for the most petty expenditure, the most serious disciplinary action and much in between. Bureaucratic scrutiny was never merely bureaucratic. Though often retroactive, the paper trail was disciplining, reminding the agent of London's ultimate authority and diminishing any impression that the depot's isolation increased his autonomy. Confirmation that expected standards of discipline, diet, hygiene and comfort were met also attested to the suitability of Dartmoor as a location for an institution of this sort. If standards could be maintained within budget, Whitehall logic dictated that the prison could be judged a success and the decision to locate it on the moor justified.

Once the depot was up and running, diet was a source of constant concern. When American POWs demanded improved rations, including tobacco, they were told that they must wait until reports were returned from agents in America regarding the treatment of British prisoners.[44] The loyal press discussed the treatment of pris-oners in terms of how humane British standards were when compared to those of enemy, and being seen to maintain standards by deploying a rudimentary nutritional science was part of the propaganda war informing the larger military conflict. Official documentation suggested the prisoners were fed a balanced diet, and the press was happy to confirm that this represented the prisoners' reality. Each could expect a pound and a half of bread each day, half a pound of fresh beef five days a week, a pound of herring on Wednesdays and dry salted cod on Fridays. A pound of potatoes was to accompany the fish and half a pound of cabbage or turnips the beef. An ounce of Scotch barley, a quarter-ounce of onions and a third of an ounce of salt was also to be provided five days a week. Precise instructions were issued regarding the quality of food, where certain items might be sourced – Newfoundland or the coast of Labrador for cod – and what substitutes could be made when sourcing proved difficult. Leeks might be a substitute for onions, Scotch barley for greens or turnips, and so on.[45] Given these benchmarks, much discussion between the Transport Board and its agents was framed in contractual rather than humanitarian terms, with Whitehall wanting to know if the depot's suppliers should be paid. Given that snow periodically cut the depot off altogether, instructions were repeatedly issued regarding the requirement to hold a plentiful emergency supply of biscuit and how it should be stored – in casks rather than bags. The agent was ticked off for his reluctance to distribute biscuit when the quality of the bread was poor.[46]

The agent's hesitancy was understandable. When biscuit was issued in September 1812 because in a violent incident the bakehouse had been burned down the previous week, 'a serious commotion' developed, making the local and national press.[47] The Times described 7,500 French prisoners in such a 'pitch of rage' that the Cheshire Militia and South Gloucester Regiment had to face them down with loaded muskets. Gallic passions were such that some 'bared their breasts to the troops' and so 'menacing was their conduct' that an express rider was sent to Plymouth for assistance. Only when the prison was surrounded by artillery the following morning were the prisoners subdued and order restored. The Times sententiously observed that 'the allowance of bread which these men have so indignantly spurned, is precisely the same as that which is served our own sailors and marines'.[48] This incident must have been on the board's mind when that winter it made inquiries about the amount of meal the contractors kept on the moor and the need to stockpile bread – was storage space for 500 extra bags available? Just two days later the board requested the prison be supplied with salted beef and biscuit in case of an emergency.[49] Food samples were sometimes sent to London for inspection, but this did not always prove satisfactory: on one occasion the bread was indeed judged insufficiently baked, but on another occasion a beef sample could not be checked because it had spoiled by the time it reached London.[50]

Some of the difficulties faced by the prisoners, shared to an extent by the guards and the agent's permanent staff, reflected the improvised nature of the depot. Letters between the board in London and Isaac Cotgrave, the agent, detailed the rather chaotic and haphazard way the state brought these relatively novel and complex institutions into being and showed officials mired in numerous petty details.

Where was the seamstress to live? Permission was given for one of the hospital wards to be partitioned to allow her accommodation. Between October and April, she and the matron were allowed each month two bushels of coal and one pound of candles, but between May and September this was to be reduced to only one bushel of coal and half a pound of candles. Where were the dead to be buried? The board approved the selection of a small burial ground, stipulating that no more than a small wall of loose stones should be built around it. How was water quality to be ensured? Who would collect the rubbish? From where would medicines be got? How could messages be conveyed? Over several months in late 1808 monies were released so someone could be paid to super-intend the leats and collecting rubbish could be added to the respon-sibilities of a local contractor. The dispensary was instructed to purchase from the Royal Hospital at Plymouth, a clerk was appointed, and instructions were given 'to purchase a small horse, and to have a Boy, as messenger, for conveying letters'.[51]

The agent's difficulties were intensified by the multinational identity of the prisoners. Predominantly French until the first batch of 250 American prisoners arrived in March 1813, the progress of the wars in Europe and America saw the Transport Board obliged to identify other nationals among their number. Foreign ambassadors had to ensure that their countrymen were not wrongly imprisoned, and the government was keen to mobilise whatever manpower was at their disposal at any given time. Successive letters from the board comprised an oblique military history of the time, requests successively made for the release or transfer of Spanish, American, German, Portuguese, Italian, Swedish, Dutch, Mauritian, Prussian, Danish and Sicilian prisoners. The Spanish ambassador was particularly exercised that justice be done. In July 1814 the Transport Board threw a number of questions at the agent. Were there still fourteen Portuguese prisoners on Dartmoor? What about the four prisoners claiming to be Prussian? And what about Wassily E. Koslofsky, who came to the notice of the authorities that August? He was stated to be Prussian but was detained as an American and was known as Clark: was he in fact Russian? Or what about six named prisoners: were they Portuguese? And was it so that there were five Sicilian prisoners still on Dartmoor? How many Italians were there altogether and from which Italian states had they come? And what was known about the six Swedes their ambassador had been asking about?[52]

American prisoners, arriving at Dartmoor for the first time, later recalled their amazement at what they found, astonished by the social hierarchies maintained within the prison. Charles Andrews described them as 'little towns' in which every man had his separate occupation, be it in 'his work-shop, his store-house, his coffee-house, his eating house, &c. &c'.[53] In Josiah Cobb's account the prisons were characterised as the 'complete epitome of an over-crowded city':

Here were trades and occupations of every kind carried on – a mixed population, made up from all the nations of the earth – and every grade of society was here as distinctly marked as in towns and cities. None of the better classes mixed with those beneath them, who in their turn let no opportunity slip whereby they could vent their scorn at those who aimed at gentility, without having the means to carrying it out; and there again had to submit to the taunts of those of still a lower grade, for trying to ape their betters.

Cobb recollected 'the stiff and measured step of the aristocrat', the mincing of the former dandy with his 'bland smile, well cultivated teeth, and ease of salutation', and the swagger of the blackguard, who 'glories in his vulgarity'.[54]

This enforced mixing reflected other ways in which the Napoleonic Wars marked 'a watershed in the treatment of prisoners of war'.[55] The emperor had little time for the old gentlemanly practice of allowing captured officers to return home if they pledged not to return to battle during the ongoing conflict. The British, however, allowed French officers to live freely in parole towns on condition that they undertook not to escape, but between 1803 and 1814 some 889 officers – including nine generals and nineteen colonels – did just this, scandalising the British authorities. The situation was thought particularly deplorable because the French government appeared to encourage this poor behaviour.[56] Recaptured officers were often sent to Dartmoor as punishment. For instance, five French officers, permitted to live in Moretonhampstead, recaptured in an open boat near Exmouth and initially incarcerated at Exeter Gaol, were eventually sent to Dartmoor. They had to make good the five guineas paid out as a reward for their recapture.[57] General Simon, also on parole, was confined to Dartmoor when he was discovered corresponding with France through Frenchmen resident in London. A period in solitary followed.[58] Seven French officers who broke their parole at Okehampton, during whose recapture a civilian was killed, were also imprisoned at the depot. Poor information meant criminal proceedings against them were dropped, though they were put on short allowances until they had made good the £44.16s.10d. their recapture cost.[59] Thus, the breakdown of aristocratic codes of honour saw the depot used as a penal institution.

Just as the prisoner community reproduced the social complexity of the outside world, so too did it function according to recognisable political norms. Robin Fabel suggests that the American prisoners developed a 'crudely democratic' form of self-government that was informed by American republican ideas and sustained by a relatively brutal but effective jury-based system of justice and punishment.[60] There is much evidence of this in prisoner memoirs. The brief entries in Joseph Valpey's semi-literate diary comment on his social life and health, the weather and, most vividly, the punishments meted out by the prisoners. Among many such recipients were two black men flogged for stealing, a cook who received eighteen lashes for 'Robbing there Fellow Prisoner's of there Small Allowance and Skimming the fat from the Soup', and the prisoner who received thirty-five lashes for stealing a watch. The young man sentenced to four dozen lashes for stealing a greatcoat fainted after twenty-six and was 'Released for another Opportunity'.[61] The savagery of these punishments greatly exceeded those permitted on Royal Navy ships, a reminder that much of the brutality suffered by prisoners on Dartmoor was self-inflicted. The board was certainly aware that this system of justice existed and largely accepted its necessity, but when an especially severe beating came to its notice a note to the agent would demand that the guards be more interventionist. In December 1815, just months before the depot was wound up, reports reached London that a prisoner was admitted to the hospital with a 'lacerated back from severe punishment'. Cotgrave's successor Shortland was instructed to announce that only he had the authority to inflict punishment.[62] A futile gesture.

Although the prison's champions maintained that its architecture and regimen adhered to the latest and most humane thinking, the prison clearly was not the 'total institution' of modern fantasy.[63] The principal role of the militia was not to control or even supervise the behaviour of the prisoners but to ensure that they were contained within the prison boundaries. Interactions between inmates and the authorities were largely limited to meal times, supervision of the daily market and daily roll call, recalled by one American memoirist as a 'cruelty' exceeding 'murder', and abandoned by Shortland following complaints that it was inhumane.[64] Escapes, most obviously, saw prisoners break a boundary, whereas riots tested the capacity of the authorities to contain the prisoners without resort to lethal force.

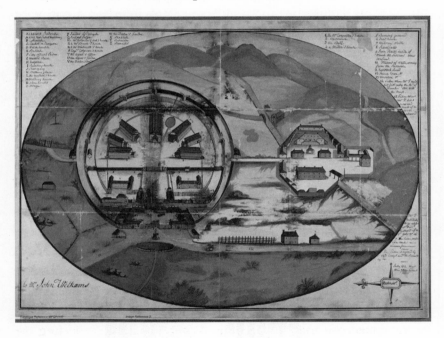

John Wethams' 1812 watercolour impression of the depot gives a good sense of where these boundaries lay. To the left of the picture are the main prison buildings and, to the right, through a walled passageway, are the guards' barracks. The outer and inner boundary walls, between which the guards patrolled, and the triangular viewing platforms can be seen; also visible are the metal palings intended to prevent fraternisation between prisoners and guards. The seven prisons, arranged in a semicircle, are divided by a wall from the hospital buildings below on the left and the petty officers' prison below on the right. In front of this can be seen the main entrance – it bore the motto *Parce subjectis* ('Spare the vanquished') – to the left of which was accommodation for the surgeons and hospital staff and to the right the agent's house. Absent is the notorious cachot, an underground punishment cell, open through a grate to the sky above.

What makes Wethams' image unusual was its portrayal of the surrounding landscape. Tyrwhitt's road can be seen, along with some fencing and other hints of improvement, but this did little to mitigate the bleak depiction of the depot's location. This was a characteristically early-nineteenth-century Dartmoor, a forbidding mountain landscape, a place of heavy skies, represented through a palate of purples and washed-out blues.

The most notorious prisoner space in the depot was occupied by 'Les Romains'. These men, their numbers running to several hundred, were so called because they lived in the prison cocklofts – 'Le Capitole'. According to one American prisoner, they were 'the most abject and outcast wretches that were ever beheld'. Judged 'too wicked and malicious to live with their other unfortunate coun-trymen', they were gathered together and confined to prison No. 4 in 1813. 'Literally and emphatically naked', they had 'neither clothing nor shoes' and were 'as poor and meagre in flesh as the human frame could bear'.[65] Visitors were scandalised by what they saw, repelled by the nakedness and the gambling. When Dr Baird, inspecting the prison in April 1812, wrote forcefully to the board about the 'vast number of the prisoners in a dirty, naked and miserable state', Cotgrave defended himself against accusations of neglect by telling the board that the state of the Romans was a consequence of their readiness to sell or exchange their food, clothing and bedding in order to buy tobacco and finance their inveterate gambling.[66] Trading of this sort had been a problem from the moment the first prisoners arrived. The board instructed that infringements be punished with short allowances and fresh bedding supplied only if the health of the prisoners was endangered, a cost which was to be borne collectively. But even if the agent was inclined to improve their condition, the board's relentless scrutiny of the depot's accounts gave him little scope to intervene.[67] When the Romans began to be removed from the prison in August 1813, new clothes were supplied at the last minute to prevent trading; a note sent by the superintendent of the prison hulks still complained they had been allowed to leave Dartmoor infected with the 'itch'.[68] Whether or not the degraded state of the Romans was largely self-inflicted, they provided shockingly tangible evidence of the limited reach of the prison authorities.

Informal interaction between the guards and the prisoners posed the authorities another challenge. The variety of life within the prison contrasted with the boredom of barrack life and offered opportunities for those out to make some easy money. Most notorious were the Nottingham militiamen who aided an attempted escape by conspiring to sell prisoners civilian clothing and pistols, but more typical was illegal trading. The daily market helped these contacts. Opening each morning between nine and twelve, it allowed prisoners to buy food and other goods at regulated prices from local traders keen to take advantage of a captive market.[69] Alongside the official market was a semi-clandestine culture of entrepreneurship within the prison – a rather innocent trade in candles between militiamen and prisoners was suspected at one point.[70] The authorities tolerated much of this activity, though they were disturbed when one prisoner's relationship with an outside potato trader allowed him to monopolise the prison's internal potato market. Reluctant to suffocate a market that allowed the prisoners to supplement their diet, the board decided the prisoners should nominate two men to act as middle men for up to two months at a time. Nominees would be ineligible to hold the position for the next twelve months.[71] By breaking the monopoly of one man, the board recognised that the prison was a complex trading environment needing external regulation.[72] A similar compromise implemented on the recommendation of the prison doctor permitted the prisoners to purchase beer in order to minimise drunkenness and undermine the illicit trade between prisoners and soldiers in smuggled spirits.[73] It's unknown whether beer proved a sufficiently appealing alternative to hard liquor.

Cancelling the market and putting all inmates on short allowances was the most effective sanction at the agent's disposal. Often used to goad prisoners into giving up a guilty man, closures were common in the early years as the prison regime was established, as when Cotgrave discovered that a prisoner was making daggers, when prisoners refused to give up pistols thought to be in their possession and as punishment for damage caused during disputes.[74] Much the most serious disciplinary issue he faced concerned the forgery of banknotes and coins by skilled French artisans. When forged notes received by a local bank were traced back to the depot in the winter of 1809–10, the sheriff officer of Plymouth was brought in and criminal proceedings ensued over the following year.[75] From the board's point of view, this threat to the national economy was more important than anything that happened within the prison or, indeed, than any prisoner escape. Henceforth, notes used in the market were marked with the prisoner's name, and restrictions were placed on the type of paper allowed into the prison. When the problem surfaced again in January 1812, the employment of French clerks by the prison was discontinued and the market was closed until the culprits were flushed out. Punished by a long stint in the cachot, they were released in June and issued notices in French claiming that civilians convicted of forgery had been executed. Despite the problem, the board insisted that prisoners could not be denied access to paper, pen and ink. The last that was heard of the problem concerned the French prisoner D'Orange. A sophisticated forger – the board was sent the plate he allegedly engraved – he spent four months in the cachot over the winter of 1813–14. The board rescinded a decision to release him after two months because this was not 'sufficient punishment for a crime, which in a British Subject, would have been punished with Death'.[76]

If cancelling the market reminded the prisoners where, in extremis, ultimate authority lay, collective punishment equally acknowledged that the authorities were reluctant to disrupt the delicate ecology of the prison by forcibly invading those parts controlled by the inmates. Exceptional interventions were made during the harsh winter of 1813–14 to protect American prisoners who had volunteered to serve in the British military. 'We often, on discovering the intention of any one to enlist into their service,' an American prisoner wrote, 'fastened him up to the grating and flogged him severely, and threatened to despatch them secretly, if they did not desist.'[77] On one occasion Shortland intervened to prevent five volunteers being tattooed with the sobriquet 'US T[raitor]'. So serious was the crime that the accused faced the Exeter assizes.[78] Like the criminal proceedings taken against the forgers, here again actions internal to the prison became subject to external justice when they challenged the national interest.

Humanitarian concern about the treatment of POWs periodically surfaced during the wars. Radical politicians and journalists fiercely contested the picture of the prison derived from the information gathered by the Transport Board and presented by government spokesmen to parliament. Sharply conflicting judgements about the implementation of policy quickly segued into disagreement about Dartmoor's natural characteristics. Polemicists divided over whether the moorscape was an exceptionally hostile or a rather more ordinary environment. Questions concerning the treatment of prisoners quickly became secondary to polemics about whether human life could be sustained on Dartmoor at all. What was Dartmoor? How should its most characteristic physical and climatic features be named? Were those hills or mountains, clouds or fog? Was Dartmoor a fixed entity, its apparently uninhabitable and unproductive state determined by nature, or was the moorscape

as open as other landscapes to improvement by human intervention? Allusions to new scientific ideas underpinned one set of
answers; fatalistic accounts of natural Dartmoor another. As the
moorscape of wind, rain, snow and ice pressed in on the depot,
this strengthened the sense that the routines of supply and discipline
that preoccupied the board were an attempt to channel the forces
of unruly nature. Whether the depot could be maintained as a
distinct place – civilisation within, exceptionally unpropitious nature
without – became a question of national significance. The front-line
troops in this struggle were the horse-drawn carts struggling
through blizzards along the moorland roads, the shivering militiamen patrolling the depot perimeter walls, and the physical and
psychological resilience of the prisoners themselves.

In late winter 1810 Dr Andrew Baird, the prison physician,
reported that 800 prisoners were sick. Cotgrave contested Baird's
figures, and although the board accepted his revised figure of 430,
it was clearly rattled and demanded daily updates. Dartmoor's
climatic conditions were also becoming a concern and Cotgrave
was instructed to keep a weekly weather journal; later that year
he was permitted to buy a 'small' thermometer.[79] In the spring
press stories referred to conditions in the depot. Picking up on
Baird's complaints about terrible ventilation in the prisons, the
Morning Chronicle helpfully reported that conditions had been
'greatly ameliorated' by allowing prisoners more time outside in
the fresh air.[80] That October a fuller account of the prison appeared
in the press. It opened with an extraordinary vote of confidence
in the enterprise as a whole:

> The Royal Prison of War at Dartmoor, in Devonshire, is one of the
> most extensive establishments of the kind in this kingdom; and, at
> the same time, demands a just tribute of applause to the judicious
> regulations which ensure kind treatment and humane attention to
> the unfortunate victims of war. Here, under the humane arrange-
> ment and control of the Transport Board, ably seconded by the

resident agent, Isaac Cotgrave, Esq. an old Post Captain, every comfort is administered to alleviate the prisoners' unhappy lot, as far as the nature of circumstances will allow. Unbiassed by motives foreign to their duty, and to the innate liberality and feelings of their heart, these gentlemen (some of whom are well-acquainted with French prisons, and have personally experienced what *they* are) pursue an undeviating system of philanthropy, honourable to them-selves, and beneficial to the objects of their care and exertions.[81]

Tyrwhitt surely inspired the report. The reader was introduced to Dartmoor, 'one of the wildest and most barren wastes of England', and the prison, located on a 'gentle declivity', was described in laudatory terms. It combined 'solidity of fabric with security of convenience'; it had an excellent hospital and medicines were 'furnished unsparingly'; representatives of the prisoners were able to inspect the food; and fair prices were maintained in the market. Most telling was the way the prison was linked to improve-ment. Though acknowledging that 'a more healthy spot might have been selected', the location of the prison was justified on account of 'the great national projects' Tyrwhitt, 'the projector', contem-plated. When 'the hand of cultivation shall have reclaimed this vast tract of moor', the report supposed, 'the existing insalubrity of the air will cease'.[82]

The importance of this last point, so casually mentioned by the journalist, should not be missed. It alluded to one of the chief claims made by Dartmoor's improvers: improvement would give Dartmoor and the surrounding areas better weather. As Vancouver explained, Dartmoor's low temperatures and high rainfall were not simply functions of its relative elevation and proximity to the Atlantic Ocean, but reflected the dynamic relationship between the heat of the sun and the cold of the earth. Dartmoor's high elevation was only one factor affecting this process, and chief among the others was the existence of high moisture levels. Dartmoor's natural characteristics inhibited efficient drainage, meaning high rainfall saturated the land

and could only be dispersed through Dartmoor's many rivers and, more importantly, through evaporation. This 'acqueous vapour' lowered the temperature of the moor and accounted for 'all those cold and blighting vapours carried by the moor-winds through all the country below'. Vancouver believed the problem would get worse because these conditions generated the 'luxuriant growth' of purple melick grass, rush cotton grass, flags, rushes and other 'aquatic plants', which when decayed added further layers to the bog, increasing its depth and absorbent capacity. In some places, Vancouver thought, the bog was already fifty feet above the granite plain on which it lay, making for a treacherous landscape that as time wore on would become more dangerous and have a more negative effect on the environment of the whole region.[83] Such thinking made Dartmoor more lake than terra firma and made improvement more than merely desirable.

More audacious still, the newspaper article praising the depot also claimed the government contemplated turning the depot into a convict prison at the end of the war, and that convict labour would be 'devoted to the cultivation of the immense waste which now surrounds the prison'. The benefits were clear to the journalist. Rather than transporting convicts to Botany Bay, where their labour was of no use to the 'mother country', these 'outcasts' could be rendered 'useful' on the moor by bringing into cultivation '80,000 acres of desolate and barren tract', thereby adding to the grain supply and diminishing concerns about national food security. Apparently, this was 'one of the most laudable and economical plans for promoting the interests of agriculture, and the benefit of the public' that the newspaper had ever known, and it 'most heartily' gave the 'applause' the project merited.[84]

Towards the end of 1810 stories were run attesting to the fair treatment of the POWs in both the French and British press, suggesting that both nations were motivated by civilised values even if the 'frippery . . . of the ancient regime had been somewhat diminished'.[85] Radical opinion was less easily pacified. William

Cobbett ran reports condemning the 'deplorable situation of the French prisoners in England' and particularly conditions on Dartmoor, 'the most unwholesome spot in England'.[86] The issue faded from view until the following summer, when reports in the *Independent Whig* prompted the radical peer Lord Cochrane to claim that POWs on Dartmoor were dying at a rate of thirty to forty a week. A government statement immediately refuted these claims. Of 45,531 French prisoners in the country, only 321 were sick. Cochrane's figures were out of date. They referred to the previous winter, when prisoners arriving from the West Indies 'in a most filthy state' had brought typhus fever with them. Cotgrave had dutifully reported the high mortality rates to the board.[87]

Cochrane addressed the House of Commons on 14 June, explaining his previous statement. He had visited Dartmoor in response to the many letters he had received describing the 'truly deplorable' conditions the POWs had to endure. Refused entry, which he believed contravened his rights as an M.P., what he could see through the grating of the inner courtyard confirmed what he had been told. He described the depot as 'exposed on the summit of the highest and most bleak range of mountains in Devonshire, where the winter winds pierce with all the keenness possible, increased by constant fogs and sleet and rain'. No vegetables grew, and such was the 'state of obscurity' caused by the fog and rain that he needed a guide the whole time he was there. Why, it was known that even Scottish men had refused to settle on Dartmoor! Certain that attracting inhabitants to the moor cannot have been the motive for establishing the depot, Cochrane described how in search of further clues he examined the prison plans held at Plymouth. They confirmed to him that the whole enterprise was fundamentally ill conceived. For instance, food was issued to the men from a single doorway, which necessitated them standing in line, a thousand at a time, in the rain and without any prospect of a change of clothing. Fever, Cochrane maintained, was bound to arise from 'wet clothes in close prison'. Moreover, economy could not explain the location, given that provisions, including coal, had to be transported from Plymouth. How was it, he pondered, that 'this prison should, by accident, have been placed on the only spot in Devon, whence the stercoraceous matter of the depot [faeces] could, by the power of gravity alone, descend on a neighbouring and elevated estate belonging to the Secretary of his royal highness the Prince Regent'? An insinuation that displeased the House. In short, Cochrane perorated, the 'Dartmoor depot ought not to have been placed on the top of the highest and most barren range of mountains in Devonshire, where it is involved in constant fog, and deluged with perpetual rain.'[88]

On 18 June further reports refuting Cochrane's allegations were made to the Commons. George Rose, treasurer of the navy, reached for hyperbole, claiming that 'human ingenuity could not contrive a building of the kind better calculated to preserve the health of its inhabitants'. He praised the 'stream of pure water running through the middle of it' and described how 'every care' had been taken 'to make the accommodation as comfortable as possible'. Tyrwhitt also spoke up, asserting that there was 'no part of this country more healthy' and, only partly answering Cochrane's insinuation, informing the House that he paid for the soil to be carried to his land. Eyebrows must have been raised when Mr Whitbread, speaking for the Admiralty, added that Tyrwhitt said he had exerted no influence over the building of the prison and did not benefit from its establishment.[89] Still, once the statistics are digested, it is hard to accept the validity of Cochrane's basic charge. Yes, official figures revealed a terrible peak in mortality over the winter of 1809–10, but this had been a consequence of exceptional circumstances. Between July and September 1809 the number of inmates almost doubled to 6,031 and deaths that September had totalled 15. From November 1809 to May 1810, the period when the board, Baird and Cotgrave had discussed the level of sickness at the depot, the monthly death rate had been respectively 63, 131, 87, 63, 28 and 25. Since then the monthly figure had not risen above 20.[90] Figures produced for Chatham showed Dartmoor's typical level of sickness was unexceptional,[91] and although the depot's aggregate mortality rate, 4 per cent, was the highest of the land prisons, a third of those 1,455 deaths were caused by the 1809–10 epidemic.[92] Explaining the epidemic, the lord chancellor said French prisoners arriving from the West Indies 'in a very dirty state' had brought the fever into the prison, but it was soon brought under control. At that moment, he crowed, the prisoners were 'in a state of comfort unequalled in any military prison, and in a state of health which was not exceeded by that of the healthiest district in England'. Whitbread endorsed these implausible claims, and, alluding to the 'Romans', argued that some prisoners were in a bad state because they had gambled away their clothes.[93] Their degraded state was self-inflicted and did not deserve the sympathy of parliament.

Deploying statistics did not see off the controversy. Later that July the *Independent Whig* renewed its criticisms, claiming not only that 'the air of Dartmoor prison is considered most detrimental to health' but it was also 'pretty well understood that the prison was built for the convenience of the town – not the town for the prison'.[94] Developing this serious charge, Princetown was described as a 'speculative project' that had proved 'unprofitable'. The towns-people – 'solitary, insulated, absorbed and buried in their own bogs' – needed a market for their goods if their town was to survive, and 'thence arose the dreary, deadly walls of Dartmoor Prison'.[95] Some £200,000 of public money, the *Independent Whig* implied, had been spent to sustain Tyrwhitt's failing community. However hard the government printing presses got to work vindicating the healthi-ness of Dartmoor's climate, and however libellous the original allegation regarding the rate of mortality in the depot, it was hard to refute the basic insinuation that the decision to locate the prison on Dartmoor was corrupt.

With the attorney general bearing down on the *Independent Whig*,[96] debate shifted from scrutiny of government decision-making towards the nature of Dartmoor itself. Between August

and October, *Humanitus* and T.H. slugged it out in the pages of
the *Examiner*. *Humanitus* began by attributing the evils the prisoners
faced to where they were imprisoned rather than the treatment
they received. A tree was not to be 'seen perhaps for more than a
dozen miles on these stony mountains, and the gutters or valleys
which nature has made between them, are full of the most horrible
swamps'. The error was to have spent vast sums erecting a prison
'in this Siberian desert', and the problems could only be remedied
'by abandoning a place never designed by nature for the abode of
man'.[97] T.H. ridiculed the extreme picture painted by *Humanitus*.
In his hands, 'every hill is quickly converted into a snow-top'd
mountain, and every agreeable stream (of which there are many
on this moor) into a frightful bog'. A native of Dartmoor, T.H.
wrote to familiarise his readers with the moorscape, explaining that
this 'circular common, with a diametrical road through it', though
less cultivated, was 'like some parts of the West Riding of Yorkshire'.
It was not, as *Humanitus* had ludicrously stated, a range of mountains but a place of 'hills and valleys', few of which were 'much
higher than Highgate or Hampstead Hill'. The clouds *Humanitus*
believed capped the mountains were merely fogs, and naming them
clouds was like translating 'the smoke of an iron-foundry or a
glass-house into the rolling volumes of Mount Etna'.[98] A letter
reprinted from the *Plymouth Chronicle* saying Dartmoor was no
different to other hilly tracts was similarly domesticating.[99]
Humanitus, his polemical powers recharged, needled T.H. for his
'happy knack of levelling mountains, and fertilizing deserts; at his
presto, the snows of Siberia may melt into murmuring rills, or the
sands of Arabia change their arid face into fruitful verdure'. This
was all good knockabout stuff, designed to expose the foolishness
of the other party, but *Humanitus* then alleged that the high
mortality in the depot had not been caused by prisoners newly
arrived with infection 'but that the *fogs and damp* of the moor had
produced a species of asthma and consumption throughout the
prison'.[100] *Veritas*, entering the fray, asked how this could be squared

with the state of the soldiers who stood guard for four hours at a
time, day and night, and 'are always healthy and well'. So should
the prisoners be if properly provided for.[101] Perhaps so, though
three months later the cold was so severe the soldiers were relieved
of their guard duties every half an hour.[102] *Humanitus*, sharpening
his quill one last time, reasserted his belief that if the prisoners
now enjoyed good health it was in spite of the 'salubrity of a spot,
exposed to fogs and surrounded by swamps'. The violent sickness
that had reportedly broken out that September strengthened his
case. It was well known, he claimed, 'that the mornings and evenings
of that month often exhale more than others the pestilential vapours
of bogs putrified by the Summer'.[103]

Not surprisingly, a uniformly negative impression of the moor is
to be found in prisoners' memoirs. These compelling accounts were
written and published during the peace and after the heat had gone
out of these debates in Britain. Where the poets valued Dartmoor
as a landscape of the sublime and the improvers as a productive
land of the future, former prisoners represented the regime within
the prison as barbaric and the moorscape as fundamentally malign,
a lifeless place profoundly unsuited to human habitation. No appreciation of the moor's natural beauty was allowed to leaven their
accounts.

Charles Andrews, an American prisoner, wrote that on hearing
they were being transferred to Dartmoor, 'the very name . . . made
the mind of every prisoner "shrink back with dread, and startle at
the thought," for fame had made them well acquainted with the
horrors of that infernal abode, which was by far the most dreadful
prison in all England, and in which it was next to impossible for
human beings long to survive'.[104] This is a startling claim. How had
such knowledge become so firmly established in the American
prisoners' minds? Had their British guards tormented them with
terrible stories? Had knowledge of the moors been communicated
to them via French prisoners? Had the public debate of 1811 been

covered in the American press? What is certain is that prisoner memoirs presented a very consistent picture of Dartmoor and the depot. Accounts often began with the march from Lopwell Quay, Andrews recalling how they arrived at the prison at nightfall, following a march 'through heavy rain, and over a bad road', to find the ground covered with snow. 'Nothing could form a more dreary prospect than that which presented itself to our hopeless view,' he wrote. 'Death itself, with hopes of an hereafter, seemed less terrible than this gloomy prison.' The tone set, he moved into a more formally descriptive mode, explaining to his American readers that the prison was situated 'on the east side of one of the highest and most barren mountains in England'. It was surrounded on all sides 'by the gloomy features of a black moor, uncultivated and uninhabited, except by one or two miserable cottages, just discernible in an eastern view' – that was Tyrwhitt's pioneering little town – 'the tenants of which live by cutting turf on the moor, and selling it to the prison'. Unintentionally or not, Andrews echoed the claims made by Tyrwhitt's critics. The moor was 'deprived of everything that is pleasant or agreeable, and is productive of nothing but human woe and misery. Even riches, pleasant friends and liberty could not make it agreeable.'[105] His detailed description of the prison itself made it a place of '"living death"', a 'seminary of misery' and a 'gloomy mansion' inhabited by 'fluttering, ghastly skeletons'.[106] The awful conditions were partly caused by the inadequacies of the buildings, for water constantly dropped 'from the cold stone walls on every side, which kept the floor (made of stone) constantly wet, and cold as ice'.[107] Above all, the depot's location lay at the root of the problem:

> During the month of April there was scarce a day but more or less rain fell. The weather here is almost constantly wet and foggy, on account of the prison being situated on top of a mountain, whose elevation is two thousand feet above the level of the sea. This height is equal to the place on which the clouds generally float in a storm,

the atmosphere not being dense enough to support heavy clouds
much above that height; almost every one that passes that way finds
the top of the mountain enveloped in thick fog and heavy torrent of
rain. In winter the same cause makes as frequent snows as rain in
summer. It is also some degrees colder during the whole year than
in the adjacent country below. This too is occasioned by the great
elevation of the top of the mountain, which is above the atmosphere
heated by the reflected rays of the sun upon the common surface of
the earth, and being small of itself, reflects but little heat. These two
causes combined, produce constant cold and wet weather.[108]

Josiah Cobb, author of the most crafted of the memoirs,
preferred Gothic expressiveness over pseudo-science:

The turnkey swung open the portal, in we entered, and the
ponderous door of bars and rivets was slammed in our rear, with
a hollow sepulchral sound, that was only equalled in dolefulness
by the harsh grating of the key and the snapping of the bolts, as
they shot into their deep-sunk sockets in the granite jam[b]s. I stood
for a moment or more, before I could collect my sight and senses,
from the glare of light and the hum of many voices which burst
upon me; and the only conclusion I came to, was, that I had suddenly
awakened from a disturbed dream, which had left me where I was
in reality, in Pandemonium.[109]

Cobb's description of the moorscape was as sinister, but more
revealing is the rare insight he offered into the ways of local people
and their relationship with the depot. Blending together a little
antiquarian and folkloric knowledge, he implied that the market
traders were doing well:

By the small irregular mounds and ridges of raised earth, that were
here and there discovered scattered over the plains, with evident
signs of great antiquity, it is thought that battles must have been

fought on these heaths in ancient times. And so fully were the country people's minds convinced, that the departed dead were yet restless, and that their spirits still danced over the moor in that witching hour, when ghosts love to riot in transparent forms, and lightly float in drapery of swathing clothes, that it was difficult for those rustics to bring their minds, to pass this extensive waste after night-fall. They would always time their startings from the little village of Princeton, so as to cross the waste ere dark. It was a picturesque sight, from an upper window, to follow those who attended the markets, each mounted upon a donkey, driving before him or her fifteen or twenty more, one strung before the other, in single files, with panniers swung across their backs, wending the way to their homes. When at a distance, and following the crooked sheep paths, I could liken them to nothing but the sinuous move-ments of the snake, as it winds itself along, without any observable motion, except a snail pace progressing.[110]

Although adhering to certain established descriptive conventions, these were specifically American narratives, each culminating in the event that gave their writing purpose. On 6 May 1815, a few months before the depot was emptied of prisoners, the 1st Somerset Regiment of Militia opened fire on a crowd of prisoners. The most notorious act in the history of the depot left six prisoners dead, another to later die of his injuries and many more maimed, some requiring amputations. Establishing for certain what happened that day is difficult. Was the shooting a response to an attempted escape or was it an attempt to quell a threatening riot, constituting the 'justifiable homicide' of the official report? Or did the militiamen lose control, shooting wildly following months of taunting and stone-throwing by prisoners frustrated by not having been freed despite the end of the war? Or was the massacre, as some memoir-ists claimed, premeditated murder instigated by Shortland, who, determined to revenge various challenges to his authority, person-ally ordered the firing?

What seems certain is that a major confrontation had become likely. American prisoners appeared to settle into prison life and accept the authority of their captors less readily than the French. This was particularly evident from the autumn of 1814, when the departure of the French and their replacement by further Americans saw the militiamen more vulnerable to violent assault, their authority within the depot ever more fragile. The prisoners became more restless still when the Treaty of Ghent in December 1814 concluded the British–American war but left the date of their release uncertain. A nasty smallpox outbreak that January, blamed on poor ventilation, took some 200 lives, enraging inmates already angered by the condition of men kept in solitary.[111] Napoleon's return from Elba and the 'Hundred Days' delayed their release further still, leaving prisoners mutinous as the best time to cross the Atlantic fast approached.

Anger focused on Mr Beasley, their official government representative and long considered neglectful of his responsibilities. He was tried in effigy, convicted of depriving Americans of their lives 'by the most wanton and most cruel deaths, by nakedness, starvation, and exposure to pestilence', and executed, the effigy paraded, abused, placarded and hung from a lamp hook.[112]

Deteriorating material conditions made matters much worse. Astonishingly, with the ratification of the peace, the American government stopped paying the prisoners their allowance. Debts were called in, and as credit dried up trading ceased altogether. Things got fractious, as Josiah Cobb recalled: 'Meetings were called, propositions propounded, backed by resolutions of a nature to destroy the propositions, memorials were drawn, adopted, cancelled, torn up and afterwards reconsidered, all in the same half hour, for not being sufficiently strong in their wording, or for expressing more than could be understood by their intended meaning.'[113] More dependent on the British than ever, a few days before the massacre a dispute over the bread ration caused a near riot. Shortland, absent in Plymouth, had ordered that one pound of hard bread be served in place of the usual pound and a half of soft bread. Refusing this ration, the prisoners forced their way to the store, and the officers of the garrison, fearful of their anger, relented and issued the standard ration. The memoirists were convinced that Shortland, 'tyrannical' but fearful (he now only appeared in public with an armed guard), returned to the depot determined to reassert his authority.[114]

HORRID MASSACRE AT DARTMOOR PRISON. ENGLAND.

Where the unarmed American Prisoners of War were wantonly fired upon by the guard, under the command of the Prison Turn-key, the blood thirsty SHORLAND; Seven were killed, and about Fifty wounded, (several mortally,) without any provocation on the part of our unfortunate American Citizens!——" Blood has a voice to pierce the Skies!"

Drainage, Manure and Redemption

In 1854 Henry Tanner, an Exeter solicitor, published *The Cultivation of Dartmoor*, recipient of the Western Literary and Scientific Union prize for 'the best Essay on the Cultivation of Dartmoor, as a source of employment for the unemployed population of the district'. Though opening with a Romantic description of the moor, Tanner diminished the threatening aspect of Dartmoor's outward appearance by explaining that the 'billowy surface' of the moor's interior, typically 1,300 feet above sea level, was less formidable than its exterior escarpments or the tors implied.[115]

Insisting it was no more intimidating a prospect to the farmer than County Durham, Tanner explained that Dartmoor's problem was neither its elevation nor high rainfall but inadequate drainage and the negative effect this had on the moor's climate. Repeating Vancouver's arguments, Tanner described how the 'atmosphere resting on it [the moor] becomes charged with vapour, and *by this evaporation a large amount of heat is abstracted from the air*'.[116] Radical interventions were needed. If drainage channels could not be established, open ditches should be dug, and the 'Moor-pan', a compact layer of ferruginous earth, be broken up. Once the red loamy subsoil was accessed, it could be fertilised by applying lime and a strict system of crop rotation instigated – first rape, then oats, and then grass, a cycle that might need repeating. The land would then be suitable for growing oats, rye and flax, as well as spring and winter vetches, grasses, clover, rape, turnips and swedes. Of Dartmoor's vast acreage, Tanner reckoned 10,000 acres were probably suitable for arable and 50,000 acres as permanent grassland for livestock nearest 'their wild condition', like Cheviot and Cotswold sheep.[117] Converting heath into grassland seemed the simplest way of making the moorscape more productive.

Providing shelter was essential and plantations were profitable, so Tanner assessed the qualities of different trees, differentiating sycamore and beech, which could take strong westerly and south-west winds, from ash, which could face bitter north and easterly winds. Larch and pine could be cultivated in between, though in the lower districts ash, Bedford willow, horse chestnut and birch were more suitable. If grazing land were mainly used in the milder months, the plantations would provide moor men with work in the winter. Profits were assured. Some 110,000 acres, reclaimed and leased, would see the per annum rental value of the common increase from £200 to £66,250.

Significant investment was needed to realise this profit, and

Tanner revived the periodically mooted idea that Okehampton should be connected by road to Princetown. The proposed route through the northern plateau followed the West Okemont, then cut between Ammicombe Hill and Great Kneeset to the junction of Rattle Brook and Tavy Cleave, thence direct to Princetown. More generally, either the proprietors, including the Duchy of Cornwall authorities, should invest heavily in preparing the land for new lessees, or new tenants with a modicum of investment capital should be induced to take on long tenures with the promise of low rents.

Modern authorities go some way towards vindicating Tanner's thinking. Studies exploring the history of drainage recognise that in England 'the removal of excess water . . . has always been the most important water management practice'.[118] Effective drainage enables soils to warm up more rapidly in the spring, reduces the amount of water that has to be removed by evaporation – evaporation *does* cool the surface of the soil – and improves aeration, allowing for the greater diffusion of oxygen into and carbon dioxide out of the soil. Drainage also causes the soil to shrink and crack, creating channels for roots and fissures for water.[119] By the middle of the nineteenth century, agricultural commentators regarded schemes of drainage as fundamental to progress. Some £27.5 million, a significant portion lent by the government, was spent on drainage projects between 1845 and 1899, with 4.5 million acres drained, about 35 per cent of the total modern authorities judge needed doing.[120] Given that drainage schemes were usually the work of tenants spending capital lent to them by their landlords, Dartmoor, as a common, could not benefit significantly from this form of capital investment: once again, the nineteenth-century perspective suggested that progress could only be made if common rights were dissolved and the upland subdivided into individually worked holdings.

Dartmoor's history is thrown into sharp relief by the development of Exmoor, where catchwork meadows were used to irrigate sloping terrain, allowing crops to be cultivated. Here, notwith-standing more benign geographical characteristics, the crucial differ-ence was that in 1815 Exmoor was purchased from the Crown for £50,000 by John Knight, a wealthy Midlands ironmaster. His son, Frederick Winn Knight, inherited the land in 1841 and built several farmsteads, which he let cheaply, encouraging his tenants to under-take the reclamation of the moorland.[121] Given that Knight behaved almost exactly as Tanner and others hoped the duchy and other landlords would, it is odd how rarely the Exmoor model crops up in their schemes. Tanner did though celebrate the efforts of G. W. Fowler, who had purchased Prince Hall in 1846 and continued work begun there by Judge Buller, but his efforts hardly compared to Winn Knight's. Tanner had come to believe that no individual landlord was equal to the task and the most important development that gave him hope for Dartmoor's future was the revival of the prison:

> I look upon this proceeding as introducing a new era in the history of Dartmoor, for the means at the disposal of a Government enable them to prosecute works of such extended character, as quite eclipse the most zealous and praiseworthy operations of individuals. I consider this portion of the Moor, as the centre from which cultiva-tion is to radiate on all sides, and although it is 'an oasis in the desert', yet eventually, luxuriance and fertility will pervade the face of Dartmoor. Not only does this central position recommend it as a desirable spot for commencing any general system of improve-ment, but such important advances have been made in the establish-ment of Prince Town, with its Church and Public Schools, and the improvement of large portions of the surrounding Moor, confirm my conviction of its future history.[122]

Had Tyrwhitt found a worthy successor? Certainly so in the emphasis Tanner placed on Princetown as 'the centre from which cultivation is to radiate on all sides', but in placing the state rather than the visionary individual centre stage, Tanner intimated a very different future for Dartmoor. Just how plausible were his hopes? Why had Princetown once again become a place of incarceration and how was this linked to improvement? To answer these questions we have to return to the 1810s and the early years of the POW depot's long afterlife.

In G. A. Cooke's *Topography of Great Britain* readers were provided with skilful but conventional descriptions of the moorscape. The 'summits of several of the higher swells of Dartmoor', he wrote, were 'truly savage', 'rendered finely picturesque, by reason of immense piles of stones, or huge fragments of rocks, thrown confusedly together, in the most grotesque manner: sometimes crowning knolls, but oftener hanging in their flows'.[123] Descriptively on message, admitting Dartmoor to the company of more exalted landscapes, Cooke also joined the chorus anticipating its improved future. Dartmoor's soils, 'greatly above the par of mountain soils' elsewhere in Britain, were 'superior' to those of Scotland and the north of England. True, parts of the moorscape were difficult, whether the more elevated spots 'covered with black moory earth' – Cooke's prose was not very technical – or the peat bogs in between, but a more 'genial climature, and a proper supply of manure' was all that was needed 'to render them valuable as arable lands'.[124] Many people had been saying the same thing, what with the government's plans to build an 'extensive' prisoner of war depot and the efforts of the Prince of Wales 'to encourage the settlement of industrious people in this hitherto uncultivated district'. Cooke was optimistic. Of Devon's 320,000 acres of 'waste', only a small part of Dartmoor could not be improved, meaning there was potentially £150,000 available in additional rents. The upland afforded 'innumerable opportunities for the beneficial exercise of industry and capital', not just to the agriculturalist, but also, thanks to the ready water supply, the promoter of 'every species of manufacture requiring the use of industry'.[125]

These urgings, taken from the 1817 edition of Cooke's book, were only a little out of date. The POW depot had been built, but it now stood empty, its reputation, always shaky, wrecked by the notorious massacre of 1815 that scarred its last days. Princetown was quickly falling into decay, once again looking like a failed experiment, only now it boasted the monstrous relic of the prison. Cooke, though, was not that wide of the mark. During the decades after the peace, the notion that Dartmoor could be dramatically improved activated the imaginations of many people, and the empty depot brought their speculations into focus. In the event, nothing on a grand scale was done until the 1850s, when the dilapidated depot was converted into a convict prison. This recycling was only tangentially related to the particular characteristics of Dartmoor, but the reasons offered for the revival of the site in the preceding years made much of Dartmoor's agricultural and industrial potential.

Cooke had probably picked up the importance of manure to Dartmoor's future from Vancouver, and the same need formed the central plank in Thomas Tyrwhitt's proposal that Princetown be connected to Plymouth's lime quarries by horse-drawn railway. Presented to Plymouth Chamber of Commerce on 3 November 1818, Tyrwhitt promoted his projected railway as the keystone element in a comprehensive scheme for the improvement of his favourite part of the moorscape. His new project was surely shadowed by lessons learned from the experience of the POW depot, when problems of supply and transportation had under-mined its effective functioning from the beginning. Tyrwhitt's imaginings could only be fulfilled if Princetown were properly connected to the outside world. With the estimated cost of the railway placed at £45,000, which Tyrwhitt hoped to raise by public subscription, his case had to be good.

Tyrwhitt's opening statement of aims is worth quoting in full:

To reclaim and clothe with grain and grasses a specious tract of land, now lying barren, desolate, and neglected; to fill this unoccupied

region with an industrious and hardy population; to create a profitable interchange of useful commodities between an improvable and
extensive line of back country and a commercial sea-port of the first
capabilities, both natural and artificial; to provide employment and
subsistence for the poor of several parishes; and to alleviate the pressure of parochial burthens, by a method, at once simply ingenious
and comparatively unexpensive, form altogether such a stimulus to
adventure and such a scope for exertion, especially to a wealthy
Company, as must dilate the benevolent heart of the patriot, while
it emboldens the capitalist gladly to lend his assistance in carrying
the plan into execution.[126]

Who could disagree? In Tyrwhitt's clever hands, distinctions
collapsed, conflicts of interest dissolved. Patriot or capitalist, ratepayer
or pauper, all would benefit. Income would be generated through the
transport of importable and exportable commodities. Up to fifteen
tons of produce could be drawn by horse from Princetown to Crabtree
in five hours – the downhill journey – and four tons of material
returned uphill in six hours. Wind, rain and frost would no longer
impede progress, the railroad making an often dangerous journey
safe. And the scheme gave 'no occasion to demolish houses, encroach
on pleasure grounds, or remove any material obstructions'.[127] What,
then, would the railroad carry? Lime, sea sand, timber, bricks, slate,
tiles, and laths, all materials needed if the land was to be cultivated
and new farmsteads built. Tyrwhitt was particularly exercised about
the necessity of bringing large supplies of lime to the moor, which
he judged the most effective manure. Culm would also be needed
– to fuel the limekilns – as would coal, an essential of domestic life,
though demand might be diminished by the abundance of peat. People
had to eat, and Tyrwhitt's settlers would desire tea, sugar, wine, spirits,
beer and porter, as well as other commodities like furniture. Improving
the moor would bring new populations to the region, stimulating the
growth of Plymouth businesses, which Tyrwhitt thought had been
held back by the absence of a 'back country'.[128]

This traffic would not be one way. Tyrwhitt rightly predicted that the local and national market for Dartmoor granite would grow, whether it was used for paving stones, gateposts, in road building or as gravel: even he did not imagine the large-scale demand for granite that major building projects in London would generate. Then there was flax, his particular *idée fixe*. Tyrwhitt's own experiments at Tor Royal convinced him there was 'scarcely a part of the Dartmoor district' which, sown to flax, would 'not amply reward the cultivator'. He imagined farms of thirty or forty acres, let on ninety-nine-year leases, supplemented by smaller plots of four or five acres let to poor but industrious men, who would provide the labour needed by the larger farmers.[129] Later schemes for the settlement of the moorscape followed a similar pattern, looking to nurture communities that embodied the proper relationship between the classes. Tyrwhitt now hit his stride, drawing out the further benefits of this scheme from 'a *moral* point of view'. Cultivating flax would provide employment for 'indigent' and 'unemployed persons, of all ages', and he imagined 'the *art of spinning*' being introduced to the poor houses. This would give opportunity to young people of a 'virtuous' and 'active' disposition, for once taught to spin, they could 'always procure . . . a comfortable livelihood' and ensure 'the preservation of their independence'. Such, he claimed, had been the case when flax spinning was introduced in Exeter and parts of London.[130]

As with all well formed speeches, Tyrwhitt's peroration returned to the beginning, in this case recapitulating the universal appeal of the scheme. It would see 'a tract of land, at present locked up in barrenness, reduced to penury by the want of culture, and wanting population . . . rapidly inclosed, cultivated, planted, built upon, inhabited, and at last rendered a most valuable integral part of the kingdom'. The scheme would reduce parochial burdens, and small investors, facing the 'present dearth of good mortgage securities', would have an opportunity to employ their capital in a 'beneficial manner'. Improving Dartmoor would halt emigration, for there was no reason why people should leave for Canada when there was so much scope for the 'colonisation of Dartmoor'. Contributing towards the development of this 'commercial interchange' would 'add a large quantity of improved land, strength, and population to the kingdom', 'yielding new incentives to industrious emulation, local prosperity, and public improvement'.[131]

Just two months later the 'Prospectus of the Plymouth & Dartmoor Rail Road' was issued. Briefly reiterating Tyrwhitt's case for the twenty-two-mile line, it added that engineers had now established that the route ensured a fall of only half an inch every three feet, suggesting a shallower decline than sceptical observers had predicted. Impressive figures detailing the value of goods the line would carry were included, ranging from the £12.10s. that might be paid for 100 tons of turf to the £1,000 that 4,000 tons of the best-quality granite would fetch.[132] A year later, William Bailey of Bankside, Surrey agreed to supply cast iron at the price of six pounds and nineteen shillings a ton.[133] Right on cue, the *Morning Post* puffed Tyrwhitt's endeavours, printing a letter from 'Juvenis Devoniensis'. Praising Tyrwhitt and, to a lesser extent, the Reverend Bray for their efforts to cultivate Dartmoor, the correspondent mocked the hubris of 'others (delighted with the idea of possessing an estate of so many hundred acres)' who, having carried 'a line of circumvallation around their extensive domains', grew 'tired of their bargain' and left the moorscape 'rather more a forest then if they have not begun it'. Had they factored in the acquisition of lime, which could only be got from Plymouth and could now be supplied by Tyrwhitt's railroad, they might have had more luck.[134]

On 2 July 1819, 8 July 1820 and 2 July 1821 parliamentary acts permitted the construction of the railway, two extensions and a share issue – other rail projects of the time had been financed by loans from the Office of Public Works.[135] Archival evidence that the railway was soon up and running comes in the form of a surviving notice dated 30 October 1824 warning trespassers that they must not use 'improperly constructed' carts, wagons and carriages on the railway. Liabilities amounted to a fine of up to ten pounds or in the case of anyone caught 'wilfully Breaking, Throwing Down, Damaging, Destroying, Stealing, or Taking Away any Part of the Railway, or other Works' transportation for seven years.[136]

It cannot be said if the activities prompting the issue of this notice reflected local opposition to the enterprise. Tyrwhitt's railroad certainly brought job opportunities to the area, as the disused quarries at Foggintor and Swelltor indicate, but local carters must have suffered, particularly given the poor quality of the road between Princetown and Yelverton.

Tyrwhitt did not forget his other pet project, and he discussed the disused POW depot in his original statement at Plymouth. One of the contingent benefits of the scheme, he argued, was it provided for the easy transfer of convicts to his putative Dartmoor Prison. Tyrwhitt keenly quoted an 1818 government report that considered Dartmoor as a means of relieving pressure on prison places in the capital. The buildings were thought suitable for conversion, and the estimated cost of rendering them appropriate accommodation for 2,000 prisoners put at £5,000; the location was judged healthy, provisions were cheap, and fuel was easily available. Practicalities were not all. Prisoners sent to Dartmoor would be 'removed from the contagious vices of the metropolis', where committing a crime was 'mostly the precursor of another more iniquitous', and 'youth delinquent', through compulsory labour, would not develop those 'habits of idleness' that 'engender depravity'. They could cut the granite needed for public works, which would 'cleanse the land and render it fit for cultivation', cut turf, which could be conveyed by railway to Plymouth where it would be sold, and they could make their own clothing from the coarse wool available on the spot. Dartmoor Prison would be morally improving and self-financing.[137]

Tyrwhitt accurately reflected the conclusions of the report, though the evidence of expert witnesses had been less conclusive. Their testimony rehearsed old debates, seeing the controversies provoked by the POW depot revisited by the original protagonists. Captain James Bowen, commissioner of the Transport Board, Daniel Alexander, Tyrwhitt's original architect, and Colonel Wood, whose regiment, the East Middlesex, had been posted to the depot, talked up the site's potential. Sir Willam Elford, living in nearby Bickham, was less positive, reckoning that the cost of improvement was prohibitive: were it not, he reasoned, the opportunity would already have been taken. Medical opinion was mixed. Baird, the depot physician, explained that it was unresolved whether conditions on Dartmoor aggravated pulmonary complaints such as consumption, but there could be no doubt that the depot's location and the 'humid atmosphere' exacerbated the difficulty of treating a major outbreak of illness. No medic gave an unambiguous thumbs up.[138]

Others thinking about possible uses for the disused depot shared Tyrwhitt's view that time spent on Dartmoor might be reformatory. A society aiming to convert the depot into the 'Metropolitan School of Industry' was established under the patronage of the Prince Regent, the presidency of the Archbishop of Canterbury and fifty vice presidents chosen from among the merchants of the City of London at the Mansion House in January 1820. Pauper children were to be removed from London, thereby reducing parochial burdens, and housed in the new school, where they would be taught the skills needed to successfully farm on the moor. Newspaper reports noting that similar schemes prospected thirty years earlier had been dropped as impractical, admitted new agricultural techniques gave new life to the idea, particularly the use of lime to fertilise land for the cultivation of flax. 'It is, we understand, upon this manure,' drily commented the *Gentleman's Magazine*, 'that the present hope of redeeming Dartmoor from sterility is founded.'[139] The society aimed to eventually settle the pauper children onto small farmsteads where they could live independently.[140]

Not all were convinced. In a great stack of solicitous notes addressed to Lord Sidmouth, the notoriously draconian home secretary, was a letter from Henry Wilson.[141] Written in anticipation of the Mansion House meeting, Wilson questioned whether the prison was a 'fit place' for this new purpose.[142] He developed his ideas at

greater length in a pamphlet. Dartmoor's distance from London
was an obvious disadvantage, for it would be difficult for the society
to provide appropriate supervision, but, more than this, Wilson
objected to relocating a child innocent of a crime to a place so
suited to a penitentiary. Particularly troubling was the idea that any
child's future would be determined by this placing. Not only would
this deny him or her the opportunity to rise according to their
merits, but the suggestion that such children would be paired off
for marriage as adults was 'contrary to the established rules of
Providence and cannot be gravely entertained'.[143]

Wilson had his own agenda. The evident problem of London's
pauper children should be addressed by establishing a school closer
to London, allowing the Dartmoor depot to be reopened as a place
of punishment for juvenile delinquents, a staging post on the way
to transportation. Convicts made to cut roads, drain the land, form
enclosures, would be 'atoning, in some degree, for the injury they
had done to society'. Like Tyrwhitt, Wilson saw the advantage of
removing the juvenile offender from 'the theatre of his early delin-
quency', and he emphasised how his ideas aligned with the improvement agenda: there were minerals to be extracted, granite to be
quarried and dressed, and flax to be spun.[144] Nurturing self-discipline
and a capacity for hard work was only a part of Dartmoor's appeal.
Its natural characteristics could have a morally improving influence
too. Wilson wrote: 'Situation and other local circumstances may,
however, have a powerful influence on the most obdurate; and
there is a greater hope of bringing the mind of the hardened
delinquent to sober reflection, in a prison in the middle of a Moor,
unsurrounded by tree or shrub, to relieve the wearied eye, than in
a Penitentiary within the dulcet sound of the marrow-bones and
cleavers of the independent electors of Westminster.'[145] For the
prisoner of war, innocent of any crime, Dartmoor's natural characteristics turned a necessity of war into a form of punishment;
for the criminally guilty those same characteristics might bring
redemption. Transformed by the experience, Wilson hoped the

convict might then look on New South Wales 'as a *land of promise*, not a place of *banishment*', transportation experienced as '*reward*'.[146]

Paradoxically, just this kind of thinking was placing the government under pressure to end transportation. Some 4,000 adult and juvenile convicts were transported each year to the penal colonies of New South Wales and Van Dieman's Land. Labour was a central part of this punishment regime, both for its redemptive and disciplining qualities. Sensational press reports, however, made much of transported convicts who earned their 'ticket-of-leave' following short periods of good behaviour and went on to acquire profitable estates in the Antipodes. Such negative publicity forced successive governments to question the deterrent value of transportation, reinforcing worries about the effect the penal colonies had on the fiscal, political and moral life of the colonies. Few were surprised when in 1837 the Molesworth Committee recommended that transportation should cease as soon as possible and convicts be confined at home and sentenced to hard labour for terms of two to fifteen years.[147]

The Dartmoor depot often cropped up as a minor theme in these discussions. An 1826 investigation into overcrowding in London's prisons examined whether it was suitable for criminals either convicted to serve prison terms in the UK or under sentence of transportation. In its favour was the railway, the ready availability of supplies from Plymouth, the fact that the Plymouth garrison could provide a guard and the possibility of 'renumerating labour . . . preparing and breaking granite for Macadamized roads' and cutting turf. Less positive was the depot's state of disrepair.[148] In 1829 Robert Peel sanctioned an investigation into whether the depot at Dartmoor could be converted 'into an establishment for the reception of various descriptions of prisoners'. Particularly concerning were the disciplinary problems created by the growing number of imprisoned juvenile delinquents. When Tyrwhitt appeared before the gaol committee of the Corporation of London to outline the depot's advantages, he prefaced his remarks by explaining that parliament was keen on the idea that some establishments develop farms 'with a view to the correction and reformation of this class of offenders'. His case was familiar. Within the boundaries of the 394-acre site, granite could be worked, peat cut, and flax and vegetables grown; the depot had a hospital, offices, a chapel and suitable accommodation for officers; the individual prisons meant prisoners could be classified and kept separately; and an iron railway stretched to Plymouth.[149]

Reportedly, the governors of Newgate and Giltspur Street were quite taken by Tyrwhitt's case. Imprisonment at a distance could be an advantage, for at present it was 'impossible to prevent communications between the depredators in confinement in the metropolis, and the depredators at large'. Crimes were even planned on board the hulks.[150] Two years later, a government report recommending that convicts serve a spell of imprisonment in the home country before transportation as a deterrent urged the government to investigate converting the POW depot. Making the case along broadly Tyrwhittian lines, the report noted that additional prison space would also obviate the need to transport at 'improper seasons of the year', which had led to unacceptably high mortality rates.[151]

If Tyrwhitt's case chimed with emerging sensibilities, his claim that the Board of Ordnance had kept up repair of the buildings was unconvincing. Girders, lintels and door jambs needed replacing to prevent walls collapsing; much re-slating was required; chimneys needed to be reinforced; yard walls needed coping; much work was needed on the twelve cottages.[152] Parliament kept alive the possibility that the building might be put to some future use by voting £7,000 for repairs, and in 1835 a committee discussed the bold suggestion that Millbank Penitentiary be moved to Dartmoor. A figure of £100,000 was bandied around as the likely cost, though J. H. Capper, superintendent of the prison hulks, thought it would be much less because the site could be prepared using prisoner labour.[153] Some momentum was developing. In 1836 Capper reported that the number of boys kept on hulks awaiting transportation was finally coming down, and if the depot was refitted the remainder could be transferred and HMS *Euryalus* decommissioned.[154] Later that year, however, Lord John Russell pre-empted yet another report, announcing that the scheme would not be taken up on grounds of cost:[155] £72,659 was just too much for a conversion that would only accommodate 726 boys. Similar conclusions were reached regarding the King's House at Winchester, Porchester Castle, Waltham Abbey and Enfield Lock.

George T. Bullar's survey provided a particularly damning assessment of the Dartmoor depot. 'Inferior' materials had been used in the original construction, especially the masonry, which was of a 'sandy porous nature', while the choice of location was dubious: 'The general decay has been accelerated by the locality and nature of the climate; the country for some miles round being subject to dense fogs and rain during the greatest part of the year. In consequence, the walls are frequently saturated with damp, which has occasioned a gradual rotting of the wood-work and timbers attached to the walls.'

Few doubted that new prisons were needed, built according to 'the most-approved principles' and located near London.[156] Pentonville and Parkhurst, respectively intended for adults and juveniles, were the result.[157] Just five years before the Dartmoor depot was converted into a convict prison, a damning Admiralty report described the buildings as 'in such a state of decay' they could not be put into 'habitable condition by repair' and needed rebuilding. Better heating and ventilation were required if they were ever to be used again. Perhaps, ruminated the authors of the report, it was time to dispose of the site. Shortly afterwards it was returned to the Duchy of Cornwall.[158]

A change in policy was expected when Lord John Russell came to power in July 1846. Opposed to transportation, he commissioned Joshua Jebb to consider the employment of convicts on public works. Jebb thought sentences of seven to ten years transportation could be commuted to periods of three, four or five years imprisonment in Britain. Of a four-year sentence, the first might be spent in separate confinement, the remaining three in public works. Convict labour would be of continuing use on navy works, particularly Palmerston's great defensive undertakings at Chatham, Portland and Plymouth, but Jebb reported that the Admiralty did not favour extending the use of hulks, which in 1844 still held 70 per cent of this category of convicts and were thought 'demoralising'.[159] Dartmoor's reprieve was granted. Under the Convict Prisons Act of 1850, along with Millbank, Parkhurst, Pentonville and the hulks *Justitia*, *Leviathan*, *Stirling Castle*, *Warrior* and *York*, the decaying depot was integrated into a new system of transportation. Henceforth, the condemned were subject to a period of separate confinement, followed by labour on public works in Britain, transportation on a ticket-of-leave to Van Dieman's Land or Western Australia and, eventually, a conditional or absolute pardon. Under pressure from the colony, transportation to New South Wales had been brought to an end in 1840; the same saw transportation to Van Dieman's Land cease in 1853, effectively ending this system of punishment.[160] Following the passage of primary legislation in 1853, Palmerston signed warrants naming Pentonville, Portland, Millbank and Dartmoor as prisons where sentences of penal servitude would henceforth be served: convicts faced a programme of separate confinement, associated labour on public works and release on licence.[161]

Condemned by the Admiralty in 1845, a concatenation of factors saved Tyrwhitt's old POW depot. First, the near breakdown in the system of transportation highlighted the need for solutions to the severe problem of under-capacity. Faced with immediate political pressures, renovating the depot suddenly made a great deal of sense. Second, hard labour on naval projects was not suitable for some convicts, and despite its fierce reputation Dartmoor was opened as an invalid prison. Third, as Tanner's long essay suggests, the old hope that Dartmoor could be improved had not been abandoned. It must be acknowledged, however, that this flew in the face of much official opinion. In 1827 the third of several reports by select committees on emigration from the United Kingdom roundly rejected encouraging paupers to settle Dartmoor and other wastes because it would prove more expensive than emigration. Cultivating poor land merely for the purpose of employing people would aggravate the problem of rising population, creating a false sense of independence and self-sufficiency.[162]

Nor did Dartmoor come off too well in 1843 when the government sought to establish a comprehensive picture of the effect of enclosure since 1800. Devon, England's largest county, was found to have the largest acreage of land classified as common or waste. At 240,872 of 952,198 acres, it was comparable as a proportion to Cornwall (131,752 of 578,695 acres) but proportionately smaller than in several northern counties. Only Elsden in Northumberland, at 75,858 acres, was a larger single common than the Forest of Dartmoor, which was reckoned at 54,241 acres, of which 50,421 were common or waste.[163] Further probing saw one leading witness, Charles Bailey, land agent, surveyor and auditor to several large landed estates, describe Dartmoor's soil substratum as 'very bad', and when pressed he said that 'for agricultural purposes' Dartmoor was 'the worst common' he had 'ever seen in the United Kingdom'. There were 'veins' of better land, but the corn grown there was not 'renumerating', and when questioned about whether ash plantations could be established he simply replied, 'I thought Sir Thomas Tyrwhitt's plantation was a failure.' Dartmoor's real value was to be found in its minerals and the granite.[164]

Ralph Cole was a more optimistic witness, arguing that hay had been successfully cultivated on the moor, and enclosure would arrest the continuing decline of the sheep population, not least by making farmers less fearful of theft.[165] The committee concluded that new developments in drainage and new forms of artificial manure, plus the fact that the Tithe Act (1836) freed land from increased liabilities if improved, made it a propitious moment to turn the wastes over to 'profitable cultivation'. No concrete recommendations or any particular conclusions about Dartmoor were made. It is hard to read the evidence without concluding that the moorscape remained a place apart, dutifully considered but of marginal importance.

Not all were so easily cowed. On 29 November 1845 a group of speculators registered the Dartmoor Improvement Company with the Board of Trade as a joint stock company. Their preliminary prospectus framed their ambitions according to now familiar precepts: 'At the present time when general attention is directed to the necessity of every possible extension and improvement of agriculture as a means of providing for the wants of our large and rapidly increasingly population the Project of the Dartmoor Improvement Company for converting the wild wastes of Dartmoor into valuable arable and pasture cannot fail to be received with considerable interest by the Public.'

This was an ambitious scheme. Capital of £200,000 would be raised through a share offer, and negotiations with the duchy for the purchase of 10,000 acres of 'the best part of the moor', including the remains of the POW depot, were expected to go well. The 'work of improvement' would see the land drained and fenced, roads established and farm buildings erected. Then the land would be turned over to 'the ordinary cause of husbandry', the soil needing only 'the application of lime or other alkaline manure'. Though the company's aim to rapidly convert the waste into a 'populous and productive agricultural district' cannot but seem hubristic, and though the Council of the Duchy of Cornwall offered the site of the prison back to the government in 1849, the basic assumption that bringing the moor into cultivation would require more than 'individual exertion and capital' seemed sound enough.[166]

Punishment and Reformation

And so Dartmoor once again became a site of incarceration. And so it would remain. It is hard today to encounter those sombre granite prison buildings, often only greyly seen through misty drizzle, without feeling that this is a peculiarly penal place. The most cynical tabloid hack would struggle to convince his readership that this apparently perfect manifestation of Victorian rigour and certainty is the 'holiday camp' desired by the reactionary imagination. Built from the clitter that lay scattered for thousands of years on the site itself, the prison's long roofs are nave-like, the great gable ends rising up, blank as though subject to a wild act of iconoclasm. The buildings are of the moor but less a natural outgrowth than a cancerous corpuscle. At night this impression is strengthened, for then to overlook the prison enclosure from the cold dark road is to encounter a great toothy-yellow light. This electric effusion, floodlights intermingled with the glow emitted by individual cell windows, is coldly beautiful. Nobody can be seen; only the wind can be heard. It's a little unnerving to know of the invisible thousands below, screws and lags differently confined.

Piecing together an internal history of HMP Dartmoor and locating this according to a carefully delineated analysis of the development of penal policy could keep an historian busy for many years. Inspiration might be sought in Michel Foucault's seminal thesis on the continuities between the disciplining regimes of the modern prison and the disciplinary foundations of modern society;[167] information would be found in Seán McConville's useful work on English prison administration;[168] attention might be given to high-profile prisoners like the Tichborne Claimant, Irish republicans like Michael Davitt and Eamon de Valera, and the habitual offender known as the Dartmoor Shepherd; and much picaresque detail could be gleaned from newspapers and memoirs, allowing anecdotes of eccentrics, escapees and violence within the prison to enliven the account. In the 1850s alone the press had a steady trickle of stories. What of the two escapees heading towards Ashburton whose convict dress gave them away to a local farmer? What of James Taylor, a Lancashire collier of 'fresh complexion', apprehended at Devonport trying to get to Australia; his companion, John Gray, a man of 'scowling . . . countenance', appears to have got away; or George Woodcock alias Massey alias Matthew Williamson alias George Johnson alias Alexander Sigismond alias convict No. 2753, who absconded in August 1855; or a prisoner named Stewart who subsequently broke into three houses, stole cash, coats, hats and two silver saltspoons; or William Dixon, captured at Ottery St Mary when trying to sell stolen silver spoons; or the two recaptured prisoners who made a run for it, one managing to evade capture for several days despite racing off cuffed and without hat or shoes; or the prisoner 'hotly pursued over the Moor' who made it the fourteen miles to Buckfastleigh?[169] Like the old war prison, HMP Dartmoor kept Dartmoor in the news, defining the meaning of the moorscape to countless outsiders. Pre-existing conceptions were reinforced by the image of the escaped desperado pursued across the boggy, foggy moor, raiding isolated farmhouses for food or clothes, securing Dartmoor a place in a Victorian imagination fascinated with the criminal and criminality.

For all the fascinating detail, underlining this imagined history would surely be an uneasy sense of the preparedness of the state to sustain an institution few ever thought fit for purpose. Dartmoor's peculiarly penal aspect might give the location of the prison some sense of inevitability, but such thoughts are fundamentally retrospective. The long interlude between the departure of the last French and American POWs and the arrival of the first convicts saw only a few favourably disposed enthusiasts fully convinced of the depot's continuing utility. If a somewhat dubious decision made in the early 1800s established Dartmoor in the public mind as a penal landscape, it is hard to imagine that this would have been sustained had the decaying buildings of the war prison not exercised the imaginations of a generation of improvers and penal reformers; it is equally hard to imagine that the governors of London's prisons would have been quite so attracted to the idea that convicts might be geographically isolated had Tyrwhitt and others not been so keen that they were.

In the gaslit Victorian world the nighttime voyeur would have seen a less intense, more flickering light, more differentiated by source, more intimate; the voices of patrolling guards and the scent of their tobacco smoke might have been caught on the breeze. This view was rarely contemplated, for it is a car with petrol in the tank that emboldens the casual observer to pass by so late, but surely then as now the contrast between the glowing prison and the dark moorscape encouraged reflection on the tremendous energy required to make modern life possible. It's a thought that takes on more substance if the history of the early years of the prison is read as an ultimately failed attempt to domesticate a very difficult natural environment. Initially, the guiding principles were largely familiar – improvement – as were the limiting factors, which were less the natural environment than the degree to which it was thought worth investing in attempts to make the moorscape habitable for a larger number of people.

Following the establishment of the convict prison, prison directors were required to submit an annual report to Joshua Jebb, surveyor-general of prisons and chair of the prison directors. These remarkable printed reports, each running to several hundred pages, comprise the statements of individual prison officers, the summary of each prison director and the conclusions of the surveyor-general. Inevitably repetitive, they nonetheless constitute an extraordinarily full administrative record of the prison establishment. Though replete with tables providing statistical evidence measuring the success of the institution, these reports no more told the singular truth about the prison system than any other bureaucratic record might. What they do provide, however, is a comprehensive picture of how the prison establishment represented itself to the home secretary and what connoted success. They are exceptionally important as a record of the penal establishment's official mind, that set of perspectives that evolved according to government policy and thinking.

HMP Dartmoor had two peculiar characteristics. First, it was an invalid prison, and the reason given for the revival of the war prison was that it was a particularly suited to this purpose. Second, written into the lease agreed between the state and the Duchy of Cornwall were obligations to improve the land. Commencing in autumn 1850, the ninety-nine-year lease applied to a 660-acre site comprising the prison buildings, the barracks, the prison farm and a number of houses, and a further 200 acres of waste to the east of the original site. Turf and peat could be cut from the waste and sold on the open market, the duchy taking a cut; stone could be freely used to develop the site but it could not be sold for fear of creating unfair competition for local quarrying interests; the prison leat and reservoir were to be used and maintained; and £15,000 was to be spent during the first five years of the lease on repairing and altering the buildings. A tramway linking the prison to the Dartmoor and Plymouth Railway was to be laid and, most importantly, it was 'stipulated that all land capable of being improved shall be reclaimed and cultivated'. Either party could terminate the lease if they gave twelve months notice, and if the duchy initiated proceedings it would compensate the prison authorities to the value of their improvements.[170] This rendered the lease more secure: increasing the value of the land through improvement made the state less likely to suddenly give up the asset and more expensive for the duchy to terminate the lease. Mr Gambier, Dartmoor's first governor, had a clear programme of work.

Progress over the course of the first year was promising. In October and November 185 men under sentence of transportation were transferred from Millbank and set to work converting five of the old prisons to their new use. A capacity of 1,300 was the initial target. Two buildings would be converted so they could hold 700 invalids in large open dormitories, sleeping in hospital beds and hammocks; two buildings were to be modelled on Portland and converted using corrugated-iron sheeting to allow single-cell occupancy for a further 500–600 able-bodied convicts; a portion of a fifth building was converted into a chapel, the remainder made suitable for kitchens and offices. By the end of 1851, 500 invalids were to have been transferred from the hulks and other prisons. Once the remaining buildings were converted along similar lines, overall capacity could be raised to 2,000.[171]

Future reports emphasised the role played by prisoner labour. Not only were they chiefly responsible for the conversion of the prison buildings but they built the parade ground, laid roads, excavated the foundations of the old walls, dug drains, quarried stones for the buildings, and broke stones for the roads; carpenters among them produced fences, doors and other fixtures and fittings; the skills of blacksmiths, painters, coopers, and wheelwrights were drawn on; a kitchen garden was established and linen, boots, shoes and clothing were repaired. So far, so good. Agricultural experiments were also begun. A three-acre patch of ground was enclosed, trenches were dug and formed into drains, three and a half tons of lime and a light dressing of manure were applied, and nine bushels of flax seed were sown in early May. The experiment succeeded. In August the flax was pulled, steeped, grassed and dressed, producing 384 pounds of fine flax and 452 pounds of tow or coarse fibre. A second experiment saw four acres of unimproved land ploughed and sown with oats. This produced a good quantity of straw but the crop of oats was 'very light'. A similar experiment with mangold-wurzel seed (a type of turnip used as animal feed) also gave disappointing results, suggesting that the land would only prove productive if trenched and drained. A fourth experiment confirmed this. A four-acre site, fully trenched, drained and manured, was planted with turnips. The portents were thought good for the crops were 'growing vigorously' until 'checked' by November frosts.[172] Later reports suggest these lessons were applied, particularly to land to the west of the prison and beneath North Hessary Tor.

Notwithstanding 'the stormy spring, ungenial summer, wet harvest' in 1853 another 25 acres of waste was added to the 95 acres reclaimed since the prison was revived. Oats, barley, swedes and turnips were grown reasonably successfully, while carrots, cabbages, parsnips and early turnips were raised in the prison garden. In all, some 59 tons of produce were supplied to the prison.[173] Keenly promoting the good work, Jebb wrote a report for *The Times* in which the statistics were allowed to speak for themselves. Belts of trees had been planted to protect the 98 acres that had been brought into cultivation; artificial manure (ingredients: peat charcoal, night soil and crushed bones) was manufactured on site; 30 tons of hay had been produced, 40 tons of heath and rushes had been cut and gathered for litter, and 1,920 tons of peat had been cut; 8,000 yards of old walls had been prepared, 3,000 yards of new walls had been built, 1,584 yards of new roads had been formed and 2,420 yards of old ones had been repaired; 6 horses, 10 cows, 6 calves and 100 pigs were now kept, supplying heavy labour, milk, butter and pork.[174]

Future reports told a story of steady reclamation and productive crops, though the shift in reliance from able-bodied to invalid convicts meant declining productivity. Governor Morrish explained in January 1857 that reclamation 'requires the co-operation of a few stout, sturdy fellows in each gang to give their weaker companions heart and energy', and he hoped that more would be sent his way: in effect, he asked that the pattern of sentencing reflect Dartmoor's needs.[175] Repeating the same request the following year, he observed that the work of reclamation in the winter, even for able-bodied prisoners, was 'severe and trying', a hint of problems to come. Despite this, another 20 acres were reclaimed and Morrish boasted that the prison farm now supported 114 head of cattle and was in need of new farm buildings.[176]

At the same time Morrish reminded his superiors that the 'good effects' resulting from convict labour should not be 'estimated merely by the money value of the work performed'. More important was the 'material improvement' of the character of invalid convicts employed reclaiming land. They had 'acquired habits of industry' that would 'prevent their relapsing into crime after they have obtained their discharge'.[177] This version of the prison's purpose had little to do with leases and productivity and much to do with a reformatory agenda predicated on evangelical Christianity, whose real value would be felt when convicts returned to civilian life. 'It will be borne in mind that they are undergoing a probationary period of discipline previous to the execution of their sentences of transportation, and that the object is not only to create habits of continuous and persevering industry, and to render them more useful to a colony, but to reduce the expense of their detention under penal discipline in this country by a judicious application of their labour.'[178]

This stock paragraph, repeated in a number of Dartmoor reports, reminded the minister of the purpose and economic justification of the penal regime he oversaw. To become an economically self-sustaining individual, whether in the colonies or in Britain, was integral to the moral reformation the prison regime sought to complete. Criminality, according to this mode of thinking, stemmed from irregular habits and indiscipline caused by an inadequate upbringing and the influence of immoral associates. Convicts resorted to theft and other forms of criminality because they were temperamentally incapable of sustained productive labour. Effective penal rigour would strengthen weak character. Or so the argument went. Central to the functioning of this regime was a system of reward and punishment that allowed privileges to be earned and could culminate in a prisoner being granted a ticket-of-leave or an unconditional release. Late-stage prisoners were allowed to wear 'their liberty clothes on Sundays', 'let their hair and whiskers grow to a moderate length' and write more letters, all allowances that give a sense of normal restrictions.[179] When the system worked, as the reports claimed it did, 'coercion' was no longer required and the convicts laboured voluntarily, completing 'a day's work equal in quantity and value to that of any ordinary body of labourers'.[180] It is striking, therefore, that the governor highlighted the importance of the distinction between those prisoners facing transportation, who could expect remission of their sentence on the basis of good behaviour, and those who were serving sentences of penal servitude, who could not.[181] Good behaviour was incentivised on the part of the former but not the latter.

These attempts to 'reclaim' the prisoner, as the prison chaplain described the process, contained a religious as well as a secular dimension. The criminal body was not simply an automaton that was to be gradually re-engineered by enforcing good habits; it was the vessel of a soul that had come to function without the influence of the divine. The reformed criminal character became morally autonomous when, 'led in sorrow of heart to the cross of Christ', he was 'taught by the Holy Spirit' to lead a new life.[182] This evangelical purpose, so integral to mid-Victorian liberalism, resounded through the chaplain's use of the word 'reclaim', neatly associating the mission wrought on the prisoner with the reclamation of waste, highlighting the harmony of institutional place and geographical space. Capturing the flavour of the chaplain's thinking is this long passage:

> You are of course aware that about one half of the prisoners in this prison are confined in cells, in which separation from companionship takes place during night and at meal time, association being only allowed during the hours of labour. In the other half of the prison, which consists of large and small wards, association exists, but a classification has been attempted, which promises well; exemplary prisoners are associated by their own choice and the evenings passed either in mutual improvement, in private reading, or in silent exercises of devotion. Very little management and oversight on the part of the Chaplain is sufficient to keep them in good working order; facilities for occasional Bible classes exist better in these classes than in the cells; so that the practice appears to be a good set off against the evils of general association. Punishment and reformation appear to keep pace with each other in Dartmoor. The utter isolation of the moor has little in common with previous evil associations in our great towns; the increased bodily health gained from the

bracing mountain air; the regular habits of living, hard labour from
morning till night; habits of industry; abstinence from intoxicating
drinks and excitement of mind; obligatory education and attendance
on divine worship, are all advantages on the part of the prisoner.
Many of them have admitted these various benefits, and promised
to continue the greater part in their families in time to come. Some
have written to say that they are doing well at home; indeed the
healthy and improved look and expression of face has struck many
a visitor to the prison.[183]

A healthy spirit required a healthy body, and the revival of the
prison saw the old question of whether Dartmoor was a healthy
place addressed once more. John Campbell, the prison physician,
examined this at length in his contribution to the 1853 report.
Marshalling statistics, he echoed the observations of the war prison
physicians: life on Dartmoor benefitted convicts with pulmonary
complaints, particularly young adults and the middle-aged. And
like his predecessors, Campbell was also struck by the absence of
tubercular illness among the locals, observing that the same could
be said of the 'moist, foggy atmosphere' of the Hebrides. Catarrhal
illness, however, took its toll in the summer, while the climate
didn't seem to benefit elderly convicts with bronchial illness. Typhus
made an appearance in September, both outside and inside the
prison, recalling the bouts of September sickness typical of the war
prison. Scrofulous patients might benefit from joining work parties
out of doors, but the 'damp atmosphere' aggravated rheumatism.
Bowel complaints, mild but not uncommon, were attributed to
'sudden atmospheric changes' and, more plausibly, to 'drinking too
freely of cold water when employed on the turf ground'; an
outbreak of diarrhoea afflicting 254 inmates and lasting for up to
four days was evidently caused by food poisoning, though the
doctor was reluctant to blame the kitchens. Finally, 'phlegmorous
inflammation of the extremities' had been 'rather common',
resulting in 'the loss of the last phalanx in some instances', which

was the doctor's way of saying he had had to amputate infected little fingers, possibly as a result of insect bites.[184]

Campbell's judgements, chiming with earlier medical knowledge of the moorscape, helped justify the decision to reopen the prison. By emphasising the climatic factors behind disease – the microbial and viral basis of infection were not yet understood – he identified the ideal type of Dartmoor invalid prisoner: reasonably young, suffering from a pulmonary complaint but capable of moderately demanding work. A steady supply of such convicts, augmented by able-bodied men who could do the heavy lifting, would ensure the prison thrived according to its founding purposes, benefitting 'youthful and middle-aged invalids' transferred from closed prisons. 'Many of them on reception are pale, spare and delicate looking, with impaired or capricious appetites; but the out-door labour has a powerful effect in invigorating the system and increasing the desire for food, and it is really surprising how little inconvenience is experienced in bad or foggy weather.'

Dartmoor, however, would not become the ideal the chaplain and the doctor imagined. The prison quickly became a dumping ground for elderly prisoners, often infirm or suffering from fatal illnesses, who could not be returned to society as productive citizens, as well as men suffering from mental illness. Typically classified as 'weak-minded' (some were thought 'insane'), a designation that blurred the distinction between moral weakness and a medical condition, these men undermined the regime. In 1856 they were isolated in Prison no. 1, an admission that they were not susceptible to the reformatory regime. This decision was later justified as follows: 'It was found that, when mixed with other prisoners, they obstructed the regular course of discipline, occupying an undue share of the officers in watching them, and when badly disposed, or irritable, proved themselves to be exceedingly dangerous, not only to the officers, but to their sound-minded fellow-prisoners, amongst whom there are always some who will take pleasure in teasing and provoking the unfortunate class of convicts alluded to.'[185]

The relatively benign regime outlined in the prison reports did not last. When the three dozen lashes meted out on the men accused of attacking the prison governor and deputy governor provoked widespread rioting in February 1862, the prison's armed civil guard were called out to quell the unrest.[186] Although an isolated event, this reinforced the wider turn against the reformatory optimism that had dictated penal policy in the 1830s and 40s and had shaped the first decade or so of the new regime at Dartmoor. The 1860s saw a significant shake-up of prisons policy throughout the United Kingdom, leading to a more centrally directed and professionalised prison service that imposed national standards and shifted the focus from reformatory towards punishment regimes.

Although sometimes retarded by severe weather or insufficiently robust labour, the pattern of gradual expansion in the prison farm's activities continued. *The Times* dutifully reported these efforts. Notices recording bad weather that caused the prison farm's crops

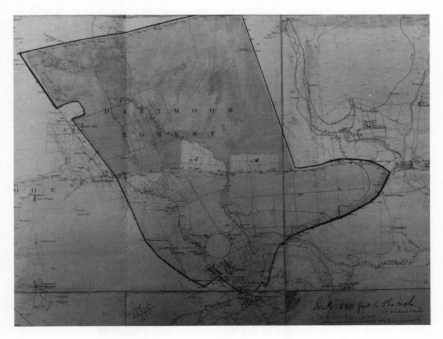

to fail or pieces puffing its achievements were proof that the news-paper considered the prison farm of national interest.[187] Progress was such that in September 1870 the prison authorities chose to lease from the duchy a further 1,010 acres of moorland. With the lease came new demands: the whole was to be fenced immediately and 125 acres of the new land was to be reclaimed every five years, an ambition which relied on the prison farm sustaining its rate of expansion for a further forty years.

In 1890 the duchy complained that these conditions were not being met, and a government probe established that only half of what was required had been done. Building a wall around the whole at the outset had been an impossible condition of the lease, the bailiff explained, for it would have required the labour of 155 able-bodied convicts, 20 horses and two carts over a two-year period: the cost of this was the same as the freehold value of the land, and the new road would be practically useless to the farm. Moreover, to catch up would take the labour of 147 men ten years while another 47 men would be needed to keep up the required annual expansion rate of 25 acres. Clearly the demands of the 1870 lease were unreasonable, but even the renegotiated terms proved hard to meet: by 1900 the prison was 306 acres behind. James Gourlay, the farm bailiff, explained that second-class labour was sufficient for general farm work – the 738 acres provided work for 160–70 men – but first-class labour was needed if reclamation was to continue. His overall assessment of the farm's effectiveness and Dartmoor as an agricultural landscape was downbeat. The reclaimed land tended to 'relapse into its original condition' and had to be regularly broken up to keep in check 'boggy and rank growths'; the grass crops might be suitable as grazing or feed but the cattle still had to be 'fattened off with a small supply of arti-ficial food' before they could be sold; the cost of keeping livestock on Dartmoor was 'considerably above the average' because 'the severity of the winter' meant they had to be fed more, which was only possible if food was bought in; grain couldn't be ripened and harvested with any confidence; and potatoes as a field crop were a failure.[188]

Gourlay's was the weary voice of experience, and, whether he realised it or not, with these few words he tossed onto the grate the speculative writings of Vancouver, Marshall, Tyrwhitt and Tanner. He effectively declared the greatest experiment in Dartmoor farming to have come to an end, even if the primitive practices it established, only gradually modernised, would continue for another century. To have 160 men working fewer than 1,000 acres was grotesquely inefficient; to have men carrying rather than horses hauling stones was a reminder that Dartmoor was a penal regime. Moreover, the prison farm had taught that draining and trenching the land was not a task that once completed was for all time. Maintaining improvements was a continuous and costly operation that required a lot of muscle. Similarly, livestock could be kept on the moor and, within reason, most crops could be grown, but neither to a degree that was self-sustaining. Farming the moorscape according to traditional arable and pastoral aspirations just wasn't achievable.

Twenty years earlier, reporting in 1881, the Howard Association, forerunner of the Howard League for Penal Reform, had offered a damning assessment of the farm:

> Thus, at Dartmoor, 2,000 acres are 'farmed' in a certain sense. But the results when compared with 2,000 acres of free land are curiously inadequate. The convicts in these prisons appear to accomplish far more than they really perform. But year after year they cost the nation from £30 to £40 each net in hard cash, notwithstanding the 'profits' alleged. They are not, as a class, trained to voluntary habits of industry, or incited thereto, adequately, by hope. Hence a large portion of them leave the prisons unreformed, and return thither speedily undeterred. Thus the present system is a very indifferent 'apprenticeship,' as tested by result.[189]

Emblematic of Dartmoor's reformatory failure was the case of David Davies, whose multiple convictions, severe sentences and peaceable demeanour made him a minor cause célèbre in the first years of the twentieth century. Known as the Dartmoor Shepherd on account of his work on the prison farm, his case made the headlines when the Prevention of Crime Act (1908) came under scrutiny. This illiberal measure permitted judges to extend sentences of penal servitude passed on the 'habitual criminal' by between five to ten years. Winston Churchill, a highly energetic home secretary known for his sudden enthusiasms, was concerned that 'preventative detention' was being applied unevenly. And the experience of David Davies, which Churchill first encountered in Basil Thompson's *The Story of Dartmoor Prison* (1907), seemed to prove this.[190]

Thompson, a former governor of Dartmoor, dryly described the Shepherd as possessing 'a full measure of the Christian virtues', his 'only fault' being his 'habit of breaking into houses when he got past the bounds of strict sobriety'.[191] It is hard to read the record of his crimes and his sentences without a mounting sense of disbelief. First convicted at Montgomery Petty Sessions on 23 July 1870, he received a month's penal servitude for larceny. Exactly two months later he received a further eight months at the Ruthin Assize for stealing a gun. On 24 July 1871 seven years penal servitude was handed down at the Manchester Assize for burglary, and he was sent to Dartmoor for the first time. Released early on account of his good behaviour, it was for stealing a watch and chain that he came before the Salop County Assizes on 16 October 1877; ten years later the same sentenced him to fifteen years penal servitude for burglary. On 18 April 1899 he got a month's imprisonment from Liverpool County Session for 'sacrilege' (he had stolen four pence from a Roman Catholic collecting box); on 14 October 1903 it was another five years penal servitude, this time thanks to the Knutsford Sessions and again for theft. Released on licence on 16 April 1907, on 6 June 1907 he earned a further three months penal servitude

at Manchester Petty Sessions for theft (sacrilege again) and, at
Shrewsbury, on 3 November 1908, another three months hard
labour. On 19 October 1909 his old friends at the Salop Quarter
Sessions sentenced Davies to three years penal servitude and ten
years preventative detention after he confessed to stealing two shillings from Whitchurch Parish Church in Shropshire.[192]

Churchill argued that the crimes of no other convict sentenced
as a 'habitual' were 'so petty' and only two habituals had been
more severely punished. Davies, he observed, 'enjoyed a melancholy
celebrity for the prodigious sentences he had endured, for his good
behaviour and docility in prison, and for his unusual gift of calling
individual sheep by name'. His record was 'not less terrible for its
punishments than for his crimes'; his crimes were neither 'daring'
nor 'professional'; he was a 'nuisance' rather than a 'danger to
society'.[193] In the House of Commons Churchill stated that 'employment for him as a shepherd' was being sought in the hope that it
would keep him on the straight and narrow. This provoked a characteristic exchange with Kier Hardie, Labour leader and Scot:

CHURCHILL: The man possesses the faculty – possessed I believe,
by no one else in this country – of being able to call individual
sheep in the flock of which he has charge, to him by their names.
HARDIE: As to the last part of the question, may I ask the Home
Secretary if he is aware that every Scotch shepherd possesses a
similar qualification?
CHURCHILL: I carefully said 'in this country'.[194]

Churchill and Lloyd George had visited Dartmoor on 24 October
– the two Davids chatted in Welsh – and two months later work
was found for Davies as a cowman with a farmer near Ruthin.
Released on 6 January 1911, four days later the Home Office wrote
to the local chief constable saying Davies had absconded and
requested that discreet inquiries be made. The press knew something was up. On 8 February the matter was raised in parliament;

on 28 February Churchill said he had no further information for the House; on 1 April Davies was apprehended when he forcibly entered the cellar of Morton Hall in Shropshire and stole four bottles of whisky.[195] To follow his career through to the end is to find more church boxes violated and more sentences served. Davies retired to Llanfyllin workhouse in 1925, where he died of heart failure in 1929.[196] At first glance, it is an endearing story, this gentle man and his flock of Dartmoor sheep, though to contemplate the whole of Davies' life suggests the Shepherd was not quite the benign figure Churchill required. Still, his case served the home secretary's purpose, ensuring that sentencing guidelines were tightened up and proportionality more rigorously applied.[197]

Bogs and Fogs

The fantasies of the improvers went unfulfilled. Dartmoor did not become patchworked with fields tilled by resilient householders whose hard work and thriftiness would realise the potential of the landscape. To examine the OS today is to see just how little enclosure there has been. Some encroachment on the edges of the high moor is evident where existing farms have been extended onto the common, but on the high moor itself what little there is hugs the roads. To the Victorians, the idea that roads brought the future to backward districts was an article of faith, which Dartmoor could be used to prove. For example, in 1867 Samuel Smiles, famous for his Victorian manual *Self-Help* (1859), wrote a life of Thomas Telford, the civil engineer who built the great roads of north Wales and the Scottish Highlands. To Smiles, roads were essential to the nation-building enterprise. They connected people, allowing insular communities to realise their potential as part of the nation; they facilitated the easy transportation of goods, ensuring commodities were available to all people at a fair price; they enhanced personal liberty by allowing people to move easily from place to place; and they allowed new forms of knowledge to penetrate isolated districts. If knowledge was a form of freedom, isolation perpetuated 'local dialects, local prejudices, and local customs', undermining the coherence and prosperity of the national community.[198]

And despite his biography of Telford's focus on Scottish development, it was to Dartmoor Smiles repeatedly turned for examples of the backwardness maintained by underdevelopment. His syntax oscillating between past and present tenses, Smiles wrote that such were the 'difficulties of road-engineering in that quarter, as well as the sterility of a large proportion of the moor' that it preserved 'much of its old manners, customs, traditions, and language'. 'It looks like a piece of England of the Middle Ages,' he wrote. Witches held 'their sway', pack horses were used, a post-chaise operated in Chagford because the roads were too steep and rugged for modern sprung vehicles, and 'the patriarchs of the hills' were to be seen in the 'straight-breasted blue coat . . . fastened with buckle and strap' that dated back to the days of George III. Agricultural implements and methods, partly in consequence of the primitive roads, were very old-fashioned: 'The slide or sledge is seen in the fields; the flail, with its monotonous strokes, resounds from the barn-floors; the corn is sifted by the windstow – the wind merely blowing away the chaff from the grain when shaken out of sieves by the motion of the hand on some elevated spot; the old wooden plough is still at work, and the goad is still used to urge the yoke of oxen in dragging it along.'[199]

It was characteristic of mid-Victorian optimism that one of its great prophets would write the biography of an engineer. It was equally of the time that Dartmoor should come to mind when an example of primitive backwardness was needed. Thanks to the attention the establishment of the POW depot attracted to the moor and to the mapping of the county by the Ordnance Survey, enthusiastic modernisers could not contemplate the great open moor without imagining its tremendous potential. In the eighteenth century Dartmoor had been an unknown space on the map, a blank given shape only by the representation of the snaking rivers mapmakers knew originated somewhere up in those hills; in early-nineteenth-century OS maps it was represented as a dark mass of forbidding hills, great hairy caterpillars of print emanating from a mysterious centre that contrasted with the paler, less threatening, representations of the lowlands. That unknown darkness fired the imagination, the enthusiasms of Georgian Tyrwhitt and Victorian Smiles being of a piece. Theirs was a civilising mission, turned not outwards towards empire, but inwards towards the mother country. New communication and travel technologies, whether in the form of a horse-drawn railroad or a tarmacadam road, would herald the future, facilitating a new regional and national prosperity, bringing light to that cartographic darkness. As Dartmoor and other forbidding landscapes became better known, the Ordnance Survey played a crucial role in this familiarising process. More sophisticated systems of representation allowed a more differentiated impression of the high moor, which was itself distinguished more subtly from the surrounding country. As the nineteenth century advanced, those first encountering Dartmoor through maps found a place less apart.

Smiles surely would have been pleased to know that as he was writing, this newly known Dartmoor was beginning to attract the attention of a new generation of speculators. Although the records of the Companies Register Office in London catalogue much failure, from the late 1860s numerous attempts were made to extract granite, minerals, peat and china clay from the moorscape. And extraction was not all. Three further companies, though not much more successful, started to tell a new story about Dartmoor. The prospectus of the Dartmoor Electric Supply Company, Ltd was issued in July 1910, and the company had some success, not being wound up until 1928. This 'Electric Light Undertaking', the prospectus explained, would supply Bovey Tracey, 'one of the most

popular of the moorland health resorts'. It was a reasonable prospect, if the success of electric light companies at other Devon and
Cornwall towns popular with tourists was anything to go by.[200]

The Torquay and Dartmoor Touring Company responded to
similar demands. It planned to a buy 'a certain char-a-banc manufactured by the Daimler Motor Co. Limited of Coventry'. This would
be the beginning of a company intent on 'running motor char-a-bancs
and motor omnibuses of all kinds' and carrying 'on the business or
businesses of garage proprietors, dealers in motor cars, motor spirit,
and tyres of all kinds, and all kinds of motor accessories'. Issuing its
articles of association in March 1914, not a propitious year, it was
dissolved in August 1919. Finally, Dartmoor Restaurants, Ltd issued
its prospectus in March 1922. It was fiercely ambitious, planning to

> carry on the business of proprietors and managers of restaurants,
> hotels, lodging houses, taverns, places of amusement, recreation,
> sport, and entertainments, livery stables, and motor garages, refresh
> ment caterers and contractors in all its respective branches, and
> vendors of and dealers in provisions of all kinds, and tobacco, beer,
> wines, spirits, minerals and aerated waters, and other drinks, refresh
> ments and confectionary of all kinds.

Everything a modern tourist could wish for.

The company's *pièce de résistance* would be a 'modern and up-to-
date Pavilion Restaurant', to be located on the 'high ground overlooking the Two Bridges Hotel 600 yards away, and the West Dart
Salmon and Trout river close by'. 'In addition to the natural attractions of Dartmoor,' the prospectus explained, 'a considerable
amount of Sport is available in the way of Race Meetings, Hunting,
(Fox Harriers, and other Hounds) extensive Salmon and Trout
Fishing, Shooting, Golf, &c.' Moreover, part of the site would be
developed 'to facilitate the putting down and taking up of passengers from Motor Coaches, Motor Cars, &c., and to provide up-to-
date and adequate accommodation for their various comforts.'

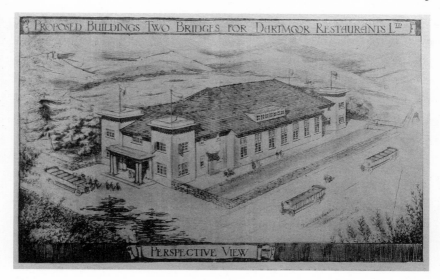

Somerset House was informed in February 1924 that the 'Company was not proceeded with in consequence of the necessary Capital not being forthcoming'.[201]

If this catalogue of failure demonstrates that the Dartmoor of opportunity continued to exist in counterpoint to the Dartmoor of the waste, the mid-Victorian period also began to see Dartmoor's unimproved physical characteristics take on new value. In January 1873 Plymouth Town Council and the Corporation of Exeter decided to coordinate a lobbying effort to persuade the War Office to select Dartmoor for its autumn manoeuvres. Inspired by Prussia's alarmingly easy defeat of French forces in 1870–1, large-scale demonstrative manoeuvres had become established as part of the army's routine. The army already had some knowledge of the moor, having used it to conduct experiments with field artillery,[202] and the case made by the burghers of Exeter and Plymouth depended on Dartmoor's undeveloped state. The absence of crops on the moor meant the army would not have to account for harvest time, and their manoeuvres would cost little in the way of compensation for damage to local farmers; the absence of game reserves gave similar freedoms. The plentiful supply of fresh water and access to local beef supplies was also advantageous, as was the proximity of the moor to troops stationed in Plymouth, Bristol, south Wales and Ireland. The location was surveyed in January, and in March the announcement was made. Manoeuvres would take place that summer at Cannock Chase in Staffordshire and on Dartmoor. Dartmoor was reportedly selected because the landscape would pose the generals challenges that contrasted sharply with those posed by the chalk ridges of Salisbury Plain and already familiar land around Aldershot.[203]

The weather that August was dismal. During three weeks of manoeuvres the man from *The Times*, filing lengthy reports, developed a mordant line in commentary on bad weather. 'Another day of rain, fog, and mist,' opened one report. 'Another night of pouring rain,' opened another. 'The weather, vile to the last, has been doing its best to spoil the March Past, fixed for tomorrow at 12,' opened a third. Occasionally, he expressed his frustration through irony, developing an extended metaphor:

> It seemed quite a strange piece of good fortune that everything was not marred by the weather, but it must not be imagined for a moment that we have had a fine day. It is true that the day did begin well, but Dartmoor weather trying to keep fine is like a drunkard trying to keep sober. It may stay itself from rain as he from brandy, but only for a time. The fatal fit must come. With a wretched countenance the morning resisted a lowering sky, but in the afternoon, yielding itself a willing victim, put the rain-cup to its lips, and shows as yet on sign of laying it down.

All the pageantry of the manoeuvres – the fluttering pennants, the smart uniforms, the polished boots, the fine horses – and much that was militarily useful was ruined by the weather. The whole enterprise could be confidently declared a failure, the journalist explained, because Dartmoor, in contrast to Cannock Chase, suffered from poor drainage. Here rain, which could be reasonably expected anywhere in England, turned solid ground into bog.[204] The army's summer on Dartmoor had been exactly as the agricultural improver might have predicted. And not for the first time a metropolitan outsider encountering Dartmoor came away baffled by the phlegmatic outlook of the locals. 'Nothing strikes one as more extraordinary,' the journalist wrote, 'than the modest pride with which the inhabitants speak of this howling waste. They are proud of its bogs, proud of its fogs, proud of its prison, and proudest of all of its rain.'[205]

When the War Office reported on the manoeuvres the following year, reviving memories of the previous summer's washout, it was inevitable *The Times* should receive a letter from a Dartmoor reader. 'The weather is now hot and brilliant,' the correspondent wrote, 'and it looks as if it would remain so.'[206] Nor had the War Office given up on Dartmoor. It remained an important site for field artillery experiments, continuing a practice began in the 1860s, and temporary camps established at Okehampton began to take on a permanent air. In the summer of 1875 *The Times* wrote enthusiastically about the location. Immediately south of the camp were the high north moors, on which were clustered the moor's highest peaks – a 'fine wild piece of scenery' – and to the north lay interesting if 'much less wild' land. 'In a word, the Camp is most delightfully situated, no position in the immediate district commanding a prettier or more varied view.' The air was also thought 'agreeable', being 'neither too hot nor too cold', though at times it was 'apt to be a little too moist, requiring trenches to be cut around the tents'. Abandoning euphemism, the newspaper exclaimed 'when it rains in Devonshire it does so in earnest'.[207]

Indeed it does, as has been the War Office and its successors' defence of its right to conduct military manoeuvres on the moor. On the OS today large parts of the northern reaches of the high moor, designated the Okehampton Range, are marked in pink ink DANGER AREA. Closer scrutiny shows the large number of military roads that have been opened up into this area of the moor. Walkers

making the long hike from Postbridge or Two Bridges to Belstone
or Okehampton might encounter no other person and feel utterly
alone for a few short hours, but only a wilful refusal to acknowledge
the flags, noticeboards, observation posts, huts and firing ranges
could lead anyone to suppose that this is a landscape untouched
by human intervention. The material precipitates of army life on
the moor can offend the eye and are an affront to preservationist
sensibilities, but it is hard to deny that they lend the region a peculiar fascination. And as the army track is picked up at Hangingstone
Hill, guiding the hiker securely north, it is equally hard to deny
that the army's presence has made this most challenging part of
the moor more accessible. Still, in the twentieth century this didn't
prevent the army's presence on the moor proving highly contentious, the controversy rivalled only by opposition to the Forestry
Commission and local water authorities.

If few yet objected to the army's use of the moor, the last decades
of the nineteenth century nonetheless saw new sensibilities emerge
that marked a break with the attitudes of the early improvers and
would also profoundly affect the politics of twentieth-century
Dartmoor. In January 1876 the Liberal MP Sir Charles Dilke
addressed the electors of Chelsea:

> It was over the New Forest that they had fought last Session, and,
> on the committee, he had been in the majority – the smallest possible
> – which had carried the report which would preserve the forest as
> it stood. Next year the struggle would be over a matter on which
> he had often addressed them – illegal enclosures of the Duchy of
> Cornwall authorities at Dartmoor. The way in which public estates
> would disappear and become merged in private ones was frightful.[208]

A year earlier *The Times* had reported that the duchy had let part
of Belstone Parish as a warren to Mr Fewins of Sticklepath. Fewins
had since prosecuted a case of game trespass against a commoner.[209]
Fifty years earlier this would have roused little comment; in the

1870s, with men like Dilke making such matters the stuff of national politics, common rights were fast becoming a significant political issue. In November 1877 *The Times* received a letter from Robert Dymond announcing the formation of the Devonshire Association. Lamenting what he regarded as the imminent loss of Exmoor 'as an invigorating breathing space', he wrote of the urgency with which concerned people must ensure the same did not happen to Dartmoor.

> Those who have traversed Dartmoor – not hastily as mere 'cheap trippers', but as intelligent tourists – well know the peculiar interest that attached to this playground of the forces of Nature, and the solemn grandeur of its granite tors, the charm of its folklore and tradition, and the music of its many waters. One of the objects of the Committee is to preserve these features of Dartmoor from disfigurement by vain attempts at improvement by enclosure. Many a fortune has been wasted in fruitless efforts to subject it to ordinary rules of cultivation. Such efforts, however promising they may have seemed at first, have, one after another, ended in discomfiture. The soil is fairly good, but invaders are invariably conquered by this climate.[210]

Much that would characterise twentieth-century preservationism can be heard here: the distinction between good and bad tourists, the celebration of the primitive cultures Smiles and his ilk wanted confined to history, the opposition to any form of enclosure and the tendency to see the natural characteristics of the moor as rendering it unconquerable according to conventional aspirations. The optimism of the early improvers is here superseded by a fatalistic naturalism revolted by the ugly remains of failed attempts to tame the moorscape.

Six years later the Dartmoor Preservation Association was launched at a meeting chaired by Lord Morley in Plymouth. From the outset, the DPA regarded the perpetuation of the landscape as intimately bound up with the preservation of the traditional grazing regimes that maintained the moorscape's physical characteristics. These had to be protected against 'encroachments', which took on many forms. Like the army, the DPA became a powerful Dartmoor interest group, as often coming into conflict with the commoners as it did with the government. For if the prison farm did not inspire transformative change, profound and dramatic change nonetheless shaped the moorscape from the late nineteenth century onwards. Large parts were militarised, valleys were flooded to form reservoirs, thousands of acres were cultivated with uniform rows of non-indigenous conifers, and Dartmoor became a site of institutionalised conservation and leisure subject to external regulation and planning. Twentieth-century Dartmoor did not become a place of 'English hedgerows', as early improvers had hoped, but largely thanks to the interventions of the state, it became profoundly modern.

3

Preservation and Amenity

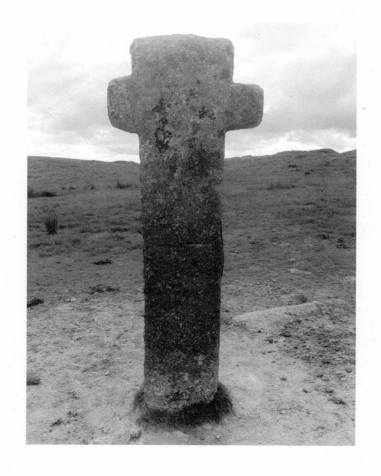

Enter William Crossing

On 24 May 1909 the *Western Morning News* announced its publication of William Crossing's *Guide to Dartmoor*. Based on obsessive note-taking, conversations with local people and wider reading, and drawing on toponymy, archival research and hearsay, the *Guide*'s 500 densely printed pages constitute an astonishingly detailed and at times overwhelming topographical description of the moor. Foreshadowing the work of Nikolaus Pevsner (1902–91) and Alfred Wainwright (1907–91), Crossing's name achieved a comparable form of brand recognition, albeit on a local rather than a national scale. An immediate commercial success, the *Guide* was lauded as the most important book ever written about the moorscape, quickly accruing cult value, and when Crossing's works were brought back into print in the 1960s and 70s they were titled using the possessive: *Crossing's Amid Devonia's Alps*, *Crossing's Hundred Years on Dartmoor*, *Crossing's Dartmoor Worker* and, of course, *Crossing's Guide to Dartmoor*. To the initiate, *Crossing's Guide to Dartmoor* was simply 'Crossing', just as others refer to Pevsner or Wisden.

Crossing was born in Plymouth on 14 November 1847 to parents interested in antiquarianism and rural traditions who holidayed on Roborough Down. He became fascinated with the theatre and the moor, and hindsight suggests his father was unwise to entrust to his son the management of the family mill at South Brent.[1] As the mill went slowly out of business, its manager wrote plays, formed a local dramatic society and 'tramped' the moor. This was not idleness, and Crossing's first major work, *The Ancient Stone Crosses of Dartmoor*, was published in serial form in the *Western Antiquary* and then as a book in 1887. A substantial work of topographical, antiquarian and historical research presented in the loco-descriptive mode, it came leavened with extracts from evocative verse and personal reflections on the moorscape and its history. Crossing's sources included pioneering works of early-nineteenth-century geology[2] and familiar books like Eliza Bray's *Borders of the Tamar and Tavy*, but he eschewed Samuel Rowe's discredited *Perambulation* and instead made extensive use of Dartmoor Preservation Association reports and the published lectures of the Plymouth Institution and the Devonshire Association. Before looking at Crossing's complex relationship with the preservationist milieu and how his unorthodox perspective prophesied twentieth-century national park thinking, something of his achievement can be grasped by taking the *Guide* on a 'tramp' through the moorscape.

Crossing lived for many years at South Brent, and much of his early loco-descriptive writing took this as his starting point. Tramping westwards across the southern quarter of the moor to Princetown is to encounter Dartmoor as the young Crossing most often did. Diligently consulting the *Guide* along the way transforms the undramatic grassy moorscape of the area into a landscape saturated with evidence of past human activity.

At the Newbridge Marsh car park on the edge of the Dart Valley Nature Reserve, the Two Moors Way can be joined on foot in the direction of the 'small inn called The Tradesman's Arms' in Scorriton. Crossing knew the pub but not the car park, the Two Moors Way and the nature reserve, three changes neatly illustrative of how the moorscape is now differently encountered.

The Two Moors Way offers a particular contrast to the Dartmoor of Crossing's imagination. Established in 1976, it traces a 102-mile route from northern Exmoor to southern Dartmoor. Marked on the OS in green diamonds and on the ground with MW signs, the Way provides hikers with a set-piece challenge and reassures more wary visitors that this is a safe way to experience the moor. Passing into the national park at South Tawton, the Way skirts the eastern uplands until Chagford Common, from where it takes an appealing route over Hookney, Hamel and Dunstone Downs before joining Dr Blackall's Drive above the Dart – a picturesque carriageway laid in the 1880s by the owner of Spitchwick Manor. From Scorriton, the Way follows the road for a short distance before entering the moor at Chalk Ford. The longest stretch of open moorland followed by the Way – an eight-mile trek over Quickbeam Hill, through Ugborough Moor and around Butterdon Hill – safely follows the dismantled tramway that once served the disused Red Lake china clay works. The Way simplifies Dartmoor for the visitor, following a route through a more complicated network of tracks, bridle paths and tramlines, warning against other routes through the moorscape. The effect is paradoxical. Imagined Dartmoor, already strongly determined by the cartographer's art, is shaped subtly by the Way's markers. Domesticating a stretch of the eastern upland makes the great central tract all the more intimidating. Facilitating access, the Way exercises control.

By contrast, Crossing was comprehensive and didactic rather than prescriptive. The *Guide*'s formidable detail was intended to give the visitor the confidence to access the whole moor, providing a corrective to earlier guides, which were 'really brief descriptions of the moor . . . full of mistakes' which led the reader to 'well-known objects . . . described times out of number . . . easy of access'. Crossing believed a directory was needed to lead the visitor to 'the hundred and one interesting objects hidden away among the hills and far from the beaten track . . . *never named*'.[3] To this end, the *Guide* opened with a detailed description and historical commentary on no less than eighty-one 'packhorse tracks and other old paths' that criss-cross the moor. Impressive enough, this served only as a prelude to the main body of the work, which picked out hundreds of possible routes or excursions through these intersecting tracks and paths. Throughout, Crossing steered his readers towards points of interest, helping them see the moor as he did.

Crossing judged the road from Scorriton to Chalk Ford as just 'another track' and less interesting than Sandy Way to the north, which connected the villages of Holne and Scorriton to the now disused Whiteworks tin mine on Holne Moor and was probably taken by traders visiting the war prison at Princetown.[4] Since its designation as part of the Two Moors Way the road is now well travelled and Sandy Way neglected, a simple illustration of how leisure is now more prominent in the moorscape than trade and industry.

Continuing west with Crossing reinforces the point. The 'by no means . . . clearly defined track' beyond Chalk Ford that heads due west and skirts the southern side of Pupers Hill leads to Huntingdon Warren. The warren, a sizeable tract of enclosed upland, is located in an area peppered with evidence of ancient, medieval and modern human activity. Crossing thought Pupers was a corruption of Pipers and attached to the three rocks on the hill's peak 'the usual story of men being turned into rocks for playing and dancing on a Sunday'.

According to Crossing, the warren had a relatively short life. It was established in 1809 when Thomas Michaelmore acquired a ninety-nine-year lease from the duchy and undertook to enclose 600 acres of open moorland. When the rabbit trade went into steep decline in the late nineteenth century the residue of the lease was sold to Mr E. Fearnly Tanner, who kept a pack of foxhounds and was keen to maintain the warren as a mean of attracting foxes for the hunt. This proved a brief interlude, and the Michaelmores soon returned, continuing trapping until the 1930s and maintaining the lease until the 1970s. Crossing says the warrener's house (now ruined) was built because workmen cutting turf at nearby Red Lake Mires resorted to poaching 'when their supplies of food were running short, or when they desired a change of diet', though Elisabeth Stanbrook's meticulous research suggests the small granite lookout on the western slopes of Huntingdon Hill was built for this purpose.[5]

Crossing has the walker circumnavigate the warren northwards, in order to take in the remains of the Gibby Beam and T Gert tin workings and the blowing house on the River Avon. His glossary explained that 'Beam' and 'Gert' indicated places that were once deep open tin workings and that a blowing house was a small tin smelter. Mixing archaeological and folk knowledge, Crossing also noted the cairn on the broad summit of the hill enclosed by the warren 'usually known as Huntingdon Barrow, but . . . sometimes referred to as the Heap o' Sinners'. The southern boundary of the warren towards the clapper bridge ('not boasting of any antiquity') passes through an area where there is further evidence of tin mining and ancient and medieval settlement, including the sixteenth-century Huntingdon Cross.[6]

Crossing might also have encountered a happy dog and its owners, but had his gaze followed the Avon's snaking way through Bush Meads to the south he would not have seen a reservoir. In the mid-1950s the South Devon Water Board dammed the valley, and in the Le Messurier edition of the *Guide* a thick black line is scored into the margin of paragraphs no longer accurate. What Crossing did find, however, were reaves (ancient boundaries formed of 'banks of earth and stone'), hut circles and 'abundant evidence' of tin streaming, some of which he dated back to Avena, the fourteenth-century settlement mentioned in the records of Edward III's bailiff of Dartmoor. Crossing pointed out another corruption in local usage, citing an unidentified sixteenth-century source that suggested Bush Meads should read Bishop's Mead, a correction adopted by the OS.[7]

Crossing the clapper bridge and rejoining the Two Moors Way leads into an area of the southern quarter where the traces left by an older history are overlaid with modern industrial archaeology. Amid plentiful evidence of ancient medieval settlement in the Erme valley, including the stone row leading from the stone circle on Stall Moor to Green Hill, is the dismantled Zeal Tor horse-drawn tramway, which once connected the industrial peat cutting at Red Lake Mire with the naphtha works at Shipley on the south-east edge of the moor. The imprint left on the surface of the moor by the tramway and the peat cuttings, known as ties, proved more lasting than the enterprise itself, which folded after just three years. When the tramway was dismantled and sold off in 1850 – Crossing remembered the wooden rails attached to granite sleepers – it left behind a new way of accessing the southern quarter by foot or on horseback.

The Way soon meets a more dramatic example of the same phenomenon. A small steam engine once pulled a three-coach train along eight miles of three-foot-gauge track connecting Ivybridge to the Red Lake china clay works deep in the moor's interior. In 1910, a year after the *Guide* was published, the Dartmoor Preservation Association feared the new track might be fenced, cutting the southern quarter in two and severely impeding the free movement of livestock and people across the moor. Their fears proved groundless, and they were equally content to report that no antiquity was 'interfered with'.[8] For some twenty years the works were profitable, and the site was not abandoned until 1932. Left behind were three artificial lakes, ruined buildings and a volcano-shaped spoil heap visible for miles around, all faithfully inscribed on the OS. When Crossing wrote it was unnecessary to explain that 'lake' on Dartmoor indicates a river and does not refer to the pits left by the mining company.

To head west from here, leaving the Two Moors Way, is to penetrate the more isolated parts of the southern quarter. The bridleway passes through Red Lake Mire, crosses Red Lake Ford, Dry Lake Ford and Blackland Brook Ford and leads to Erme Pits Hill, another site with extensive evidence of mining, the names themselves connoting past activity. Lonely desolation is the dominant note, telling of lives once lived rather than geographical isolation.

The OS tells a similar a story, for between here and Ringmoor Down to the south-west the map is densely inscribed with the locations of cairns, cists, tin workings, blowing houses, warrens,

pillow mounds, stone rows and settlements; intriguing names prolif-
erate like Deadmans Bottom, Giant's Basin, Evil Combe, Edward's
Path and Grime's Grave. Crossing did not explain these, though he
grumbled that the OS had adopted Drizzle Combe despite 'an old
map' suggesting the correct name of the 'little lateral valley' was
Thrushel Combe. 'It is easy to understand,' he wrote ruefully, 'how
it would become Drizzle Combe in the Dartmoor vernacular.'
These dense cartographical markings, which transform this section
of moorscape into a palimpsest of the ancient, medieval and
modern, depended on the detailed descriptive work R. Hansford
Worth presented to the Devonshire Association in 1889.[9]

The poorly defined track from Broad Rock skirts the southern
slopes of Great Gnats Hill and joins the rather better path at Plym
Ford that traces the gentle decline from Crane Hill to Nun's Cross
Farm. Crossing wrote little of the farm, simply mentioning that
the small enclosure dated from 1870 and that in recent years the
'modern dwelling-house' had taken the place of 'the quaint thatched
cottage'.[10]

Although this grim farmhouse exerts a peculiar fascination,
Crossing was more interested in the cross the farm was named after.
Siward's or Nun's Head Cross, a photograph of which heads this
section, featured prominently in Crossing's first book as Dartmoor's
oldest 'stone cross', and he reiterated his earlier interpretation of
the two names in the *Guide*. Sywardi was mentioned in Henry III's
1240 Perambulation of the Forest of Dartmoor, and Crossing
concluded that it must date back to the time of Edward the
Confessor, when Siward, Earl of Northumberland, held the manors
of Tavei (Mary Tavy) and Wifleurde (Willsworthy). He argued that
the puzzling inscription on the cross suggests it originally marked
the boundary between lands held by Northumberland and the
monks of Buckland Abbey. Nun's Cross, the alternative name,
reflected the location of the cross at the head of the Swincombe
valley and derived from the Cornu-Celtic word *nans*, which meant
valley, dale or ravine, a usage possibly of seventeenth-century

origin.[11] Thus, as the original legalistic purpose of the cross became redundant after the Reformation, it accrued a meaning that displaced the logic of property with the logic of the moor man's encounter with the moorscape.

From Nun's Cross a reave can be followed into Princetown, passing Tyrwhitt's plantations. The Whiteworks mine just to the east was disused by Crossing's day, though then as now 'evidences of it are abundant'. Crossing remembered an earlier Dartmoor soundscape 'when two large waterwheels were to be seen revolving here, and when the blacksmith's hammer was constantly heard ringing on the anvil'. Peat Cot, the small farm located a little further on, stands in counterpoint to the relics of Tyrwhitt's 'dream', exemplifying what 'the nineteenth century settler' could actually 'accomplish on Dartmoor'.[12]

Following a night spent in Princetown, the Dartmoor Way can be followed to Dartmeet. It passes through the bleak abandoned newtakes in the Swincombe valley and into the cosy hamlet of Hexworthy, where Crossing's favourite inn can be found. Passing over Huccaby Bridge – a dreamscape for the wild swimmer – and then by St Raphael's Anglican chapel leads to the short steep stretch of road to Dartmeet. From there the fun scramble along the west bank of the Dart – an area of exquisite natural beauty – completes the circle begun the day before.

Crossing followed up the *Ancient Stone Crosses of Dartmoor* with the breezy *Amid Devonia's Alps* of 1889. Early readers who mocked the title's grandiosity missed its irony, for this lively personal account of nine walks or journeys across the moor was told with a lightness of touch and a self-deprecatory humour generally absent from its predecessor or the exhaustive loco-descriptive content of the *Guide*.[13] Nonetheless, the all's-well-that-ends-well nonchalance, enthusiasm for comestibles and air of irresponsibility that gives *Amid Devonia's Alps* its puckish charm should not obscure the polemical gesturing that surfaces throughout. In 'A brief chat about Dartmoor', his apparently whimsical opener, Crossing emphasised the failures of the improvers, accentuating how little of Dartmoor was cultivated and the exceptional way it remained a place of nature: he dismissed the newtakes along the Tavistock to Moretonhampstead road as merely enclosures for cattle and sheep; he belittled Tyrwhitt's misguided efforts as part of a wider 'mania for cultivating the moor'; said it was possible to walk in a straight line for a day and see no evidence 'of recent occupation or cultivation'; and claimed that what cultivation there had been, including the establishment of conifer plantations, had not 'materially' changed Dartmoor's 'natural aspect'. Crossing scorned attempts to extract naphtha from peat and mocked the old thought that it was 'actually feasible' to build a railway across the moor. Capital's power had its limits, and human speculation and ingenuity had proved unequal to the 'soil and climate' of this 'primeval region'.[14]

Dartmoor's natural state was emphasised in Crossing's other writings. He took issue with John Northcote (1746–1831), the Plymouth-born painter who claimed the moorscape 'was not worth painting'.[15] Northcote missed what an unnamed artist who chiefly painted 'under the sunny skies of Italy' had realised: he could only paint Dartmoor if first he were to 'tramp over it for a least six months'. It took a long time to 'know' the moor because 'Nature' sometimes 'refused to be interpreted'.[16] Another instance of Dartmoor special pleading? On the moor Crossing could *feel* 'the influence of an older day' and *see* 'the face of the country almost as it was centuries ago'. This inclination to contrast 'progress' with the past and to equate the past with 'the handiwork of Nature' and the 'primeval' confused human and natural history, but what should not be missed is the importance of his claim that on Dartmoor was found 'uncultured Nature without a sign that man has ever intruded upon her domain'.[17]

Textual evidence suggests that Crossing derived much of his Dartmoor lexicon from the fervently polemical preservationist writing of W. F. Collier, founding member of the Dartmoor Committee of the Devonshire Association. Collier came from a

long-established Plymouth family of Quaker radicals. His father
cropped his hair in sympathy with the French Revolution, objected
to corporal punishment, was critical of tree-clearing by the city
authorities and taught his son that he had rights of access to
Dartmoor. Collier's aunt Kate, 'a glorious stickler for woman's
rights', was outraged by the continuing use of the ducking stool.[18]
His first major intervention, read to the Devonshire Association at
Ashburton in July 1876, polemicised against all that disturbed
Dartmoor's 'state of Nature'. Characterising the moorscape as
'fresh from Nature's workshop', Collier found it somewhere people
could find 'the solitude, the quiet, and the grandeur of the works
of primaeval nature'. Dartmoor's 'ever-varying contrasts' led him
to observe a paradox: civilised people admired 'nature apart from
civilisation' and were touched by the 'sublime grandeur' of 'nature
in a wild state'; uncivilised people reduced nature to a 'severe task-
mistress' who gave no 'delight', 'relief', 'rest' or 'pure unalloyed
enjoyment'. On Dartmoor the materialist striving of modern civilisation met a 'higher' civilisation that sought to make of nature a
'friend, a mother of her, placing her at the head of our affairs,
choosing her for our queen'.[19]

With nature identified as Dartmoor's essential characteristic,
Collier's sweeping polemic found enemies among friends. He took
the archaeologists to task for methods that damaged the moor's
surface and findings that served only to highlight the insignificance
of Dartmoor's human history. The 'only tale' told by the 'circular
rows of stone, which some people delight in calling villages', he
exclaimed, is that man has always 'defaced the features of Nature'.
Much more threatening than the scrabbling about of the archaeologists was the working of metropolitan capital. Angered by the
damage caused by granite quarrying and fruitless gold speculation,
Collier shook a provincial fist at the metropolis, insinuating a simi-
larity between its treatment of Britain's marginal landscapes and
Britain's colonies. Why should London, he asked, hungry for
granite, 'be enriched by the spoils of Dartmoor'? Hadn't the capital

'spoils enough from all countries and people that it must take from us even our tors'? The much-touted Dartmoor railway would be yet 'another of the numerous channels through which the hard-earned wealth of the country is drawn to the coffers of London capitalists'. Collier's radicalism extended to the politics of enclosure and access. Acknowledging the rights of private property, he reminded his listeners that 'rights of common and rights of way' were also properties 'distinctly protected by the law'. More significantly, there were 'rights of free foot on the face of the earth, rights of visiting the high places and worshipping the powers of nature, rights of disporting oneself (I do not allude to sporting), not distinctly recognised by the law, which are, however, nevertheless legal'. Collier warned landlords that they should be wary of driving 'these questions to an issue' for this may 'rouse the whole nation to claim the wild uncultivated tracts of land to which they have for ages resorted for the air and the exercise that can alone restore the health and vigour sacrificed to modern forms of industry'.[20]

When less sanguine about the long-term survival of Dartmoor's natural state, Crossing also framed his fears in terms of national priorities. He admitted that the district lying around Princetown had 'ceased to be as it once was', conceding that an apparently uncompromising natural environment could be made subject to human desires. Extending enclosure or the quarrying operations into the northern and southern uplands, presently 'as ever they were', should be prevented. 'No amount of profit, even supposing they could be made to yield such would compensate for the loss of their primeval character, and it behoves those who believe there is something more to a nation than money to aid in the preservation of these stretches of wild moorland, which have come down to us untouched, and in which we have a glimpse of the world as it was.'[21]

The improvers, foolish yet persistent, threatened still, and Crossing was unusually direct in asserting Dartmoor's national importance as 'a domain of Nature altogether unlike any other that England can show'.[22] In those few words Crossing's certainties gave way, and an existing English past, particular and inviolate, was reclassified as perishable.

Just as Collier believed civilised people felt compelled to seek a compact with nature, by placing his plea to preserve Dartmoor's natural state in the national context Crossing placed universal human needs at the heart of his preservationist agenda. He saw that the prison had changed what the name 'Dartmoor' signified, but shared with his Romantic predecessors an idea of the moor as a place of freedom. Freedom was strongly associated in his writings with descriptions of the moor's apparently natural state and, as in Carrington's Dartmoor poem, the opportunity the common provided for free pedestrian movement over long distances. Like Collier, he regarded enclosure as antithetical to national freedom, complaining that a century earlier 'many open spaces were lost to the nation'. And like Collier he deployed the classic anti-enclosure image of the poor man seeing 'the common stolen from the goose by the greedy landlords'. Dartmoor freely provided a rare commodity – 'solitude' – that 'those who would preserve Dartmoor should particularly guard from invasion'.[23] Society itself was a form of bondage from which the free were periodically liberated.

Collier's polemic – and by extension Crossing's – chimed with the demands of lobbyists representing an emergent national and international culture of landscape preservation based in some cases on the protection of common rights and more generally on the idea that the community had a 'popular proprietorial stake in the countryside'.[24] Since 1865 the Commons Preservation Association, a small but well connected and broadly liberal organisation, had 'promoted protective legislation in Parliament, acted to prevent the passage of bills injuriously affecting scenery and open spaces, supported legal action in defence of common rights, and kept up a relentless propaganda campaign'.[25] In the year Collier delivered his Dartmoor lecture, members of the CPA successfully shepherded through parliament the Commons Act. Its novelty rested on the fact that it was primarily aimed at regulating and preserving commons rather than enabling enclosure. Its protective provisions were extended in acts of 1899 and 1908, while landmark legislation like the New Forest Act (1877) sought to preserve common rights and provide public access to a landscape recognised for its historical importance and natural beauty. However, as Paul Readman argues, the most significant achievement of the CPA was its defeat or amendment of a 'huge mass' of less prominent legislation that 'threatened commons, open spaces and rights of way'.[26] Collier's complaint that enclosure on Dartmoor saw the 'whole nation ousted from a wild piece of land, enjoyable only as a place of exercise, where the climate is wretched, and the soil valueless' linked a specific claim about Dartmoor to a national political agenda;[27] his intuition that forms of 'progress' believed to enhance individual freedom could impair humanity's collective capacity to be free reflected broader shifts in late nineteenth-century attitudes towards landscape.

Despite his early polemics, Crossing was too interested in the Dartmoor he encountered on his tramps to fully embrace the cult of nature. His Dartmoor was not Collier's unspoiled nature nor the improver's waste but a landscape long shaped by human civilisation. His writing populated the uplands with people who, past and present, extracted a living from the moor through the exercise of ancient rights and modern means. His earliest preservationist intervention was simply a 'plea' that stone crosses lying damaged on the moor be repaired and re-erected. When the 'Ancient Stone Crosses' articles were republished in book form he appended an account of the day in August 1885 the Dartmoor Preservation Association spent cementing four fallen crosses into place.[28] Similar work had occupied the Dartmoor Committee of the Devonshire Association since its inception in 1876. As Spence Bate explained, a fallen stone 'frequently attracts no interest; but if restored to its proper position, under the careful superintendence of the committee, it would become an *object of attention* even to those who may not study archaeology.'[29] The allusion to expertise and due process should be set against the fact that the Dartmoor Committee, like the Dartmoor Preservation Association, was not subject to external supervision or guidance by a state or professional body but carried out its chosen work on its own initiative. Later conservationists and archaeologists have sometimes lamented the crudity of their methods – and Crossing was not above pointing out their mistakes – but the sense of urgency accompanying their work was not misplaced. As Crossing light-heartedly recalled:

> we can hardly hope for much improvement if there are many on Dartmoor like the farmer whom I heard speaking on this subject the other day. 'I wouldn't tich a stone,' he said – and my heart warmed towards the worthy man on hearing his words, though I speedily found what a mistake I had made in judging him – 'I wouldn' tich a stone 'pon the moor that was sticked up – that is, if he was marked; but if there wudden no letters 'pon un, way I might as well hev'n vur a paus as any other body.'[30]

Cementing stone crosses or Neolithic monuments into place did indeed secure them against casual use by farmers and road builders.

Acts of restoration that populated the moor with 'objects of attention' gained preservationist value when recorded as part of its wider representation. Every little inscription added to the OS – a hut circle here, a cairn there – enhanced the national value of the land and strengthened the case against 'encroachment'. To this end, the Devonshire Association began in 1878 to create a map that would 'constitute a lasting record of the principal characteristics of Dartmoor as it now exists'. Initially, the plan was to record all natural features, roads, rights of way and enclosures, the purpose being to provide a record that could be used in disputes over further encroachments, but it seems the antiquarians and archaeologists also got involved, helping to fill in the map's blank spaces with new details.[31] Recording the material evidence of past human activity on Dartmoor appealed to an emergent sensibility that regarded such survivals as part of the nation's heritage.

Crossing's topography was also a form of mapping, and by adopting a largely descriptive approach for the *Guide* he produced a work of record and, as such, a valuable preservationist tool. By making the civilisation/nature dichotomy difficult to sustain – the *Guide* contains an austere form of nature writing but primarily encountered the moor through human works – Crossing transformed Collier's essentially natural landscape into a cultural landscape. Naming anthropic alongside natural features of landscape enhanced the cultural value of the moor. But this apparently innocent exercise is not a neutral process, and the forms of official knowledge mapping produces are exercises of power and authority over the landscape and the people who live off it. Recent histories of the Ordnance Survey in nineteenth-century Ireland and mapmaking in other imperial contexts have been shaped by this thinking, and something similar characterised the mapping of marginal places in Britain itself.[32]

Crossing exercised a comparable form of authority in the way he treated local usages when recording Dartmoor place names. Of sites without recognised names, he recorded the folkloric origins of locally known names and by doing so transmitted the cultural meaning of particular places to a wider audience, giving the name an almost official status. At the same time he sometimes exposed contemporary usages as 'corruptions' of earlier names, which he established using historical documents, privileging their authority over local knowledge. When sites were referred to by more than one name, he often dismissed usages derived from local superstition or folklore, sometimes taking issue with the names provided by the OS, providing corrections or additional names subsequently adopted. 'It may therefore be necessary to explain,' he boasted, 'that not only was a list of Dartmoor place-names submitted to me for revision before being engraved on the map, but that I also added several.'[33] Crossing was thus active in the process that saw local usages 'corrected' and names only known locally inscribed onto the state-sanctioned sheets of the Ordnance Survey, rendering the malleable fixed.

Like many later Dartmoor preservationists, Crossing occupied a position of privileged externality: his encounters with the people of the moor reflected his status as an educated, propertied, middle-class outsider with leisure time at his disposal. Just as he could write condescendingly about the Okehampton farmer who claimed to know 'a good deal of Dartmoor' but whose horizons did not extend beyond the ten miles or so south towards Postbridge or Hexworthy,[34] so he insouciantly recalled his encounter with a community of whortleberry gatherers: 'A day spent with the hurt [whortleberry] pickers – not as a gatherer, perhaps, for that exercise might be found rather tedious – is most enjoyable, especially if there happen to be some aged labourer among the party who has "heerd tell" of certain things that happened "years agone".'[35]

Much that Crossing came to know of the moor, which he communicated to similarly situated readers, was accumulated in this way. Whether presented as anecdote relayed through recalled conversation or as authoritative fact divorced from the local encounter that created it, the translation of this knowledge into newspaper copy commodified local knowledge and implicitly identified the moor men and women as a subaltern group unable to represent itself. Crossing's preservationist impulse, typical of his caste but as yet only half realised, was orientated less towards the preservation of nature untouched than the preservation of the seemingly harmonious relationship established between nature and the indigenous population that created the environment he celebrated.

Consequently, despite his residual admiration for Dartmoor's pioneering improvers, Crossing thought their failures vindicated local knowledge. More wisdom could be found in the amused observations of the locals – recounted by Crossing in a barely comprehensible demotic – than in the experiments of modern agricultural science. In his 'Present-day life on Dartmoor', a sequence published by the *Western Morning News* in 1903, the farmers, moor men, newtake wall-builders, peat cutters and warreners are characterised by their sensible expectations of the wealth that could be extracted from the moor.[36] Crossing had fun with the old promise that labourers would have a bright future as leaseholding cultivators of the land, claiming the labourer had 'never believed' in these schemes because 'he knew too well what Dartmoor was'. The peat cutter was similarly astute. As attempts to commercially exploit this resource failed – compacted turf burned too quickly – 'the Dartmoor man' was not 'beaten' and continued to exercise his ancient turbery rights.[37] Consistent with this was Crossing's sententious comment in the *Guide* on the current inhabitants of the Whiteworks mine near Nun's Cross: 'Those who now live at White Works look not to the bowels of the earth for their support, but to its surface. By breeding ponies and rearing other stock, and doing such labour as their hands may find for them to do, they contrive to get a living, and if the prize of wealth is not to be obtained, they have what is far more than its equivalent.'[38]

Crossing believed that Dartmoor's 'proper use' was 'as a grazing place for ponies, cattle and sheep', and so employed 'it is put to its best and truest use', but this did not prevent him from admiring the 'spirit of enterprise' that underpinned granite quarrying, attempts to revive tin mining and the establishment of china clay works, the most successful of Dartmoor's modern industries.[39] He accepted the extractive industries as modern examples of ancient practice and even insisted that 'we must not grudge to enterprise' the comparatively small portions of the moor's surface they demanded.[40] Even the greater efficiency the electrical fuse brought to quarrying, allowing great lumps of granite to be blasted out of the hill at Merrivale, excited rather than disturbed him. Crossing also accepted the necessity of swaling – the annual burning of heather and gorse to create pasturage – rejecting the view that the practice should be prohibited because heather enhanced the moor's natural beauty: 'sentiment . . . cannot expect to be listened to when it would inflict a loss upon the community'.[41] The Dartmoor Preservation Association, by contrast, deplored 'promiscuous and excessive swaling' because it destroyed herbage, made large tracts of grazing land unfit as pasturage and killed ground game, black-game or black grouse and small birds.[42] If regulation was needed, Crossing argued, this was because outsiders set the heather ablaze for fun. Crossing was equally pragmatic about the new workers brought to the moor by the revival of industry, the prison and the moor's use as a military training ground. Distinguishing these incomers from those he considered of the moor, he happily char-acterised Princetown as 'a bit of "in along" dropped "out auver"', a place of football and cricket matches, lawn tennis courts and billiard rooms, concerts, entertainments and public dinners. That which could be understood in the phlegmatic dialect of moor men seemed less threatening. Visitors were also unthreatening. If a little fun could be had at the expense of those who felt their day or two in the uplands meant they had 'done' Dartmoor, Crossing neither denied the pleasures of a coach trip nor resisted the temptation to fill column inches with detailed descriptions of the routes offered. Most amusing were the antiquarians with their ever-changing theor-ies. Druids, Vikings, Celts: what next?[43]

Dartmoor for Devonshire?

Crossing's easy-going preservationism belied a more heated political context. The needs of both the armed forces and local and national water undertakers, including the Corporation of London, raised the possibility that the duchy might sell the whole of Dartmoor, precipitating the dissolution of common rights and rights of access. Radical responses were proposed, including the idea that Devon County Council might buy Dartmoor for the people of Devonshire. Although this came to nothing, the debate it provoked announced the preservation of Dartmoor as an expressly national question. Much of this chimed with the national debate that saw the concept of amenity emerge as the cornerstone of national policy towards Britain's least developed landscapes. The history of Dartmoor preservationism thus provides an important case study of how late-Victorian pessimism about access and 'encroachment' gave way to the optimism that led to the national parks legislation of the post-war Labour governments. Once again, W. F. Collier can be found voicing opinions that would soon become orthodox.

In the 1880s and early 1890s the Corporation of London, facing a ballooning population and the intense demands of industry, was preoccupied by the need to secure an adequate supply of clean water. Water engineers considered whether water could be piped from Dartmoor, south Wales or Cumberland to London. They thought pipes could be run along Brunel's Paddington to Penzance railway line after it was converted from broad to standard gauge.[44] When in 1894 the corporation placed a private bill before the House of Commons seeking a purchase order for the whole of Dartmoor, Collier and Robert Burnard of the DPA led the charge against. Lobbying Devon County Council, they urged it to buy Dartmoor from the Duchy of Cornwall for the people of Devon. Although the water bill fell on a technicality, the Collier and Burnard proposals reflected wider national trends, as Collier explained: 'It has been of late years a maxim of public policy that waste lands – lands which have lain waste from time immemorial – shall be kept as parks or playgrounds for the people, where a dense population crowded in a town may disport themselves, refresh themselves with pure air, enjoy the beauties of the wild fauna and flora of the field, and learn to know something of the works of Nature.'[45]

Dartmoor, he argued, belonged alongside Dean Forest, the New Forest and Epping Forest as a 'wild and beautiful part of England preserved for the good of us all'. The New Forest Act, 1877, known as the Commoners' Charter, had forbidden further enclosure, guaranteed common rights and recognised the amenity value of the forest to the wider public. Epping Forest was a more instructive example still. Threatened with enclosure, popular political pressure led the Corporation of London to pursue a long legal struggle to prevent enclosure and then buy the forest. When the master of the rolls ruled in the corporation's favour in 1874, complex interests had to be identified and compensated before the transaction could be completed. Queen Victoria opened the forest to the people in 1882. Two factors were crucial in the Epping Forest case: first, its proximity to a very large population accustomed to using the forest as a leisure ground and, second, the Corporation of London, a public body able to raise the revenue to make the purchase and whose representative power could trump powerful vested interests. These criteria were not satisfied elsewhere until the Local Government Act of 1888 replaced the administrative power of resident magistrates with unitary county authorities empowered to levy rates and borrow money.

Collier's purpose was simple. Purchase would see customary rights preserved and the right of access for all citizens guaranteed. Under his proposed scheme, the sale of Dartmoor would be made in perpetuity, the new owner retaining no right to alienate any part of the property. Establishing a fair purchase price was difficult because this involved calculating the total annual yield of the moor's revenues over twenty years, including its pasturage and turbage value, the remaining lease value of enclosures, including the prison, the various mineral rights that had been sold, and the duchy's timber and forestry interests.[46] Collier was particularly exercised by the importance of Dartmoor to Devon's water security, playing the anti-metropolitan card as he alerted his audience to London's designs on the moor. 'I loathe the thought of our pure water,' he wrote, 'caught as it is by our high hills from the heavens, and held for us by those blessed bogs, conducted into pipes, taken to London by Act of Parliament, eventually to flush the sewers of the modern Babylon.'[47] An echo of earlier talk of Druids and lustral waters tinkled through this polemic alongside its clanging hostility to urban life and its ways.

> I can see Dartmoor, in my mind's eye, turned to good account by the London County Council in their own London way – Cockney villas in all directions, with railway and tramway approaches; large reservoirs in the place of our river heads, now silent spots for thoughtful men, far from the madding crowd; perhaps boats and electric launches on them, with bands of music, a superfluity of the sort of civilisation which is peculiar to this *fin-de-siècle*; tourists on every remaining Tor, the granite of which may not have been good enough for London Police Stations, and trippers staring at the reservoirs, calling them pretty, like the Serpentine.[48]

As horrified by a phantasmal Cockney as this Devonian was – Bellever Tor would become Belle Vue Tor![49] – Collier was more mystified by the apparent indifference of county councillors to his warnings despite his call having been taken up by the Devonshire Association, the Dartmoor Preservation Association and the mercantile associations of Plymouth and Tavistock.[50] Whether this was because the councillors were 'men of business' whereas the members of the Devonshire Association were 'men and women of Science, Art, and Literature', or because the location of the council in Exeter distanced them from the concerns of Dartmoor, or because the council was beholden to 'rate-fearing people', Collier insisted the Local Government Act of 1894 obliged the council to protect what he now termed the people's rights of common.[51]

Collier wrote in a more restrained register when he returned to his theme in 1896. Arguing that action was needed to protect 'our property', he adopted the emerging language of amenity, articulating his case in ways that emphasised the virtues of utility rather than the sensibilities of the sublime. Dartmoor would only remain an important source of water if its 'bogs and wastes' remained in their 'primeval state'; the preservation of common rights was of 'great value' to the industrious farmers who lived on farms adjacent to the moor; and the moor's 'infinite capacity of giving health, pleasure, and enjoyment to the people' could only be maintained if its 'scientific and aesthetic rights' were also preserved. He imagined a future Dartmoor managed by a new cadre of officers in which the relics and tors marked on the OS were subject to statutory protection, and mining, peat cutting, stone taking and sporting rights were more closely regulated.[52]

Similar thinking informed Robert Burnard's influential Plymouth Institution lecture of early 1894, which also drew on ideas of collective ownership to make the case for preventing the further 'plunder of Dartmoor'.[53] When Burnard spoke to the Exeter and District Chamber of Commerce in January 1897, the chair acknowledged the 'sentimental' reasons for preserving Dartmoor but foregrounded the 'strictly commercial' importance of the moor as a source of good water. Burnard neatly tailored his conventional preservationist points to the priorities of his audience. He not only highlighted London's interest in Dartmoor's water but also took the opportunity to criticise the prison, arguing that its acts of enclosure, reclamation and drainage reduced the moor's capacity to store water, thus increasing flooding during the 'rainy season' and drying out during the summer.[54] Civil society proved susceptible to such proselytising, if the seventeen members of the none-too-populous Exeter Literary Society Debating Club who voted in favour of acquisition by the council were anything to go by. Seven voted against.[55]

Signals from the duchy were not encouraging. Offence may have been caused when the county council's inquiry regarding a possible sale included complaints regarding recent enclosures. The duchy insisted it had not approved any recent enclosure. Quite how tense these exchanges became is unclear, though councillors were evidently concerned about water rights and hoped parliament would veto any transfer of ownership. When pushed, the duchy said it was unwilling to sell, though it transpired that it was in discussions with leaseholders seeking freeholds.[56] The *Western Morning News* recognised the difficulties posed by the myriad rights at work on the moor but was still disappointed, also sternly criticising the duchy for its failure to prevent illegal encroachment. Evidently the guilty parties were often the commoners themselves and it was unclear what authority could forcibly prevent them from building drystone walls. The duchy's apparent fear of a breach of the peace meant if the people who were 'filching slices of Dartmoor' had 'sufficient hardihood to stick to their thievings they need not fear anything more serious than a remonstrance'. Perhaps an act of parliament could give Devon County Council 'concurrent authority with the Duchy to prohibit the fencing of common land and prosecute those who do it'.[57]

If the duchy's refusal closed the issue for now, Collier's hunch that it would sell under the right circumstances was correct. While discussions regarding the purchase of Dartmoor proceeded, a greater threat to public access to the moor privately unfolded. A long report in a War Office file tells the story.[58] Ever since the duchy had permitted artillery practice on the northern quarter in 1875, commoners had been compensated for 'disturbance'. The modest scope of military operations made this straightforward until the Barracks Act of 1890 made £15,000 available to establish permanent buildings at the camp in Okehampton Park. Following agreement with the duchy and the commoners, the War Office took a twenty-seven-year lease at a cost of approximately £6,000, which was divided equally between the two interests. The need for more regular practice and the development of longer-range guns necessitated frequent clearance of stock from a larger area of common, and the War Office soon faced increased claims for compensation. When in 1896 the War Office attempted to settle the matter by agreeing to increase the annual compensation payment from twenty-five to two hundred pounds, claims brought by 'dissentient' commoners caused continued legal difficulties. The attorney general recommended that the War Office seek to buy the land outright and have parliament extinguish the rights of common or obtain statutory recognition of a single body representative of commoners. Further investigations concluded that it was best to continue dealing with those commoners on an ad hoc basis because attempts to clarify the legal position risked alerting many more locals to the existence of dormant rights. As it was, the War Office already needed the agreement of a number of local leaseholders plus the Okehampton, Okehampton Hamlets, Belstone, Bridestowe and Sourton commoners. Each required an annual payment, and the commoners demanded that the additional cost of clearing stock from the ranges on the days of practice was covered. A particular flashpoint was Watchett Hill, which the army was particularly keen to use but tended to avoid on account of opposition by the Belstone tenants. The commoners were not particularly motivated by preservationist concerns, but the money they extracted from the War Office reflected a new confidence that owed something to continued preservationist harping on the rights of common.

Purchase through parliamentary act, with common rights extinguished, offered the War Office a decisive alternative to this wearisome process of negotiation, renegotiation and spiralling costs. The Military Works Bill of 1901 made £125,000 available to the War Office for the purchase of 25,130 acres of the Dartmoor upland. Maurice Holzmann of the duchy office, admitting he would be 'extremely pleased' if the sale could be made because Dartmoor was a 'source of constant worry', highlighted the 'well nigh insuperable difficulty in dealing with Dartmoor without special legislation' and the 'impracticable' aspects of selling only the north quarter. The political ramifications had also to be considered. Any move by the War Office would prompt Devon County Council to renew its agitation to purchase while a successful sale would see the Prince of Wales 'abused in unmeasured terms all over the country' and 'hooted whenever he set foot in Devonshire'. Still, Holzmann recognised that the duchy was obliged to act in the national interest, albeit in a manner that chimed with its own:

> In the event, however, of the Secretary of State considering it indispensable for the purposes of national defence that he should become the owner of the Artillery ranges on Dartmoor, His Royal Highness, on the advice of His Council, would be prepared to sacrifice all personal considerations and would consent to negotiations being entered into for a conveyance, subject to the requisite sanction of His Majesty's Treasury, of the whole of the Duchy interests in the Forest of Dartmoor and adjoining Commons of Devon, comprising not only the unenclosed lands, but also all the inclosed premises of every description.

At a meeting with the War Office on 20 December 1901 Holzmann put the full cost to the Treasury of unburdening the duchy of its responsibilities at £200,000 for the full 68,000 acres. A laconic note by Lord Roberts, commander-in-chief of the army, closed the question: 'This is unfortunate, but is clear we cannot get the land, and must look elsewhere.'

Dartmoor: a National Park?

In the interwar years the agenda of the Dartmoor Preservation Association and fraternal organisations elsewhere began to influence national politics in new ways. Increasing suburbanisation and new land use by statutory undertakers, notably water authorities, generated anxiety regarding the uncontrolled development of rural landscapes. Of particular concern was planning law, which mainly applied to urban spaces and provided little scrutiny and control elsewhere. Such concerns were magnified by the development of new leisure cultures that made much of the physical, spiritual and moral refreshment brought by time spent in the countryside, particularly for the industrial working class. Civil society organisations like the Commons, Open Spaces and Footpaths Preservation Society (1865), the National Trust (1895), the Council for the Preservation of Rural England (1926), the Society for the Promotion of Nature Reserves (1928) and the Youth Hostel Association (1930) looked to state intervention. Thus, landscape preservation and nature conservation began to coalesce, aligning with an emerging politics of access, generating considerable pressure on government to intervene to protect and manage what was coming to be seen as the national interest. And this last point is crucial. At stake in these debates was the notion that rights of property should be made partially subordinate to the interests of the community as a whole. If such thinking had been of increasing political importance in the late nineteenth century, the advent of universal suffrage in 1928 and greater leisure time, again particularly among the industrial working class, saw political pressure on the government intensify.

Looming large over these debates was the idea that Britain should follow the example of the United States and Canada and designate national parks. Historians such as John Sheail have traced the political development of this idea and its culmination in the National Parks and Countryside Act of 1949.[59] The fine detail of this complex story can be reduced to two interweaved strands. One concerns the integration of rural landscapes into planning law in the 1930s and 40s, the other the growing political consensus favouring the creation of national parks. The 1949 act was shaped by three government reports instigated by Labour ministers that reflected the slow social democratic turn in British politics. To simplify a complex story, Addison (1931) made the case for national parks, exploring through extensive international comparisons what form might be suitable in Britain; Dower (1945) identified possible sites, moving beyond Addison's reluctance to be prescriptive; and Hobhouse (1947) addressed the question of implementation and governance. Addison concluded that the purpose of the parks was the preservation of areas of natural beauty, the protection of sites of scientific interest and the enhancement of their recreational and educational value. Though there was a presumption in favour of public access to all areas of a park, a national park designation would not connote a right of access, principally because it was expected that a significant part of each park would be enclosed farmland. It would be incumbent on the new national park authorities to seek access agreements wherever possible. Among the suggestions adopted by Addison but later rejected was the notion that all or some of the parks might be nature reserves and thus subject to a very restricted right of access. By contrast, there was never much appetite for the notion that the parks should be purchased outright, though the National Trust, having purchased vulnerable parts of the Lake District in the first decades of the century, tended to favour this solution.

Landscapes identified as suitable for national park status were often described as wilderness, though in the British context this did not carry the North American connotation of wilderness as pristine. Indeed, the national parks proposed in Britain were often valued as heritage or cultural landscapes, their national importance partly attributed to the survival of pre-industrial ways of life rather than the absence of human activity. To adopt the terminology of modern environmental history, these were generally anthropic pastoral landscapes. Tensions quickly arose in national park thinking between the curatorial impulse to preserve and the desire to see the prosperity and well-being of rural communities enhanced through the introduction of new agricultural methods. Above all, the discussions that preceded and followed Addison centred on the concept of 'amenity', which placed at the heart of national parks thinking the needs of the general public. Attempts to protect amenity, focused on preserving the natural beauty and historical features of the

parks, dominated national park politics from the 1940s through to
the 1970s.

Designating and enhancing national parks was a costly business,
and when Addison's report was published on 23 April 1931 the circumstances could not have been less propitious. Arthur Greenwood, the
minister of health, urged that £100,000 per annum be reserved for
the next five years for the purpose of implementing Addison's
proposals, but the fall of the Labour government in August amid
the escalating economic crisis saw these plans sidelined. Instead, the
incoming 'National' government passed the Town and Country
Planning Act in 1932, which empowered local authorities to protect
'beauty spots'. When the question of creating national parks was
again debated in parliament in 1936, opponents often claimed that
the 1932 act had empowered local authorities to implement Addison's
recommendations. If this stretched credibility somewhat, it nonetheless underscored parliament's reluctance to wrest powers from local
government and vest them in a new national organisation. Local
planning control would be further strengthened in 1947 by a new
Town and Country Planning Act. Consequently, when in 1949 the
decision was made to establish a National Parks Commission with
the power to designate a series of national parks, the commission's
long-term role was to represent the amenity interest to the local
planning authorities, advising on how the parks could be developed
in the interest of amenity. Nature Conservancy, established by royal
charter that year, added scientific expertise to the mix and another
new organisation local planning authorities had to accommodate.

Dartmoor was the fourth and last of the first wave of national
park designations, following the Lake District, the Peak District and
Snowdonia. During the Addison hearings, when the qualities of a
very wide range of landscapes were assessed, Dartmoor's status
was the least assured of the big four. When asked whether the
national parks should be designated as representative of particular
types of landscapes, several witnesses identified Dartmoor as a
prime example of 'moorland', one witness preferring it to the

Pennines because no valuable sporting rights existed on the moor, which would make reaching agreement with the duchy relatively straightforward.[60] That said, during the Addison hearings it became clear how much the amenity societies distrusted the duchy. Lawrence Chubb, representing the Commons, Open Spaces and Footpaths Preservation Society, could scarcely conceal his contempt. Since 1820, he told Addison, the duchy had allowed the commonable land of Dartmoor Forest, 130,000 acres in total, to be reduced to 50,000 acres. Among the encroachments was a 'solid wedge of the moor' from Princetown to Chagford, but more indicative of the current threat to Dartmoor was the duchy's decision to permit extensive china clay works at Board Down. Chubb urged that any development of the 'great central enclosure . . . out of harmony' with its 'wild and romantic scenery' should be prohibited.[61]

Patrick Abercrombie of the CPRE, one of the moving forces behind the establishment of the Addison Committee, took a less confrontational view. He thought the High Peak (the north Pennines) and the South Downs had the strongest claim to protection if the measure was the proximity of a large urban population, but if the criteria was 'national interest and intrinsic beauty', then he thought the Lake District, Snowdonia, Exmoor and Dartmoor were the leading contenders. A long and discursive submission from the National Trust placed purchase high on its agenda and lined Dartmoor up alongside the South Downs and the Lake District, though only the Lake District, to them self-evidently Britain's finest landscape, received its unambiguous support. The British Correlating Committee for the Protection of Nature also found the Lake District the most important landscape in Britain. Of sixty-six landscapes examined, only the Lakes were judged valuable under all the five headings (mammals, birds, plants, geology and insects). Dartmoor was among a small number of areas judged important under three headings, namely birds, plants and geology. The Geological Society identified Dartmoor as particularly important, calling for the protection of isolated granite outcrops like Haytor.[62]

Sir Walter Peacock, representing the Duchy of Cornwall, made the case against designating Dartmoor a national park. Peacock insisted there had been little enclosure since the Napoleonic Wars, besides which the most popular beauty spots, such as Dartmeet, were already enclosed. The duchy, he assured Addison, was 'alive' to the threat from developers having recently 'wrested an important site' up for auction 'from a jerry builder'. Peacock's remaining points were more potted. Few visitors actually made it to the heart of the moor; mining leases were granted only for a year; the artillery firing range would be an obstacle to making Dartmoor a national park; and the volume of stock carried on Dartmoor had been reduced in recent years. If the pertinence of these points was not altogether evident, Peacock's basic proposition was simple: Dartmoor was safe in the duchy's hands.[63]

Early drafts of Addison's report were strongly influenced by the duchy's lobbying. Dartmoor was recognised as 'a typical Moorland area which by its extent and beauty is well fitted to become a National Park' but access to the open moor was judged already 'liberal' and 'little' was required beyond maintaining 'present privileges' and ensuring 'that neither the Moor nor the enclosures' were 'disfigured by development out of keeping with their environment'. The duchy could be relied upon to maintain farm buildings 'in the traditional style of architecture of local granite, in keeping with the character of the neighbourhood', and new buildings of 'an unsightly design' were not allowed.[64]

For reasons that are not altogether clear, the drafting of the report saw the duchy's view rejected, and in the published version of the report Dartmoor was listed as among a dozen or so landscapes that might be designated after the Lake District. A page later, Dartmoor was edged up the scale, Addison adopting Abercrombie's judgement: 'From the population point of view the High Peak and South Downs would appear to have the first claim; the Lakes and Snowdonia, Exmoor and Dartmoor are pre-eminent from the point of view of national interest and intrinsic beauty.'[65]

Dower made Dartmoor a high priority, placing it third after the Lakes and Snowdonia and among the first six areas recommended for national park status; Hobhouse placed the Peaks above Dartmoor but included a long appreciative description of Dartmoor in his report.

Addison, Dower and Hobhouse were important steps in the revaluing of the British landscape, envisioning the transformation of Britain's most marginal and underutilised landscapes into its most controlled and managed rural areas. Dartmoor's achieved status as one of Britain's preeminent rural landscapes owed much to the efforts of its champions. Had Carrington's poetry not won the approval of George IV, had the Devonshire Association not undertaken its archaeological investigations, had Crossing not demonstrated his skill as a journalist and topographical writer, and had the Dartmoor Preservation Association not raised its noisy and insistent voice, Dartmoor might not have found itself counted alongside the Lakes, the Peaks and Snowdonia. The same, of course, can be said of those other landscapes, each of which had advocates of an equally if not more distinguished lineage.

Does this suggest that selection was simply a political process whereby the most effective lobbyists got their own way? Had the Dartmoor lobby been weaker – it was certainly weaker than the Lakes lobby – would Addison and his successors have identified Dartmoor as of national importance on the basis of its natural characteristics and particular history alone? What is certain was that the voluntarist enthusiasm of the preservationist lobbies was largely responsible for producing the knowledge that led the government to consider introducing national parks, an idea of North American origin, into the UK.

Harry Batsford, writing of 'Country and Coast' in a valuable volume commemorating the first fifty years of the National Trust, imagined the threats then faced by 'a pleasant, moderate-sized hilltop estate':

But the War Office wants to take the place for tank manoeuvring and artillery observation; the Admiralty are after it for a compass-testing station and a wireless experimental centre; the RAF think it is just the thing for a glider landing ground, or flying-bomb launching site. The Forestry Commission regard it as an ideal spot for the acreage of its legions of regimented conifers; a large town wishes to lay out a huge housing estate on the top, a still bigger and more distant city wishes to construct a reservoir on the estate, and a county council plans to set down a super-splendid madhouse, which it has been the hobby of county councils to erect between the wars – and they always want to put the beastly things on the highest hilltop for miles around, possibly to cheer the inhabitants by the sight of their future residence; presumably now the war is over, they will be avid to put up enormous institutions for TB and VD.[66]

Replace the 'super-splendid madhouse' with a prison, and this becomes remarkably prophetic of the early decades of Dartmoor National Park.

Designation and Television

The decision to grant the National Parks Commission the authority to designate Dartmoor a national park generated a rush of anxiety regarding developments that might occur in the meantime. Geoffrey Clark, director of planning for Devon County Council, met Patrick Duff of the National Parks Commission in April 1950 to offer warnings of 'untoward developments'. One imminent threat was an application to reopen a café in Widecombe 'out of keeping with the village' and the owner's plan to run 'an undesirable bus service across the moor'.[67] In early May the Devonshire Association and a host of amenity organisations passed resolutions demanding Dartmoor's early designation. Sylvia Sayer, president of the revived Dartmoor Preservation Association, enclosed these resolutions in a letter to Duff, reinforcing the sense of urgency. Various interests, Sayer argued, including the military and the china clay companies, were trying to expand the scale of their operations before the designation was made.[68]

A national park requires a boundary, and establishing where that boundary should lie is contentious because inclusion or exclusion affects the future value and development potential of the land. Few doubted that Dartmoor's boundary would be placed significantly beyond the common, taking in extensive enclosed farmland, many villages and several small towns. This 'in-country', among the most popular and most beautiful parts of the region, also hosted signifi-cant commercial and industrial activities thought to be out of keeping with a national park. In the event, the National Parks Commission exercised considerable latitude in making its recom-mendation, going beyond the strict criteria provided by the 1949 act and towards the creation of a national park that reflected the popular sense of Dartmoor as a geographical space. Despite the emphasis placed on preserving the natural environment, the most significant landscape features shaping the popular sense of Dartmoor were the main roads connecting Exeter to Okehampton, Okehampton to Tavistock, Tavistock to Plymouth and Plymouth to Exeter; parts of Brunel's great railway line between Exeter and Plymouth was suggestive in the same way. Such a maximalist designation was never likely, though evident in the designation process was the way the rough triangle of land these roads bounded influenced the map image of Dartmoor.

A case in point were the villages of South Tawton and South Zeal, both just off the northern quarter of the common and within the boundary suggested by the Exeter–Okehampton road. The initial draft boundary excluded the two villages because committee members 'could not recall any great natural beauty north of the draft boundary at this point'.[69] After visiting, they changed their mind and opted for inclusion. The claim that the parishes of Bridford, Dunsford and Christow were not typical of Dartmoor was also dismissed, and their inclusion pushed the boundary significantly east of Moretonhampstead until it fell a few miles short of Exeter.

Such decisions were influenced by representations sought from local authorities, amenity societies and interested bodies like Nature Conservancy, the National Trust, the Forestry Commission, the parliamentary Standing Committee on National Parks, and the Ministry of Agriculture and Fisheries. Sensible solutions often came easily. Setting the boundary immediately north of the railway line between Ivybridge and Bittaford Bridge, for instance, excluded the town but included the village. Most contentious was the Bovey Tracey area on the eastern side of the moor and the stretch of land between Plymouth and Tavistock comprising Whitchurch, Plaster and Roborough Downs and the small towns of Horrabridge and Yelverton. Commercial and industrial interests were worried, and Devon County Council and its Dartmoor Committee successfully urged Bovey Tracey's exclusion on commercial grounds, though committee members noted that 'those who were about to explore the district should have a pleasant place as a jumping-off ground, and Bovey Tracey would be an admirable place for a hostel'.[70] By contrast, the committee tended to favour arguments advocating the inclusion of the Plymouth–Tavistock hinterlands, though it recognised that the arguments against could not be easily dismissed. Plaster Down had been the site of a US field hospital during the war and was only connected to the western quarter of the common by a narrow aperture of open land. Whitchurch Down on the western edge of Tavistock, separated from the open moor by enclosed fields and woodland, already included a cricket ground and was ripe for development. Roborough Down was still more problematically placed. Lying mainly to the west of the main road connecting Plymouth and Tavistock, it hosted a golf course and Second World War RAF base; Horrabridge and Yelverton lay between Roborough and the enclosed farmland of Sampford Spiney and Meavy. Visiting representatives of the National Parks Commission concluded that 'on landscape grounds the area did not qualify for inclusion',[71] but local lobbying and pressure from Devon County Council outweighed local business and landlord interests and Plymouth's 'playground' was included in the designation order.[72] These inclusions saw amenity rather than the aesthetic or natural quality of the land win out, pushing the border of the new national park to the western edge of Tavistock and to Plymouth's northern suburbs, placing a limit on the expansion of both. Roborough Down would prove the focus of considerable controversy in the future: sooner or later a large stretch of unenclosed land near an expanding city was bound to attract the attention of city planners and developers.

Another contentious site near Plymouth was the bulge of moorland to the east of Shaugh Prior. North of the village is the

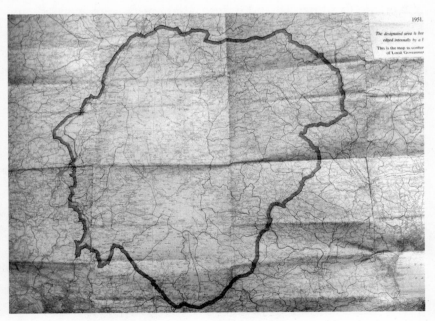

Dewerstone Rock, an iconic location much favoured by the Victorians. Sitting at the southern head of Wigford Down and overlooking the confluence of the Plym and the Meavy, it offers an unsurpassed view over Bickleigh Vale and the Plym Valley to Plymouth Sound hazily beyond. Less than a mile to the east were the china clay works of Lee Moor, the most destructive industrial development of modern Dartmoor history. By closely skirting the Wotterwaste, Shaugh Lake and Whitehill Yeo works the boundary protected the antiquities of Saddlesborough and Trowlesworthy, limiting further development.

On 15 August 1951 the designation order was made. It was one of the last acts of the pioneering Labour governments that laid the social-democratic foundations of post-war Britain. At 365 square miles, Dartmoor was the smallest of the first four national parks, coming in behind the Peak District at 542 square miles, Snowdonia at 845 square miles and the Lake District at 866 square miles. Compact and cartographically pleasing, the hexagonal shape of Dartmoor National Park contrasted with the irregular outlines of the other three, creating a landscape whose natural integrity seemed self-evident.

Although Dartmoor contained the largest open space and most impressive uplands in the densely populated region south of the line from the Bristol Channel to the Wash, as the Addison, Hobhouse and Dower reports had shown, until the designation order its status alongside the big three was not secure. Uncertainty about quite what was being preserved remained, as was evident in a letter Geoffrey Clark wrote to Duff following a celebratory lunch in London. Reiterating the pleasure the county council took in the designation, Clark reflected on the problem. 'Dartmoor is a strange place,' he wrote, 'very different from the Lake District, in the sense that its scenery is of a kind which requires individual enjoyment.' This was the old claim that Dartmoor's particular qualities could only be comprehended by experiencing the isolation and quiet of the uplands. This was a minority pursuit, and Clark reckoned Dartmoor attracted fewer walkers than other national parks. As such, he anticipated 'very little major threat to this aloof moor', adding that he did not think enjoyment of the moor was 'seriously damaged' by the 'military occupation' of some parts. More threatened was the popular 'surrounding fringe', though as yet only in a few places had the 'well-defined tourist roads' generated 'a certain amount of unpleasantness' like 'shops selling things like highly coloured gnomes'.[73]

When the *Sphere*, an illustrated magazine, celebrated the designation with a double-page spread, it chose to represent the new national park with aerial photographs of Buckfastleigh, Buckland-in-the-Moor and Widecombe-in-the-Moor. These locations gave readers a Dartmoor of pretty villages, patchwork fields and rolling hills, the idealised England of nucleated villages that would be celebrated by W. G. Hoskins in his *The Making of the English Landscape* (1955) just a few years later. Dartmoor's uplands still loomed menacingly beyond, just as they had in the nineteenth-century imagination and just as they continued to do in the mind of the journalist and the county planning officer.

On 9 August 1951 Patrick Duff wrote to tell Sylvia Sayer that the designation would be announced on 15 August. Acknowledging that it had been a slow and complicated process, he hoped they 'could now look forward with every confidence to seeing England's green and pleasant land grow more green and more pleasant, and its healing gifts find their way to the hearts of more people'. Sayer responded with a characteristic mix of gratitude and readiness for the next fight that would become the bane of Whitehall civil servants for the next forty years. 'I will now try to curb my unconscionable impatience, which you so delightfully forgive,' she wrote, 'though I fear it will still give me a little trouble until at last the local administration of the Dartmoor park is on the right footing!'[74]

Sayer's allusion was to the status of the Dartmoor committee of the Devon County Council, which the DPA insisted should cease to be a subcommittee of the County Planning Committee and thus subject to the veto of members who 'know little about Dartmoor'.[75] The CPRE feared that the county committee might not take sufficient account of the 'national' purpose of the park and that its ministrations would not come under proper public scrutiny because the press was not admitted to subcommittees.[76] The law required that the county council and the National Parks Commission agree how Dartmoor should be managed, and in the discussions that followed it became clear that the county council had no intention of allowing the Dartmoor committee full autonomy. At times its position was very obscure, and the commission found letters from Clark 'extremely difficult to understand'. Harold Abrahams, former Olympic athlete and secretary of the National Parks Commission, explained that the 1949 Act made it clear 'the Committee dealing with the National Park should be independent and not subordinate to another Planning Committee', but when clarification was sought in the Commons, Lewis Silkin, under pressure to be seen not to dictate from the centre, fudged the discussion, generating 'an extremely muddled and verbose Debate'. That November, Devon County Council proposed that the Dartmoor Standing Committee should have eighteen members, twelve members appointed by the council and six nominated by the minister. Development applications would be made to the appropriate district council, which would pass them on to the county's director of planning. Significant proposals would be passed to the Dartmoor committee for a decision, but if a dispute arose between the committee and the district council, the county council would have the final power of decision.

In November the National Parks Commission approved these arrangements. The DPA objected on two grounds. First, the requirement that the Dartmoor committee report to the planning committee made it subordinate, which they claimed was contrary to the 1949 act – in reality, the act was notoriously weak in this respect. Second, the nominated committee members – which included Sayer – responsible for protecting the national interest could not address the county council, meaning only the voices of the council nominees, many of whom were also members of the planning committee, could be heard in public. Privately, Sayer conceded defeat gracefully; publically, the DPA newsletter sounded a positive note, though when the membership of the committee was announced in January the association was less happy, concluding that Silkin had been swayed by the need to keep the county council happy.[77]

While squabbles over the precise status of the Dartmoor Standing Committee continued, the controversy that dominated Dartmoor politics for the next few years came to the attention of the National Parks Commission. On 20 August 1951 Clark informed Duff that the BBC had notified the county council of its plan to erect a 750-foot television aerial on North Hessary Tor at Princetown. As the 'first major proposal from a statutory body . . . likely to alter the landscape of the heart of the wildest part of the moor', Clark warned that this was 'bound to create a disturbance among the purists'.[78] Two days later, the National Parks Commission received a BBC memo explaining its intentions. The mast would be erected on Dartmoor as part of the corporation's statutory obligation to provide near-total television coverage throughout Britain. Five main transmitting stations were being built, and they needed to be supplemented by five relay stations. Sites had been found for four of these relay stations and North Hessary Tor was to be the fifth. The aerial would relay the signal from the transmitting station at Wenvoe in south Wales, and would serve Plymouth, Exeter, the Torbay area and much of Cornwall – a booster mast was needed in west Cornwall. The difficulty the BBC faced was less the strength of the signal from Wenvoe and more the positioning of Dartmoor in relation to Plymouth: a good signal was only possible when physical obstructions were kept to a minimum, hence the aerial's need for a high mount. The BBC argued that North Hessary Tor was a good location for technical reasons – it would bring a good-quality signal to almost a million people – and because the prison had already spoiled the site's amenity value.[79]

Clark was right. Once the DPA learned of the plan in October, it mounted its first major campaign, establishing an approach that would become familiar. It polemicised against the threat posed to Dartmoor's amenity status; it sought expert opinion in order to refute the arguments made by the proposer; it offered alternative solutions to what it acknowledged was a legitimate problem; it pushed for a public inquiry and, when successful, engaged the services of a lawyer in order to present its case effectively. Public meetings, pamphleteering, petitioning parliament and writing letters to the press were the outward signs of the campaign. As important, the DPA understood that the national parks legislation obliged the government to consult with the commission on the best way to preserve amenity when deciding on a planning application: as Duff put it, the government's difficulty was the 'task of putting into one scale of the balance the effect on amenity when a more directly apprehensible good is on the other scale'.[80] Sayer seemed to intuit that the effectiveness and determination with which the commission could resist any given encroachment was largely dependent on the activism of organisations like the DPA. This was not because the commission was necessarily inclined to comply with the needs of powerful interests but because amenity did not provide objective criteria against which a planning application might or might not be approved. Amenity existed only insofar as interested parties could persuade the government that it did.

In December the DPA submitted to the National Parks Commission a preliminary memo outlining its opposition. It was reprinted almost word for word in the DPA newsletter, beginning a campaign framed in terms of Dartmoor's provision of an escape from modernity:

> The alien presence and associations of an outsize television mast would be a perpetual reminder of those aspects of modern life which most people come to a National Park to forget. The tor itself, as we know it, would disappear. Twenty-five acres of the summit would have to be acquired for the spread of the mast's steel wires, and the buildings – of BBC functional type: flat-topped rectangles – would crown the tor in place of its natural rocks. A road and a concrete platform, fencing, pipes, and cables would have to accompany the mast on to the tor. It would be landscape-slaughter on a more than usually impressive scale.[81]

Anti-modernity carried specific aesthetic and material judgements: steel and concrete were 'alien' materials, and modernist 'functional' architecture represented values contrary to those signified by Dartmoor. The mast, Sayer argued, would be 'a sharp reminder of the 20th century to anyone contemplating the beauty of an otherwise relatively timeless scene'.[82] Scale was also important – the mast was 'outsize' – and as the commission's campaign advanced it tended to emphasise Dartmoor's small size. A government functionary adumbrated the aesthetic case, arguing that the mast was 'ugly in itself'. 'It will consist . . . of 610 feet of triangular latticed mast with 9-foot sides. This will be topped by a 100-foot cylindrical section, not much smaller and looking like a floating factory chimney, and this will be capped by a 40-foot square section mast with aerial sprays sprouting like exotic succulents from the top . . . The mast will be tied by 4 sets of stays which will spread over several acres of hill-top.'[83]

DARTMOOR NATIONAL PARK: NORTH HESSARY TOR.

THIS —

OR THIS ?

"A National Park may be defined as an area of beautiful and relatively wild country in which ... the CHARACTERISTIC LANDSCAPE BEAUTY IS STRICTLY PRESERVED." — Mr. Harold Macmillan, July 1952.

Edward Bonong, the BBC's lead engineer, responded forcefully, giving voice to a modernist sensibility that conflicted with the DPA's preservationism. A 'graceful engineering structure' would not have such a destructive effect on 'the wild and natural character of a very wide area of ancient hill-country landscape'; the prison ensured the location was 'by no means a place of beauty' anyway; the Royal Fine Art Commission had approved the building design; the alternative sites suggested would not reach as many people and, contrary to the DPA's claims, the North Hessary site would ensure a good signal reached Truro and Falmouth in Cornwall and a weaker one beyond. Moreover, Bonong found the elitism of the DPA objectionable: the 'intangible values' of their 200 members could not dictate what was good for nearly a million people in the region.[84] When the National Parks Commission sought the expert opinion of Dr Smith-Rose and Dr Saxton of the government's Radio Research Station at Whitton Park, they affirmed Bonong's argument.[85]

The Dartmoor Standing Committee voted to approve the scheme on 17 June 1952, prompting Sayer to observe angrily that it had relied on the casting vote of the chair and the absence of three members who would have voted against. Sayer could not imagine any argument justifying a television mast on Dartmoor. 'No other feature of landscape value, or such a beloved viewpoint,' she exclaimed, 'has ever been destroyed deliberately for a <u>luxury</u>.' Even as she mobilised technical arguments, including the dubious claim that future developments would see new wavelengths discovered that would allow for two separate Devon and Cornwall transmitters, she lashed out seemingly without irony at 'local interests, even materialists, even rabid all-out TV viewers', making it clear she had little truck with the Reithian case for television. When in July she failed to persuade the committee to rescind their earlier decision, she walked out, her view of the committee confirmed. Two days later Devon County Council approved the proposal by a large majority.[86]

The CPRE firmly opposed the mast, demanding that the National

Parks Commission intervene; the Commons, Open Spaces and Footpaths Preservation Society took a more measured line, saying it was unconvinced by the case for North Hessary but would fall in behind the decision of the National Parks Commission.[87] The commission itself had to balance its statutory obligation to protect the amenity value of the national park against its status as an advocacy body that could not unreasonably obstruct government policy. Having lost the argument with the planning authority, the commission turned its attention to the BBC itself and Harold Macmillan, Hugh Dalton's Conservative successor at the Ministry for Housing and Local Government. Patrick Duff now played a more active role, corresponding directly with fellow members of the political establishment. He appealed to W. J. Haley, the BBC director general, asking that no decision be made until the Wenvoe transmitter was switched on and its effectiveness assessed. He asked the same of Macmillan, whose office replied that no decision would be made until the BBC submitted a planning application. Sayer's letters to Macmillan elicited the same response.[88] Around the same time the commission began to receive notifications from local authorities that they supported the favourable decision taken by Devon County Council.[89]

Sayer was not ready to give up and continued to propose alternative solutions in the light of the apparently successful Wenvoe tests, which she claimed – a 'rather comic footnote' – had supplied the only television in Princetown with a good picture. In November she wrote apologetically of her reference during an agitated telephone conversation to 'Coronation TV hysteria'. 'The hysteria,' she explained, 'that is being worked up about televising the Coronation has nothing to do with regard for the Queen's own feelings or wishes.' She was certain 'much of the clamour is being raised by those who would quite willingly sacrifice Her Majesty's personal feelings to make a TV holiday', and the BBC 'will make the most of it to get their foot in on North Hessary Tor'.[90] The lobbying paid off that summer when the BBC conducted tests at

Modbury, an alternative site south of Dartmoor, though the results confirmed their preference for North Hessary. Sayer was quick to claim their judgement was made on the grounds of cost rather than strict technical necessity, seemingly oblivious that questions of cost must shape any such decision. The CPRE weighed in, telling the National Parks Commission that the BBC's case was not proven and that their opposition must continue. Duff had few options but to push for a public inquiry. Evelyn Sharp, a senior civil servant of formidable reputation, let him know that once the BBC made its formal planning application, Macmillan would announce an inquiry. On 6 August 1953 the BBC made its move, and Macmillan announced that a public inquiry would be held on 29 September 1953 at the Castle, Exeter, under the eye of an experienced town planner.[91]

The inquiry was the first big test of the commission's capacity to represent the amenity interest on Dartmoor when faced with a major statutory body seeking to implement government policy. All the evidence confirmed that the relay station was needed. In particular, the number of television licences taken up in Devon in anticipation of the Coronation was relatively low, providing little evidence of Wenvoe's adequacy. In the 12-month period to June 1953 (the month of Elizabeth II's coronation) the number of Devon licences increased from 343 to 7,662, of which 4,372 were in the Exeter, Exmouth, Torbay and Plymouth areas. Take-up in Somerset, which was served by Wenvoe, was proportionately greater (1,316 to 14,745), while take-up in Devon's major urban centres was low by national standards: 4.2 per cent of the population held licences in England, whereas the figure was 1 per cent in Torquay and 0.3 per cent in Plymouth.[92] The commission's task, therefore, was to make the case against locating the mast on Dartmoor rather than question the need to develop the region's broadcast infrastructure. Considerable time and resources went into assembling the case, drawing on the knowledge of technical experts and working closely with the DPA.

Sayer was to hand. She provided a raft of questions for the

experts, although the answers received were sometimes contradictory and not always welcome. Most authoritative was Smith-Rose. He conceded that placing the mast on the moor meant the signal lost strength by the time it reached densely populated areas like Plymouth and Torbay, giving some credence to the DPA's case for the alternative site at Modbury. He also recognised that the second-grade first-class service provided in Liverpool by the Holme Moss transmitter in the Peak District was satisfactory, which went some way towards conceding that Wenvoe provided a reasonable signal to a larger field than had been previously acknowledged. Nonetheless, he explained that erecting two transmitters – the DPA's favoured solution – was not feasible. Masts transmitting or relaying broadcasts on the same wavelength needed to be between 200 and 400 miles apart in order to prevent interference; other wavelengths were not available, partly thanks to agreements with the French, who were expanding their television coverage to Normandy and elsewhere. The BBC's own research, which accorded with the conclusions of international engineers, suggested they should opt for the highest available point. This was why transmitters had been placed on the Empire State Building in New York and Alpine peaks in Germany and Switzerland. To be barred from using hills would create technical difficulties and significant additional cost.[93]

When Duff pondered whether the BBC might consider locating the mast on Bodmin Moor, a colleague admitted that 'if it were a straight issue, either Bodmin or Dartmoor', he would 'save D. at the expense of B. every time' for there was 'no real comparison in their landscape quality', but raising Bodmin would place the BBC under pressure to explore other sites, and that would mean further delay. Also examined was whether the earlier decision to locate the transmitter station on Holme Moss in the Peak District had any bearing on the Dartmoor controversy. Holme Moss had been approved before the passage of the National Parks Act and the commission viewed the station and mast as a disfigurement that should have been opposed. Mervyn Osmond of the CPRE explained that they withdrew their opposition to the Holme Moss plan when it became clear there was no alternative site where the amenity objections were weaker. Osmond did not think the case for North Hessary was conclusive. The commission reckoned the size of the Peak District made Holme Moss a less significant intervention than the North Hessary plan, and it was alarmed by the profound effect the mast would have on perceptions of Dartmoor as a whole. 'The size of the mast is so great that it will not only diminish the apparent scale of the Park (which at present looks much more expansive than it is) but will actually appear to reduce it beyond its actual scale. The size of the mast will in effect make Dartmoor look small.'[94]

In a similar way, Sayer was troubled by strained attempts to capture North Hessary's particular merits on camera. Writing to Abrahams, she worried their case would be weakened because the site did not adhere to romantic notions of a great natural landscape: 'I'm a bit doubtful about photographs: Hessary Tor is deceptively unspectacular in itself, having no startling crags or beetling precipices; but if the cameras can take good long-distance photographs & the day is clear, the evidence will be good. Particularly from the ridge about Nun's Cross. Nothing however can be as good as personal observation. And everything depends on clear weather.'[95]

When Duff presented his evidence to the inquiry, he laid similar emphasis on the ineffable nature of Dartmoor's particular value and the need to understand that this was a question of scale. To this end, he quoted Hobhouse's commentary on the moor: 'The austere windswept and almost treeless plateau of Dartmoor, the largest mass of exposed granite rock in the south-western peninsula, gains by contrast with its surroundings an appearance of height, wildness, and extent, greater than the evidence of maps and contour lines suggests.'

Duff explained that the National Parks Commission opposed the mast because it would have an 'alien and formidable impact on the scenery of the moor' and more particularly because the position of North Hessary Tor, 'a commanding eminence in the landscape', ensured that the 'visual disturbance of anything done there makes itself accordingly felt over the widest possible area of country'. Dartmoor National Park, he explained, was 'very small' and 'except for down in the valleys, this mast would be visible almost from end to end of the park'.[96]

After the inquiry, Duff and Sayer exchanged mutually admiring letters; Robert Watson-Watt, their expert witness, and Ramsey Willis, their counsel, received hugely congratulatory letters from Duff.[97] However, on 25 January 1954 the director general of the BBC received a letter from the ministry. Permission was granted, though certain conditions had to be met. The county council would have to agree the precise siting of the mast, its design and the building materials; the approach road should not be fenced; the existing barrow on North Hessary was not to be disturbed until the Devonshire Association had a chance to excavate it; and all cables were to be placed underground. The *Western Morning News* reported that 'prehistoric Dartmoor was to fall before the demands of the present';[98] the DPA briskly dealt with the fallout in its newsletter. Some solace was found in the requirement that the rocks of the tor and the viewpoint would be preserved by erecting the mast just off the summit, as did the stipulation that the buildings be constructed from granite and proper procedures be followed regarding the acquisition of the land. But the DPA's anger was still palpable, and it gave vent to its most reactionary sentiments. North Hessary Tor had been thrown to the 'Philistines'; 'local interests' had trumped 'the national interest' because there were 'no VOTES or MONEY in landscape protection these days'; the mast represented the 'most recent and conspicuous step' taken towards transforming Dartmoor, 'a wild and lovely survival of ancient beauty', into 'a barren and characterless municipal recreation-ground, complete with all the appropriate notices, paths and car parks, its open spaces plastered at intervals with the latest man-made contraptions and conveniences'.[99] The North Hessary controversy threw into sharp relief the rapid disillusionment the DPA experienced regarding the national park designation. The new status had not protected Dartmoor from a development that significantly altered how the moor looked, while the requirement to enhance access to the moor added new disfigurements.

It takes a leap of the historical imagination to see beyond the virulence of the DPA's elitism or the hyperbole of its claims to appreciate its legitimacy. Dartmoor is now a highly managed space with a developed tourist infrastructure that strongly shapes how the moors are encountered. Whether this makes the moors more or less loveable is impossible to say, but there is no question that the development of this tourist infrastructure has significantly changed the moorscape. Despite this, a simple narrative of ever-increasing encroachment will not do. Since the 1950s the military presence on the moor has been significantly reduced, the management of the Forestry Commission plantations has become more aesthetically sensitive – the same can probably be said of the tourist infrastructure – and the reservoirs have created valuable new ecosystems and sites of leisure that many find pleasing. Economic change means quarrying on the moor has ceased, and the railways are long gone; popular pressure has seen the china clay companies become more environmentally responsible. As the next three chapters demonstrate, this was not a story of unmitigated preservationist success, but there is plenty of evidence that lobbying by the amenity societies has significantly restrained the development of the moorscape.

At the same time it is surely true that the Dartmoor enthusiast loves the Dartmoor they first encounter. The television mast has hovered over the moorscape for over sixty years. It might be wished gone, just as might the prison, the reservoirs and the plantations – even the roads – but just as Crossing's imagination was fired by the anthropic Dartmoor of the late nineteenth century, so these encroachments are integral to what Dartmoor now is. Imagining the moorscape denuded of these presences is necessary if a sense of Sayer's Dartmoor is to be recovered; but to so scroll back is to do damage to the imagined and experienced Dartmoors of successive generations. Moreover, although certain of Dartmoor's natural characteristics have been irreversibly damaged by encroachment – if damage is understood as irreversible environmental change that diminishes biodiversity – the degree to which the natural landscape remains an agent in its own production should be no more underestimated than its resilience exaggerated.

Before turning to the Dartmoor politics of wood, water and the armed services, it should not be missed that the revival of the Dartmoor Preservation Association brought to the fore the most exceptional personality in modern Dartmoor's history. In the following chapters Sylvia Sayer will become known through her actions, but at this point it might simply be observed that the importance of the cordial relationships Sayer cultivated with senior members of the National Parks Commission and its successor agencies can scarcely be exaggerated. Much of this reflected the ease she felt among 'the great and the good' who shaped post-war British life. Following Duff's failure to make the case against the North Hessary scheme, Sayer graciously rewarded him with a painting of the tor, 'almost the last representation of that landscape that can be made while it is still unshadowed and unspoiled'.[100] It was a characteristic gesture. And when in 1959 her husband, Vice-Admiral Guy Sayer, was knighted on his retirement as admiral commanding the Reserve Fleet, the standing and respect owed the new Lady Sayer was only enhanced.

It is telling to compare the consistently respectful treatment Sayer received in Whitehall with that of her near-neighbour Katharine Parr (1874–1955). Better known as Beatrice Chase, Parr was a Catholic convert and activist and the successful author of *Through a Dartmoor Window* (1915). Though Parr and Sayer often agreed, particularly regarding the use of Dartmoor by the armed services, Parr thought the creation of the national park a grotesque violation of private property rights and was ready to say so. A handwritten note by a civil servant described one letter Parr wrote as 'a madwoman's production, objecting to Dartmoor National Park – seizure of watersheds, sinister military plans. Anti Semitic venom. etc.'[101] The letter prompting this outburst appears to be lost, though an undated letter she sent to the under secretary of state for war a few years earlier objected that land use policy now rested with 'Mr Lewis Silkin, the Jew Minister for the upstart Town and Country Planning'. A scribbled note by a civil servant observed that 'this woman is obviously mad & no notice need be taken of any further communications from her'; another civil servant warned that she was in fact an 'authoress' and thought it 'unwise to issue a curt reply, which would probably be circulated to White Knights and others'. Chase was fobbed off with an anodyne reply.[102]

By contrast, although Sayer was prone to exaggeration and hyperbole, her letters were unfailingly civil and lucid, rarely feeling like they were dashed off in the heat of the moment. Her relationship with the National Parks Commission was sufficiently close that she could write to Abrahams shortly after the designation enclosing a press cutting of a letter from Chase calling on people to write to the commission in protest against 'the compulsory acquisition of land; compulsory provision of public rights of way; and abolition of any right to appeal'. It says much that Sayer commented, 'You know the lady: no more need be said.'[103]

Wood

From the middle of the nineteenth century a surge in tree felling and little replanting left the UK dependent on wood imports from Scandinavia, North America and Russia. Deteriorating diplomatic

relations in the years before the outbreak of the Great War gave
the government further cause to find this deeply worrying. Offering
grants for planting was one solution, but landowners were reluctant to embrace an opportunity offering a low return at a distant
point in the future – the normal period of rotation for conifers
was 80 years, for hardwoods 120 years, and for mixed conifers and
hardwoods 100 years. Large-scale felling – or 'devastation' – during
the First World War happened on Dartmoor at Brimpts, which
had been planted in 1862 with a mixture of conifers and oaks and
was felled using Portuguese labour during the war, an overhead
ropeway system carrying the logs to Princetown station.[104] Fears
of a post-war global 'timber famine' enhanced the feeling of vulner-
ability and, like governments throughout the industrialised world,
Britain recognised that timber should be 'produced as a crop, not
removed as a mineral.'[105]

The Acland Committee, tasked to consider how afforestation
could be integrated into the government's plans for post-war reconstruction, designed a scheme that aimed to ensure Britain was
self-reliant in timber for up to three years in the event of war or a
trade boycott. Acland recommended establishing an agency to either
purchase land and manage planting or encourage landowners to
plant through an augmented system of grants. More radically,
Acland hoped underused land whose owners did not take up new
financial incentives to plant would be subject to compulsory
purchase. Achieving wood-supply security was Acland's priority,
but he also hoped the policy would help regenerate impoverished
upland areas in northern England, Wales, Ireland, the Scottish
Borders and, most importantly, the Scottish Highlands. Acland
envisioned forestry generating new service infrastructures and
creating modern rural communities in deprived upland areas.
Foresters would work on the new plantations for at least 150 days
each year and on new smallholdings the rest of the time, creating
a new class of part state employees, part yeoman farmers, an early
twentieth-century spin on the ideals of the nineteenth-century

improvers.[106] And if not Acland's major preoccupation, it was generally assumed that the managers of the new state forests would
facilitate access, creating valuable recreational sites.[107]

The passage of the Forestry Act (1919) and the establishment of
the Forestry Commission proved two of the most significant landscape developments in modern British history. Thousands of
hectares have since come under the commission's direct ownership
or supervision, and significant areas of British uplands now host
large conifer plantations. By the early twenty-first century, the
Forestry Commission was responsible for over one million hectares,
making it the largest land manager in Britain. Its developments
often proved controversial, provoking opposition from lobbyists
who resent the transformative effect non-native conifers like Sitka
spruce, Norway spruce, Japanese larch, Lodgepole pine and
Corsican pine, as well as Scots pine, have on the landscape.

The argument is both aesthetic and ecological. Plantations
smooth out small differences in elevation, blunt the profiles of hills
and peaks, obscure topographical features and create a 'dark
serrated skyline'.[108] It is suggested that conifer forests are disliked
because they seem foreign, a little too German, and it is certainly
the case that modern forestry techniques are of German origin.[109]
Densely regimented evergreens also block out the light and generate
a mattress of pine needles ('a deep *mor* humus layer') that leaves
little potential for other plants and wildlife to establish themselves
on the plantation floor. As Ian Simmons explains, 'acid soils and
low ambient light inhibit soil microfauna and so little breakdown
of litter occurs, giving a forest floor which is bare except for fallen
material'. Significant ecological damage is thus caused by techniques
that James C. Scott sees as the 'radical simplification of the forest
to a single commodity'. By reducing the diversity of insect, mammal
and bird populations essential to soil-building processes and by
creating a favourable habitat for 'pests' that thrive in those tree
species, increasing quantities of fertilisers, insecticides, fungicides
or rodenticides are needed. As Oliver Rackham observes, the

'Forestry Commission induces trees to grow on blanket-peat by dint of ploughing, draining, and fertilizing; but these are heroics of technology and have nothing to do with native vegetation.' There is, however, evidence that populations of nightjars and birds of prey can benefit where there are sufficient prey species, as do deer, pine marten, fox, polecat and wildcat.[110]

When thinking about the politics of forestry on Dartmoor, it is important to recognise that the preservationists and conservationists who demonised the Forestry Commission dealt with an agency whose practices were determined by statute. Ministerial permission was needed if approaches were adopted that were less immediately profitable but more environmentally sensitive or congenial to amenity organisations. This is not to say that the commission was merely a passive tool of government, for the voice of its collective expertise was often heard in Whitehall, but it could not simply decide to manage its operations differently, sidelining its statutory obligation to produce a profitable softwood crop.

Not all Acland's recommendations were taken up. The commission was denied the power to make compulsory land purchases, and its room for manoeuvre was limited by being answerable to the Treasury rather than parliament. Despite this, its budget survived the savage cuts in public spending of 1921 (the Geddes Axe) and increased a little under Ramsay MacDonald's Labour government of 1924. By the end of 1933 the commission managed the Crown estate and owned or leased 274,166 acres of plantable land in England and Wales, and 181,054 acres in Scotland. Of this total, 207,923 acres had been newly planted with conifers and 11,663 acres with hardwoods; during the same period a further 95,228 acres of private land was 'dedicated' for planting by the commission, making it eligible for state subsidy. Acland's short-term goals had been largely met, but meeting his target of five million acres of mature conifer plus a substantial acreage of mature hardwood forest by 1980 saw forestry opinion in the 1930s begin to favour compulsory purchase powers.[111]

Acland identified moorlands close to coal mining and manufacturing areas as particularly suitable for afforestation, making Dartmoor
a relatively minor player in his overall ambitions. The Fernworthy
and Bellever sites, with a combined acreage of 3,100 acres, seem
small when compared with the 1920s Forestry Commission planting
at Kielder in Northumberland. Now covering 250 square miles,
Kielder is England's largest forest, and around half the total acreage
was planted in the interwar period. Still, Dartmoor contained land
included in the one and half million acres Acland thought suitable
for afforestation in England and Wales, and the Fernworthy and
Bellever plantations crossed the 1,000-acre threshold generally
thought economically viable.[112] Much of Dartmoor boasted another
asset: a landlord keen to be seen to act in the national interest. In
1917 the Duchy of Cornwall had bought Fernworthy, a farm of
Saxon origins, and following advice began planting conifers in 1919
as part of a larger plan to plant 5,000 acres on the moor. Two years
later the duchy replanted Brimpts.

In 1930, by which time the duchy had planted 800 acres at
Fernworthy, the Forestry Commission bought the plantation and
similar farms at Bellever and Laughter Tor. Within two years the
commission had planted a further 400 acres, 250 of which was
interplanting or replanting. Despite this expansion, O. J. Sanger,
conducting a Forestry Commission inspection in 1933, was unimpressed by what he found. A 'very lax . . . regime', he reported, in
which 'something of the Duchy atmosphere . . . prevailed'. Forestry
was subservient to sport; the keeper was little supervised and rather
too interested in what he could earn from the rabbits that overran
the plantations; fencing was inadequate, allowing damage by ponies,
sheep and rabbits; and techniques, in so far as any were deployed,
were outmoded or lazy – drainage was particularly poor. Efforts
to limit the damage done by rabbits in the first half of 1933 saw no
fewer than 703 trapped. Sanger concluded that the forester, the
district officer and the divisional officer should be reprimanded and
a new foreman appointed at Fernworthy.[113]

The duchy's poor management practices were only a part of the challenge faced by the new authority. As the Forestry Commission assumed its new responsibilities, so the Dartmoor Preservation Association attempted to rouse opposition to its plan to plant Bellever and Laughter Tor. Press reports suggested this was part of a 5,000-acre scheme that would extend the Fernworthy plantation towards the powder mills, making of Bellever and Fernworthy a single large plantation right in the centre of the moor. R. Hansford Worth took up the matter privately, appealing to the duchy authorities on behalf of the preservationists to abandon the afforestation of Bellever and avoid 'heartbreak to all lovers of the Moor'. The duchy explained that it was following government advice that the planting would be good for the country.[114] Worth then went public, giving a lecture fiercely critical of the duchy at the Athenaeum in Plymouth. He located the planned planting within a familiar preservationist history of encroachment on the rights of common by the duchy authorities.

Purchasing the ancient tenement at Fernworthy and then putting the land to new use had banished from the moorscape yet another industrious 'free tenant' expert at 'wresting from the moor a productive return'; planting Bellever and Laughter Tor farms would make these illegal newtakes permanent. Worth anticipated the slow demise of such sturdy yeomen as a 'national loss', but the old preservationist shibboleth that new developments would fail was not convincing. Instead, Worth made much of the amenity argument. He claimed that antiquities and 'natural beauty' – when contemplated a stimulus to patriotism – would be damaged by afforestation, while the threat new developments posed to access would undermine the 'health of the nation'.[115] When Worth repeated his arguments in the *Western Morning News*, Acland was quick to refute his claims, informing readers that there would be no restrictions on access to the tors and that the Ancient Monuments Department of the Office of Works was monitoring developments.[116] Civil society organisations like the Teign Naturalists' Field

Club, the Plymouth Institution, the Devon and Cornwall Natural History Society and the Devon Archaeological Exploration Society supported the preservationists, though the CPRE and the Commons, Open Spaces and Footpaths Preservation Society did not oppose afforestation on principle if public access and views were protected.[117]

The Bellever plantings went ahead, creating what remains the second-largest plantation on the moor, but the Forestry Commission did not get a free hand. They agreed that the southern boundary of the plantation would fall just north of Laughter Tor, ensuring the summit remained exposed, and they agreed not to plant the strip of land running through the middle of the plantation, leaving bare antiquities like Kraps Ring and the summits of Lakehead Hill and Bellever Tor. Cairns, hut circles, a stone row and evidence of an early field system remained uncovered, which contrasted with the duchy's less sensitive planting at Fernworthy. Although frustrated by this first encounter with the Dartmoor preservationists, the Forestry Commission had partly got its way, although the rumoured ambition to link Fernworthy and Bellever came to nothing.

Although controversial, forestry on Dartmoor in the post-war period formed but a small part of a much larger national story of state-led expansion. In 1943 Acland's target of five million acres was reaffirmed, of which three million acres should be the afforestation of 'bare land'.[118] Indicative of this increased ambition was the planting of Galloway Forest Park in Scotland in 1947. Now the largest forest in the UK, it covers some 300 square miles (97,000 hectares), an area not much smaller than Dartmoor National Park itself. Post-war developments on Dartmoor were hardly comparable. Operations were developed at Fernworthy and Bellever; Soussons Down was leased, fenced, deep-ploughed and planted in 1945–9; and the commission took over the management of the Plymouth Corporation plantings at Burrator Reservoir, originally begun in 1921. This was not the limit of the commission's ambition to expand its open-moor plantings, and during the 1960s it sought to agree an afforestation plan with the local authorities and the Countryside Commission to guide the future expansion of its activities.

Just as contentious was the commission's role in legitimising hardwood felling and softwood planting in the river valleys. Private forestry interests encouraged landowners to take advantage of tax relief and grants for felling and replanting, actions the preservationists regarded as unscrupulous but which might be interpreted as the successful implementation of government policy. The availability of grants significantly increased the value of some holdings in the national park, helping to create a new land market. In a small number of cases forestry interests were awarded compensation when they acquired land with the intention of planting but then faced prohibitions triggered by opposition aroused by amenity societies. An alternative scenario saw amenity societies like the DPA or Nature Conservancy induced to purchase privately owned land in order to protect it from planting. It is unclear whether individual landowners intentionally manipulated the market created by the availability of government subsidies, but the broader paradox is evident enough. By providing financial incentives to plant without introducing compulsory purchase or bringing forestry into the planning system, the government increased the value of the land. This created a new market that could lead to land being purchased in order to protect it permanently against planting. Landowners happily sold land previously considered almost worthless and not previously the focus of controversy regarding public access.

That the Forestry Commission was anticipating a period of rapid expansion after the war was evident during the national park designation process. After some prodding, it provided the National Parks Commission with a map of Dartmoor with its present holdings and likely future holdings marked. Areas already under its management were marked in red, areas subject to continuing acquisition negotiations were marked in pencil, and those 'dedications' likely to be submitted for approval for planting were marked in black pencil hatching. Torquay Corporation was a particularly important client, and in 1951 the Forestry Commission was looking to expand the plantation at Fernworthy by seventy-six acres and take responsibility for planting at the Kennick–Trenchford–Tottiford reservoir site. Other sites identified for planting included a significant stretch of the River Teign at Castle Drogo and a scattering of smaller sites that were a part of the Chagford House, Heatree and Dartington Hall estates.[119] Although the commission owned a number of small sites at Lydford on the western edges of the park, its immediate plans for expansion were focused on the river valleys of the east Dartmoor in-country.

Conflict over forestry policy began anew in January 1953 when the Dartmoor Preservation Association wrote a memo proposing a 'Woodlands Policy' for the new national park. It criticised the Forestry Commission's 'short-term, quick-return, "ad-hoc" policy of clear-felling and softwood planting', and urged no further additional planting anywhere in the national park and replanting only after careful consideration. Though the commission's file copy of the report is annotated with cross and exasperated marginalia, particularly regarding the claim that forestry caused soil erosion and flooding, O. J. Sangar – now Forestry Commission director England – cautioned against a brusque response. Local politicians who sympathised with the DPA had to be mollified, and consideration needed to be taken of the government's view that ancient oak woods should be preserved as a part of the national heritage.[120]

The DPA's vocal opposition also ensured the commission recognised the need to develop a more sophisticated understanding of the legal ramifications of commoners' rights. During the BBC's purchase of a small part of Walkhampton Common for the North Hessary television mast, Frank Hayman, MP for Falmouth and Camborne, had reminded the government that 'common rights over Dartmoor' were 'older than Magna Charter'. Opponents of the Forestry Commission's plan to 'coniferise' 147 acres of the Yarner Estate near Bovey Tracey cited similar arguments. And although the Forestry Commission was privately dismissive of the opposition because it was roused by the DPA, such rights were a legal impediment to planting. When the commission delayed completing the purchase and the vendor's solicitor claimed that they had entered into a binding agreement, the commission observed that the vendor had not disclosed the existence of these rights at the point of sale. By the summer of 1955 the land at Yarner had been bought by Nature Conservancy and is now part of the East Dartmoor Woods and Heaths National Nature Reserve.[121]

The Yarner dispute exposed the personal tensions that disfigured the management of the national park and the rapidity with which the relationship between the Dartmoor Standing Committee of Devon County Council and the amenity societies was breaking down. First victim would be Mr Popert, conservator of forests for south-west England and a member of the county council committee. If senior figures at the Forestry Commission could be rueful about Sayer's tendency to act on the committee's behalf without its approval or authority, Popert found her behaviour infuriating. As his intemperate scrawling on the DPA's 'Woodlands Policy' document had shown, Popert took offence at what he considered unfair or ignorant criticism, and he proved temperamentally unsuited to what was essentially a political role. Popert was succeeded by C. A. Connell in 1956, a much wilier operator who got off to a good start by refuting – correctly, as it turned out – rumours there was a plan to build a road connecting the Fernworthy and Bellever plantations through Assycombe valley.[122]

Misgivings about the situation on Dartmoor began to be felt at the highest levels in the Forestry Commission. Cuckoo Ball and Trendlebeare Down, two upland sites marked for afforestation, had to be abandoned in the face of organised opposition.[123] Cuckoo Ball was an ancient enclosure on the edge of the moor near Moorhaven village that had reverted to common land and was noted for its prehistoric antiquities; Trendlebeare Down was in the already controversial Yarner area. When plans to afforest Swine

Down, a seventy-acre site near Hound Tor, were also abandoned, the commission began to ask if agreement could be reached on the afforestation of any upland site – a reluctance to waste valuable resources on relatively modest plantings on Dartmoor when massive expansion was taking place elsewhere was a conditioning factor. Conversations between Lord Radnor, a cross-bench peer and lord warden of the stannaries, and Lord Strang, chair of the National Parks Commission and a member of the Nature Conservancy Council, signalled the beginning of a concerted effort to find a more effective modus operandi. Meetings between the Forestry Commission, the National Parks Commission and the county council committee were predicated on the hope that new upland planting might be agreed if hardwoods were included. The depth of their differences became clear when Nature Conservancy visited Dartmoor with senior representatives of the Forestry Commission in May 1958. Connell's suggestion that Dartmoor be subject to zoning, as had controversially happened on Exmoor, was not enthusiastically received. More positive was the response to his suggestion that members of the committee be taken on a tour of the commission's plantations. H. A. Turner, a senior Forestry Commission official, was pleased 'that quite a few of the members who had no idea of how beautiful a conifer plantation aged 35 years or over can look were agreeably surprised by the whole expedition'.[124]

The DPA was not party to the visit and its members were not feeling agreeable. In 1959 it was preoccupied with a timber syndicate's plan to clear-fell 300 acres of Buckland Woods and to thin another 150 acres, replanting mainly with conifers. The portents were not good. When questions were raised in the Lords in February, Earl Bathurst, speaking for the government, stated that the owners had already made 'considerable sacrifices' in response to an assessment made by the National Parks Commission.[125] Objections were registered with the Ministry of Housing and Local Government by five organisations, 180 individuals and seven petitions attracting 437 signatures. The preservationists accused the

county council committee of 'extraordinary defeatism' for agreeing
to the scheme, which gave their campaign little hope given government's reluctance to contradict local planning authorities.[126] Still,
evidence that the 'tidal wave' of protest had sloshed against
Whitehall's windows came when Henry Brook, the minister, sought
a modification of the syndicate's plans. Easily dismissed by the
preservationists as 'Ministerialese', it did indeed seem likely that a
gesture from the syndicate regarding felling in particularly sensitive
areas would secure permission to fell and plant the rest. In April
representatives of the Dartmoor Standing Committee, the Forestry
Commission and, unusually, the ministry met the owner and his
forestry adviser on the site. Devon County Council confirmed its
approval of the scheme, and permission was granted on the basis
of 25 per cent deciduous replanting, closer to the National Park
Commission's 40 per cent than the syndicate's initial proposal of
15 per cent.[127]

The preservationists were unimpressed. Some 800 acres of 'lovely
mixed woodland' would be thus exchanged for 'one enormous
conifer expanse (with a sad little fringe of beeches here and there)'.[128]
This 'catastrophe' exposed the conspiracy between the industry
experts and commercial foresters, the weakness of tree protection
orders, the overlapping interests of the major players and the
perverse incentives offered by government policy. That at least was
the view of the DPA. New disputes quickly followed. A formidable
enemy was identified in D. W. Carter, simultaneously representing
the Economic Forestry Group, the most powerful private forestry
interest in the UK, and Stephen Sebag-Montefiore, scion of the
banking dynasty and new owner of Houndtor Wood near Manaton.
Carter was also behind felling and ring-barking in Gidleigh Park,
had been consulted on an application for the clear-felling of twenty
acres of mixed hardwood near Steps Bridge in the Teign Valley and
had been involved in the Bridford and Buckland Woods controversy.
If for the preservationists Carter was fast becoming a Svengali
figure, a more immediate threat took the form of Wing Commander

Passy, purchaser of the Blachford Estate of Hawns, High House and Dendles Wastes, three adjacent sites in the parish of Cornwood on the southern tip of the moor.[129] Planting High House, Hawn and Dendles would be the most significant act of afforestation on the open moor since Soussons Down was planted in the late 1940s.

The controversy took a familiar shape. Sayer wrote to the press, Hayman asked questions in the House of Commons, the National Parks Commission expressed its reservations, the minister looked for a way out. Dendles and Hawn were ploughed in early 1960 but permission to afforest High House was withheld. Its fate became bound up with national debates about the purpose of the Forestry Commission and more localised efforts to reach agreement regarding the future management of Dartmoor forestry. Just as the commission was becoming more efficient and effective, exploiting technological advances, influential voices began to question whether it was needed at all. At the same time conservationists began to imagine a future for the commission as the custodian of Britain's most important woodlands rather than merely the harvester of a softwood crop.[130] Writing of Dartmoor in particular, the National Parks Commission recommended in May 1960 that the government acquire Dartmoor's most precious woodlands and hand them over to the Forestry Commission to be run on an amenity basis. Thirteen woodlands were thus classified as of 'high amenity value', with most urgent attention needed at four sites, namely Holne Chase, the Dart Valley woodlands below Dartmeet, the Teign Valley woodlands between Clifford Bridge and Steps Bridge, and Skaigh Woods at Belstone. The National Parks Commission placed the value of these sites at £12,000, though it lacked the expertise to estimate the management costs.[131] Skaigh Woods was instructive. Purchased by local people in 1956 to prevent its acquisition by the Forestry Commission, conventional wisdom suggested it now needed professional forestry management.

Recognising that the Forestry Commission did not have a statutory right to acquire land for amenity purposes alone, the National

Parks Commission simply asked whether it agreed in principle. Internal discussions suggested that although the commission was favourable, it resented the implication that it had not already proved capable of running its commercial operations along amenity lines, and professional sensitivities were easily irritated by any suggestion that successful woodland management was not a continual process and could be overseen by non-professionals. 'If we acquire woods for amenity purposes we should expect to manage them to <u>preserve</u> and <u>enhance</u> the amenities, not to apply a dead hand', wrote G. B. Ryle, the director of forestry (England). 'We should also quite essentially expect, as a secondary objective, to secure reasonable revenues and to let the public see a virile management which would both be pleasant for the visitor and useful for the local rural worker.'[132]

Acutely aware that its reputation had suffered thanks to the public's perception of its activities on Dartmoor and Exmoor, the commission remained wedded to a commercial forestry model and insistent on its professional autonomy. Managing its holdings according to broad agreements reached with the planning authorities was conceivable, but detailed requirements regarding the exact proportions of species to be planted or the positioning of rides and compartment boundaries were unacceptable. Ever more defensive, by April 1961 the commission was unable to see how satisfying the planning authorities and the amenity societies was commensurate with producing a reasonable return on its capital investment. The commission did, however, appoint Dame Sylvia Crowe as its first landscape architect in 1963, inaugurating a new approach to planting that showed more sensitivity to contour lines and a greater readiness to soften plantation edges with broadleaved trees and larch. The absurdity is apparent. The National Parks Commission and other official bodies with a statutory obligation to protect amenity might purchase woodland and run it on that basis, but the professionals at the Forestry Commission could not.[133] A Treasury veto halted further moves in the mid-1960s to empower the FC to purchase and manage woodland as amenity.[134]

Parallel to these national discussions were attempts to resolve the specific problem of forestry on Dartmoor. Reluctant to bring planting into the planning system, the government encouraged the Timber Growers' Association (established in 1958), the Country Landowners' Association, the Forestry Commission and the National Parks Commission to enter into a voluntary agreement. The resulting 'gentleman's agreement' of January 1961 was supposed to prevent every forestry development on Dartmoor becoming a source of controversy by requiring that each party approve a felling and/or planting scheme before the Forestry Commission made the dedication and rendered the scheme eligible for a management or planting grant.

The ineffectiveness of the agreement from the point of view of the expansionist forestry interest was quickly made evident by the case of Pudsham Down near Rippon Tor, a site comprising 173 acres of open moorland surrounded on three sides by enclosed land and abutting Buckland Common on the fourth. Purchased by Fountain Forestry Limited before the agreement, it was tipped for afforestation until the National Parks Commission refused to approve the planting. Ryle surveyed the site, concluding that its poor soil quality meant it was neither environmentally valuable nor likely to yield a particularly profitable conifer crop. 'Along the road-side, bordering the South of the area,' he wrote, 'it would be very possible to introduce in a highly irregular manner, and not so as to form a hard fringe, a small admixture of different species to give a more variegated effect to travellers along the road.' Hoping the National Parks Commission would recognise that Fountain Forestry had behaved in good faith and permit the planting, Ryle gloomily concluded that if 'pressure makes it impossible to afforest this area I would consider that there is very little hope of carrying out any proper forestry anywhere on Dartmoor'. Reports that the Dartmoor Standing Committee of Devon County Council investigated buying the site with the intention of selling it to the Ministry of Housing were not encouraging, and the commission soon abandoned its ambitions.[135]

High House Waste was a more challenging case because the Dartmoor committee made it clear that if it were planted it would consider the gentleman's agreement defunct. In June 1961, after some mud-slinging in the press, Economic Forestry recognised that the case was proving too damaging and offered to sell the 142-acre site to Devon County Council. The county's plan to buy ran aground when the disgruntled vendor imposed unreasonable conditions, but in the event the DPA raised enough money to purchase the site in 1964. Initially intending to hand it over to the National Trust, the DPA owns and manages High House to this day.[136] A year later Nature Conservancy bought Dendles Wood, protecting the seventy-two-acre site from felling and replanting.

The fate of the Dendles Wood, Dendles Waste and High House Waste complex was peculiarly illustrative of how forestry developed on Dartmoor in the twenty years after 1945. Government incentives to plant increased the value of land, and the landowner sold a mixture of open upland and ancient broadleaf woodland to an investor intent on felling, replanting and afforestation. When the buyer sought a Forestry Commission dedication and hence eligibility for government subsidy, the preservationists galvanised the opposition of the National Parks Commission, which duly took the case to the responsible government minister. Under pressure from all sides, the minister granted the forestry interest part of what it wanted, but delayed a decision on the site as a whole. Further manoeuvrings culminated in two thirds of the site changing hands again, the purpose of which was to prevent felling, replanting or afforestation. Supporters of the Dartmoor Preservation Association and the taxpayer-funded Nature Conservancy were thus compelled to buy land that only had market value because the government had introduced a flawed system of taxpayer-funded incentives. This perverse outcome, in which the costs of policy failure were partly socialised, was indicative of the mess forestry policy was in by the early 1960s.

Nobody found this more frustrating than Alan Connell. A dedicated forestry professional of notable political skills, he spent much of the 1960s grappling with the Dartmoor problem. With the backing of government and the National Parks Commission, he worked closely with the Dartmoor Standing Committee to agree an Afforestation Survey. Connell sought to have areas of the national park classified according to whether there was A) a strong presumption that afforestation would be acceptable, B) a presumption against afforestation but a proposal might be acceptable, and C) a strong presumption against afforestation. The survey was not an afforestation plan, but Connell hoped that acceptance by the National Parks Commission would make dedications aligned with its classification more easily approved. Connell's judgements were inherently political. He did not classify land according to whether it could be successfully planted, but to whether planting might be consistent with amenity considerations.

Laboriously listing all Connell's recommendations would be tedious, but some focus on the detail reveals much. The sixty areas where there was a strong presumption in favour of afforestation covered 5,100 acres, about 2.2 per cent of the park. Most of these areas were located either in the in-country, close to the park boundaries, or on enclosed land of a moorland character, including the Bellever strip left bare since the 1930s. On the whole the areas selected had already been enclosed but were no longer judged agriculturally productive. Adding these areas to the Forestry Commission's existing holdings (4,400 acres) and areas already covered by dedication covenants (7,200 acres mainly in the eastern in-country) indicates that Connell and Devon County Council thought about 7.2 per cent or 16,700 acres of the National Park should be planted.

A much larger area, about 97,000 acres, some 41.5 per cent of the park, was placed in the second category. Here Connell's larger ambitions became clear. Included were most of the unplanted in-country, the enclosed uplands around Princetown and Prince Hall, and an extensive area between the Forestry Commission plantations at Soussons Down and Bellever. The remaining 120,000 acres, some

51.3 per cent of the park, which included all common land and the remaining moorland, including small patches in the eastern in-country, were classified as unsuitable for afforestation.[137]

Connell's senior colleagues were impressed. Persuading the Dartmoor committee that 41 per cent of the National Park might be acceptable for afforestation was judged a remarkable achievement. Approval still needed to be sought from the Country Landowners' Association, the Timber Growers' Association and the National Parks Commission, the latter of whom would have to consult with the CPRE and Nature Conservancy. Trouble was anticipated. 'When the National Parks Commission begins to consult the amenity bodies,' Ryle wrote, 'there will, of course, be certain loud noises from one source but that is for the National Parks Commission or the CPRE to settle.' He was referring, of course, to the preservationists. 'We ourselves', he wrote pompously, 'do not even know what is the standing of the Dartmoor Preservation Society.'[138] That a civil society organisation capable of rousing a significant section of local and national opinion could be summarily dismissed reflected the streak of authoritarianism that characterised post-war statism. Stakeholder political culture was something for the future. Sayer did indeed shoot off angry letters to the press, and she remained on the warpath throughout the long consultation process, but the preservationists were not alone in objecting to Connell's designations.[139]

The Country Landowners' Association and the Timber Growers' Association were not particularly receptive. Already irate at Devon County Council's liberal use of tree preservation orders to frustrate forestry developments, they judged Connell's Afforestation Survey another attempt to restrict their rights as property owners pursuing a legitimate form of business. Though accepting Category C, where the presumption was strongly against afforestation, and though unambiguously respectful of common rights, they found the wording of Category B unacceptable, thinking it maintained the presumption against afforestation for land of marginal agricultural value but otherwise suitable for profitable forestry. They suggested that category be reworded as 'Areas where proposals for afforestation will be considered on their merits' and that twenty-six areas be added to Category A, taking the total from 5,122 to 8,447 acres.[140]

The National Parks Commission demanded adjustments in the opposite direction, tackling both Connell's original proposal and the CLA-TGO revisions. Connell was outraged, wrongly regarding this as an unacceptable interference in an agreement made between the Forestry Commission and Devon County Council, the designated planning authority. An unexpected source of potential opposition was also found in Plymouth City Council. Though significant landowners at Burrator they had not been party to the gentleman's agreement or consulted about the plan. Piqued, Mr Elliott, the city water engineer, implied they would stand by their rights as landowners and afforest Burrator irrespective of the opinion of the National Parks Commission. A bit panicky, Connell anticipated the collapse of his whole scheme and suggested to Ryle that the minister of health be asked to bring Plymouth to heel. A few weeks later Connell reported that Elliott would not seek to embarrass the Forestry Commission. On 8 July 1965 the National Parks Commission approved the map on the basis of minor amendments.[141]

What satisfaction Connell felt was short-lived. On 18 September *The Times* made a story of the agreement's publication by seeking Sayer's opinion. Sayer delivered a maximalist reading of the survey, claiming the Forestry Commission planned to plant 15,000 acres, including Yelverton golf course and '"people's back gardens"'.[142] A day before *The Times* published its story, Sayer and J. V. Somers Cocks had written to Fred Willey, minister of land and natural resources, complaining of the Forestry Commission's 'acquisitive guesswork'. Assembling fourteen points over four single-spaced typed pages, the gist of their case boiled down to four 'factual considerations'. Thirty-ton forestry vehicles needed wider roads and bridges; the plans threatened prehistoric and historic monuments; the removal of topsoil increased the danger of flooding;

and the government had already admitted that although forestry might serve a social good it was not economically viable. More striking than these specific claims was the absolutism of their fundamental position. 'Even if it is accepted (as our Association does not) that the right way to decide the future land use of national parks is by striking some kind of balance between opposing claims, those claims ought at least to be serious contributions, and never should be a work of inflated overstatement presumably submitted for tactical purposes.'[143]

Another voice joining the chorus of objections belonged to retired Vice-Admiral Philip Ruck Keene of Hexworthy. Ruck Keene took up the case with Dame Evelyn Sharp at Housing and Local Government, bluffly explaining that Anthony Sampson's *Anatomy of Britain* described her 'as one of the most formidable characters in Whitehall and very keen on the countryside & Green Belts etc.'. Ruck Keene's case against the survey combined concern for Dartmoor's future with a particular preoccupation with loss of grassland for grazing, which he believed should be considered in the light of the UN's estimate that half the world's population suffered from hunger and malnutrition. Despite signing off 'Please do what you can & don't bother to answer,' particular efforts were made to ensure that Ruck Keene was furnished with a considered reply, the essence of which was that the proposed planting on Dartmoor was of minimal significance and proportionately low, as shown by comparison with plantations on Snowdonia, the North York Moors and the Lake District – respectively, 70,000, 50,000 and 26,000 acres.[144]

Two years later the squabbling continued, the wording of Category B still at issue. The forestry and landed interest wanted either the negative bias removed or a further 3,000 acres moved from Category B to A. Compromise proposals were batted back and forth until the following evasive wording for B was accepted: 'Intermediate areas which in the long term may be subject to reviews and in the short term may admit afforestation proposals

of acceptable scale and character in particular localities.'[145] Tempers
among the amenity interest were now very frayed. In August 1968
C. F. J. Thurley, chair of the Devon branch of the CPRE, accused
Connell and the forestry companies of bringing 'undue and indeed
unethical pressure' to bear on the Dartmoor Standing Committee,
while he found the new wording for Category B objectionable
because it reflected the needs of an expansionist forestry industry
and had been 'made behind closed doors'. He also argued that the
whole process contradicted the recent Town and Country Act, a
claim brusquely rejected by the Ministry of Housing and Local
Government, which was determined to hold to the voluntary
principle.[146]

    Connell's tenure as conservator was coming to an end. In
September 1968 he enjoyed a pleasant farewell lunch with
members of the Dartmoor committee but was denied any final
satisfaction by the last-minute intervention of the Countryside
Commission. Created by the Countryside Act of 1968, this was
the successor body to the National Parks Commission and its new
leadership wanted to know what the fuss was all about. After
seven frustrating years of negotiation, Connell handed responsibility over to his successor, G. D. Rouse. He signed off with a
defensive letter that denigrated the DPA, 'an amateur pressure
group of no status', and insisted that he had always acted
according to ministerial instructions, particularly regarding the
exclusion of the public from a process that was always intended
to be confidential. In January 1969 the Countryside Commission
notified the Dartmoor committee that it had approved the survey,
including the new wording for Category B. In a notably effective
letter that May, Sayer registered preservationist objections one
last time. If by now little more than an act of witness, it also
seemed Connell's ambitions had been somewhat neutered by
delay. In a condescending letter of June 1969, R. B. M. Williams
of the Forestry Commission explained to Miss Ruffell at Housing
and Local Government that the purpose of the survey was only

'to give a broad indication of the possibilities of afforestation'. That phrasing hardly reflected either Connell's expansionism or preservationist alarm. The change in mood was evident in the 1972 edition of G. D. Rouse's *The New Forests of Dartmoor*, an official publication of the Forestry Commission:

> There are many demands upon the limited area of the National Park, and fears have been expressed that uncontrolled afforestation might upset the balance of land use. In consequence the Forestry Commission, in consultation with the Timber Growers' Association and Country Landowners' Association, have negotiated an agreement with the Dartmoor National Park Committee, which consulted the appropriate amenity societies.[147]

In the early 1960s the Forestry Commission had been bent on expansion; in 1972 it was beginning to reformulate its priorities in terms of conservation; in the twenty-first century there has been little prospect of a significant expansion in the commission's Dartmoor holdings for some decades. Preservationist pressure as well as the conservationist turn in British politics has played no small part in this change. At Burrator today Forestry Commission signs boast of antiquities it once strove to obscure with planting; in the 1990s Hawns and Dendles were clear-felled of conifers, bought by the National Park Authority and, thanks to help from the taxpayer and the Heritage Lottery Fund, cleared of stumps and brash, allowing the moorland to gradually redevelop. Lottery funding, as well as public donations and three legacies, allowed the Woodland Trust to buy Houndtor, Pullabrook and Hisley Woods and begin to restore an ecosystem destroyed by clear-felling and conifer planting: as conifers are thinned and oak and ash nurtured, so bluebells, dog's mercury, bryophytes and ferns are re-established.[148]

Of the upland plantations Ian Mercer offers a characteristically upbeat appraisal of their value as a wildlife habitat, emphasising the importance of the 'edge' (if not the whole) and the niche created by planting up to the drystone walls of the newtakes. Insect life is encouraged by deciduous planting on the external and internal edges of the plantations, the latter creating a habitat good for species that thrive in relatively humid environments with little wind and exploit the leats, streams and rivers for the larval stage in reproduction. Other birds that thrive include wrens, suited to life in the drystone walls, but most iconic are the crossbills, lesser redpolls and the goshawks attracted to Dartmoor by the plantations. Moreover, the plantation cycle makes this a dynamic ecosystem, in which clearing and replanting after the long maturation period creates opportunities for new varieties of flora and fauna to temporarily colonise the plantation space. If the tawny owls that follow the arrival of field voles and wood mice embarking on their familiar four-year boom-and-bust cycle are most charismatic, grasshopper warblers, spotted flycatchers and nightjars add to the liveliness of a moment of opportunity that begins with new floor growth; red, fallow and roe deer, rare on Dartmoor for 200 years, now also put in an appearance, inevitably being seen as a management problem.[149]

Water

But the worst calamity is the dearth of water at Plymouth. The open leat, 20 miles long, which Sir Francis Drake cut, has never been known to fail, but for the past week it has been filled with ice and snow, and hundreds of men were engaged night and day cleaning it. This morning they hoped to complete their work in 30 hours, but the hurricane has undone three days' work. A large force of military proceeds to the spot tomorrow. Not one house in a hundred in Plymouth has any water. All manufactories are stopped, many bakers have closed, and supplies for the 300 engines in Devonport Dockyards are deemed more serious still. The railways along the lines have exhausted supplies, and notice was issued this evening that all western traffic would close tomorrow from this cause. To-night, however, the announcement is made that some trains may run to-morrow, water being fetched from the river Teign, thirty miles distant.

Sheffield Daily Telegraph, 22 January 1881

To the poet Nicholas Carrington, Dartmoor's peculiar appeal rested upon the contrast between the static unchanging uplands, a permanence exemplified by the rocky tors, and the river water rushing outwards in all directions. Water and the poet-walker perambulate the moor's surface, one bound by the gravitational pull of ancient contours, the other free to roam but drawn the same way by water's excited chatter. Hardly a stretch of river on Dartmoor is without an accompanying path. And although these paths take on definition as gradually as water flows become rivers, the process occurs in reverse, the path questing towards the uplands, the river forming as it flows from them. Such journeying saw the heads of Dartmoor's rivers invested with a mystique that transfigured the barren waste into a life-giving force. If this myth making belies topographical actuality, for the rivers often originate not in the singular openings indicated by the symbols of mapping but in indistinct swellings of water in the boggy uplands, here nonetheless originate the waters that feed the lowlands. Water was not drawn from Dartmoor as though from a well, but was delivered by its many rivers, efficiently distributed north, south, east and west.

Investing this plenitude with providential agency, the Romantics wrote at a time when it had already long been tapped. Drake's Leat and Devonport Leat had supplied Plymouth and Devonport with fresh water since, respectively, the late-sixteenth and late-eighteenth centuries. These open canals, about six feet wide and two feet deep, were complex feats of engineering that relied on gravity alone to guarantee a strong and constant flow of water from the north quarter to Plymouth. Drake's Leat tapped the Meavy; Devonport Leat – like the prison leat – begins north of Beardown Hill and taps the West Dart, the Cowsic and the Blackabrook. To the agricultural improver confounded by excess or a Romantic revelling in abundance, water seemed omnipresent on Dartmoor. This association would not fade, but in the rapidly developing towns and cities of nineteenth- and twentieth-century Devon concerns about supply generated a new language of water scarcity. New science added to the fear, generating talk of pollutants entering the water supply through the open leats.

Anxiety about water security was not unique to Devon. Pressures were much greater in the industrial centres of London, the Midlands and the north of England. From the 1840s, the new incorporated municipalities placed water security at the centre of politics and in many cases used new powers to wrest control of supply from the private companies that dominated the market. Piping water directly from the rivers was one solution, but the riparian rights of land-owners and other users limited the quantity that could be extracted directly from the flow. Growing demand necessitated greater regulation, and calculations regarding the level of water extraction the ecosystem could bear were rudimentary exercises in sustainability. Towns and cities needing a reliable water supply made storage the new panacea, and reservoirs formed by damming river valleys or existing lakes soon became the preferred tool for ensuring sufficient downstream river flows.

Manchester Corporation responded to that restless city's increasing demand for water with a succession of highly ambitious schemes. Between 1848 and 1877 the corporation successfully sought parliamentary permission to establish a string of reservoirs at Longdendale in the Peak District, but the twenty-four million gallons of water they supplied each day were soon judged insufficient. Rather than build new reservoirs, the Waterworks Committee looked instead to the Lake District and damming Lake Thirlmere. As one of the least regarded of the Lakes, the corporation imagined its scheme would prove relatively uncontroversial and, despite the lively opposition mounted by the Thirlmere Defence Association, a campaign described by Harriet Ritvo as witnessing 'the dawn of green', the Manchester Corporation Water Act was passed in 1879. In time Longendale and Thirlmere proved inadequate, and the Waterworks Committee turned its gaze on Haweswater, the highest and deepest of the Lakes. Parliamentary authorisation to dam was granted in 1919.[150] If supplies of clean water were to be guaranteed, municipalities needed to acquire the valleys or lakes to be dammed and their watersheds; only then could they be confident of preventing pollutants from entering the water supply. The watersheds for Longdendale and Thirlmere were, respectively, 19,300 and 11,000 acres.[151] Responsibility for supplying water turned municipalities into major landowners.

In November 1848 Nathaniel Beardmore addressed a public letter to Viscount Ebrington on the question of Plymouth's water supply.[152] Beardmore (1816–72), first the pupil of a Plymouth architect, was soon articled to James Meadows Rendel, a well known engineer. Beardmore completed his pupillage in 1838, and his early professional work with Rendel included improving the service reservoirs and distributing mains fed by the Devonport Leat, and helping Edinburgh win a parliamentary act for building six reservoirs in the Pentlands. He is best remembered as the engineer who oversaw the drainage and navigation of the River Lee and for his *Manual of Hydrology*, which remained a standard textbook until the First World War.[153] Hugh Fortescue, Viscount Ebrington, later third Earl Fortescue (1818–1905), the recipient of Beardmore's letter, was MP for Plymouth 1841–52 and 'a serious Whig of religious bent' who took a Chadwickian interest in public health and sanitation schemes.[154] Beardmore could expect a sympathetic hearing.

Beardmore assumed that Plymouth should have a water supply comparable to the best being developed elsewhere in Britain. He claimed the citizens of a modern city required not far short of fifty gallons of water per day, and Plymouth's poor were particularly badly served. New needs like street cleaning, summer watering, protection against fire and cleansing the sewers also had to be met – what one historian has described as the new 'culture of flushing'[155] – and Plymouth's pioneering leat system now looked primitive and highly insanitary. Beardmore supposed the principal threat to the water quality was urban rather than rural – water gathered filth as it flowed through the town – and he proposed that the leat should run into pipes at Knackersknowle (Crown Hill) on the outskirts of Plymouth, from where the water would be delivered to a large distributing reservoir at Torr-house, an elevated spot overlooking the town. Furthermore, Beardmore urged that the system be rationalised by amalgamating supply to the county boroughs of Plymouth and Devonport and the urban district of Stonehouse, the three areas eventually formed into the county borough of Plymouth in 1914. Beardmore's seminal intervention looked to produce what Jamie Linton calls 'modern water', a product whose essence is characterised by its material containment (in pipes) and deterritorialisation (the user thinks nothing of its origins).[156]

Central to Beardmore's scheme was the case he made for establishing a store reservoir on the River Plym 'by walling or embanking up the narrow and picturesque gorge below Sheepstor Bridge'. This would create on Dartmoor a reservoir 20–25 feet deep with a surface area of some 40–50 acres. The land, he argued, was 'inexpensive, and the dam would be of trifling cost'.[157] By providing a solution for the three towns, the reservoir could be fed by both leats as well as the Plym. With a typical rainfall in the area of 42 inches a year and with wastage accounted for, Beardmore reckoned the reservoir could be used to regulate a waterflow of 966 cubic feet per minute, more than the 600 cubic feet needed to supply 36 gallons per day to the 150,000 people of the three towns.

Beardmore's conclusions emphasised Plymouth's peculiar advantages: 'Placed at the feet of mountains of rain and mists which the great laboratory of nature pours forth, at 1,200 to 2,000 feet of elevation, powerless yet powerful, it is only to guide and control, on the uplands, these silent forces which now steal down their vallies[sic], and to distribute them as agents in the multitudinous ramifications of social life.'[158]

The desire to 'guide', 'control' and 'distribute' nature's 'silent forces' is clear enough, and humanity's attempted 'conquest of nature' is an established theme of environmental history.[159] An idealist subtext was also at work here. At their most optimistic, Victorian water engineers believed technical and geological expertise allowed the divination of the perfect solution inherent in the combination of human need, technological know-how and environmental conditions. Such brilliance, however, was troubled by uncertainty over future needs, which only sophisticated guesswork could predict. As the Manchester example demonstrates, dynamic expanding urban environments made once-and-for-all solutions delusional.

Much that Beardmore recommended was adopted over the next half-century. Service reservoirs were built on the outskirts of Plymouth, and the lower stretches of the leats were replaced with piping. The lynchpin of his scheme, however, the storage reservoir on the Plym, proved slow in coming thanks to a long-running dispute with Sir Massey Lopes, the landlord. After the Duke of Cornwall, Lopes was the most powerful landlord on Dartmoor and an influential politician dedicated to protecting the agricultural interest. He was the grandson of Sir Manasseh Masseh Lopes, born in Jamaica to Mordecai Rodriguez Lopes, a slaveholder of Sephardic-Jewish background. Manasseh inherited his father's fortune and worked to establish himself in Britain as a member of the social and political elite. He bought the manors of Maristow, Buckland Monachorum, Walkhampton, Shaugh Prior, Bickleigh and Meavy, some 32,000 acres in total, converted to Christianity in 1802, and entered parliament as the Tory member for the rotten borough of New Romney. Though Pitt rewarded his loyalty with a baronetcy, he was severely punished when convicted of electoral fraud in 1819.[160] Ralph Lopes, Manasseh's son, proved surer-footed, embarking on an undistinguished political career while further consolidating the family's interests in Devon. Massey Lopes, beneficiary of a Winchester and Oriel education, followed his father and grandfather into parliament and had a successful career as an MP between 1857 and 1885, which peaked when he became Disraeli's civil lord of the Admiralty, 1874–80. Once retired from national politics, he continued to play a role in local affairs as a prominent landlord, philanthropist and alderman of Devon County Council.[161]

Following the passage of enabling legislation in 1867, it was generally assumed that the new dam would be sited at the old Head Weir in the Meavy valley and would be fed by the River Meavy and the Devonport Leat. Lopes and Plymouth Corporation were the most significant landowners but Lopes refused to sell his share because it meant a significant loss of productive land to three of his farms. He understood that Plymouth's needs were real, though doubted whether the leat was inadequate and offered to sell the corporation an alternative site to the north that was fed by Harter (Hart Tor) Brook before it joined the Meavy. He believed the Harter site would meet the corporation's needs, reasoning he should not be pressured into selling more valuable land to the south. His legal case, however, rested on a set of claims concerning the corporation's riparian rights. Lopes claimed the 1867 act prejudiced rights he had to 'foul' the Meavy's waters above Head Weir such as by establishing a china clay works. He also believed the corporation had no legal right to sell the Meavy's waters to users outside the borough and that the storage reservoir would hugely increase the volume of water available for sale. At the Court of Common Pleas on 6 May 1873, Justices Bovill, Grove and Denman found in the corporation's favour, establishing that the Elizabethan statute that had enabled Drake to cut his leat gave Plymouth the right to all the waters of the Meavy. Lopes should be compensated for the losses he might suffer as a consequence of the reservoir but he possessed no right to prevent the corporation's unlimited use of the Meavy's waters or to jeopardise the purity of the water by establishing works upstream.[162]

Lopes still refused to sell and the conflict dragged on into the 1880s, generating what R. N. Worth, a strong advocate of the Head Weir scheme, described as the 'most exciting controversy recorded in the annals of the Municipal Life of Plymouth'.[163] Activists established a water rights association, heated exchanges occurred in the letters page of the local press, public meetings were called, leaflets were distributed. At a statutory meeting of 4,000 Plymouth ratepayers in July 1883 the corporation proposed it accept the offer of the Harter site. Members recognised that Harter could only provide a temporary solution to the problem of Plymouth's water supply, but it was cheap, and with this offer on the table they doubted parliament would force Lopes to sell his Head Weir holdings. The ratepayers overwhelmingly rejected the proposal, citing engineer reports that had consistently made the case for Head Weir.[164] On 25 September 1884 John Bayly, owner of a section of the Head Weir site, gave the town the twenty-four acres it needed. Lopes would still not budge, and on the eve of further deliberations in 1886 again stated his opposition to the use of the Head Weir site: it was good farmland and had not been conclusively shown to be the best site for the reservoir. If the town selected Head Weir, Lopes threatened to contest the right of the corporation to sell water held by the storage reservoir to customers outside the borough.[165] In response, Worth reminded readers of the *Western Daily Mercury* that Lopes' claims had already been rejected by the 1873 arbitration and there was plenty of evidence to show that the town had exercised this right of sale in the past. Plymouth's riparian rights had not been dissolved when Lopes' predecessors purchased Walkhampton Common, location of the head of the Meavy. Even if Harter were the best site, Worth insisted no act by Plymouth should suggest it had surrendered its legal rights.[166]

Under pressure from all sides, Lopes finally buckled, agreeing to sell the site for £15,750. Reluctantly accepting the higher costs, in 1887 the corporation successfully sought from parliament powers to build the reservoir. But new engineer reports now overturned the Head Weir orthodoxy, recommending the valley be dammed further downstream. With the corporation refusing to swallow the higher costs, Lopes seized his opportunity, offering the Harter site for £5,000. In the face of ratepayer opposition, Plymouth now promoted a bill in parliament based on the new offer. Lopes, however, had left a hostage to fortune. Having always claimed that the defence of his private interests at Head Weir was legitimate because a compelling case for the site had not been made, he was placed in a difficult position in 1889 when the engineer James C. Inglis, commissioned by the Plymouth Water Rights Association to assess both sites, concluded that Head Weir was 'obviously the best possible site for a Storage Reservoir'. The ground surface at Harter was unstable and had not proved watertight, and with a catchment area roughly a quarter of the size of Head Weir's, 'the extra leakage would prove a serious loss of efficiency'. However, although Inglis found no 'geological reasons to doubt the existence of solid bed-rock' at Head Weir, he too thought the 'proper place' for the dam was further to the south at the 'gorge' below Sheepstor Bridge.[167] Further pressure from ratepayers saw the corporation agree to settle the issue through a local poll. Of the 9,508 votes cast in March 1890, 6,849 were against Harter. The polling costs were defrayed by the twenty-seven council members who supported the Harter scheme and the public subscription got up by their opponents, the symbolism of which can hardly be missed.[168]

Harter was off the table, but it would take yet another crisis to fully mobilise the corporation. In March 1891 the *Taunton Courier, and Western Advertiser* ran the following story:

The leat which supplies the town [Plymouth] with water from Dartmoor was choked with frozen snow on Monday night, and inhabitants were in straits for want of water. On Wednesday gangs of workmen tramped all the way to the moor, a distance of nine miles, the roads being so blocked by drifts of snow and fallen timber that vehicular traffic was impossible. They worked all night to try to clear the leat, but made little progress. By some oversight they did not carry food with them, and they were obliged to work up to their waists in snow without any food or drink. Many of the men deserted the job half-fainting with fatigue and exposure. On Thursday morning 500 soldiers and navvies were despatched by special train to Yelverton, which station is near the course of the leat. They were accompanied by the Mayor of the borough and other officials, and an efficient commissariat corps. Operations were carried on all day in the teeth of a furious gale of wind, and on Friday the leat was finally cleared.[169]

On 13 October 1891 the corporation instructed Edward Sandeman, water engineer, to consider 'what works should be executed' in order to ensure a 'constant and ample supply of water at all seasons of the year' and how to 'prevent the waste, misuse, and contamination of water'. That December Sandeman reported that a storage facility was needed to hold sufficient water for 130 days' supply, approximately 422 million gallons. Impounding this quantity at Burrator would need two dams, one across the narrow gorge below Sheepstor Bridge and the other across the low-lying ground just to the east above Burrator hamlet. Without the second dam, the storage capacity would be limited to 300 million gallons. The beauty of the scheme lay in the potential to raise both dams at a later date, increasing the reservoir's capacity to 800 million gallons. A pipe with a diameter of twenty-five inches was needed to deliver the water the four and a half miles to Roborough Reservoir and thence to Plymouth. Supporting materials and further expert opinion testified to the quality of the water and the suitability of the site's geology.[170]

In March 1892 the corporation adopted the scheme and in June 1893, forty-five years after Beardmore published his original proposals, the Plymouth Corporation Water Act gained royal assent. That September Massey Lopes made his peace with the people of Plymouth, giving them the twenty acres they sought for free.[171] September 21 1898 was declared a general holiday, and with all due civil pomp the mayor closed the valves of the great dam at Burrator Reservoir. Like many national newspapers, the *Glasgow Herald* was dazzled by the scale of the achievement. It was the largest masonry dam yet constructed in Britain, extending fifty-three feet into the bedrock and one hundred feet above. The mayor drank a last ceremonial glass of water lifted from Head Weir in memory of Sir Francis Drake. The following year it was submerged.[172]

Given that the valleys themselves were a small part of the land required to guarantee a supply of clean water, local authorities were advised to purchase the whole Dartmoor watershed. Turning the land over to carefully controlled sheep farming, plantations or deer parks was one way to ensure a return on the investment,[173] but councils were reluctant to follow the advice of government and local medical officers and use controversial purchase orders. Torquay Corporation established a cluster of small reservoirs at Kennick, Tottiford and Trenchford south-east of Bovey Tracey in 1861, 1884 and 1907, and in 1897 the *British Medical Journal* praised its decision to use its purchase order for the watershed.[174] Six years later Torquay water was found 'slightly turbid' with a 'peaty taste' but free from poisonous metals. Purchasing the 2,241-acre watershed, removing all farmhouses and other homes, and prohibiting 'objectionable fertilizers' was vindicated – tourists flocking to the 'English Riviera' could drink the water with confidence.[175] Paignton Urban District Council followed suit in 1899, announcing a plan to acquire about 600 acres of Brent Moor as a watershed for a reservoir. In the event, in 1901 the council secured 748 acres of Holne Moor as a watershed for Vennford Reservoir, opened in 1907. In both cases the Dartmoor Preservation Association had battled the enclosure. Fifty years later the Brent Moor site provided the watershed for the controversial Avon Dam Reservoir. Plymouth, perhaps chastened by the 'sites battle', did not buy the Burrator watershed until 1916. Those 5,000 acres are now marked by 'PCWW 1917' boundary posts.

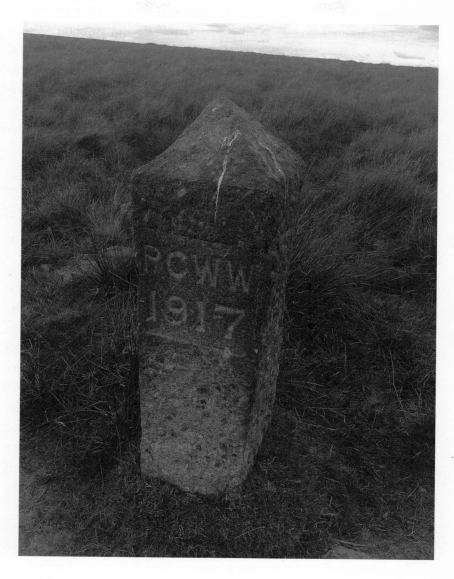

Dartmoor was targeted as a site for storage reservoirs because it was geologically suitable and the land cheap because relatively unproductive for conventional agricultural purposes. The land-owners right to sell was never in question but a transfer of owner-ship did not dissolve the existing rights of third parties. Common as well as riparian rights had to be dissolved by parliamentary act if the purchaser was to avoid complex negotiations that might require them to pay compensation for many years to come. The importance of enabling acts that dissolved rights of common could hardly be gainsaid, though additional provisions often provided statutory guar-antees that certain customary practices might continue. Another sign of the times was the tendency for customary rights of access to be guaranteed by these acts, the Vennford enabling act stating that the public 'shall be entitled to the privilege at all times of enjoying air, exercise, and recreation on the common land acquired by Paignton District Council, whether fenced or not'.[176]

Once again, the situation on Dartmoor mirrored developments elsewhere. For instance, in December 1901 the Dartmoor Preservation Association minuted its hope that Devon County Council would oppose the Devonport Water Company's bill seeking the acquisition of 383 acres of the Cowsic valley in order to ensure the insertion of the 'Birmingham Clauses'. This referred to the controversial Birmingham Corporation Water Act of 1892, which included a compulsory-purchase clause for 180 square miles of Powys in mid-Wales and allowed for the construction of several dams and a seventy-three-mile aqueduct to transport water to Frankley Reservoir just outside Birmingham. Although the act transformed common land into the Elan Valley Estate, dissolving existing common rights, addi-tional clauses provided statutory guarantees for some of these rights and additional rights of access and recreation to the general public.[177]

My region has been bedevilled by the problem of water supply. This is particularly so because a national park is a major part of our catchment area for water.

David Owen, MP for Plymouth Sutton, 1 May 1973

The small developments punctuating the interwar history of water undertakings on Dartmoor were dwarfed by the next two major initiatives. Torquay's construction of a dam at Fernworthy was enabled by legislation passed in 1934. Work began in 1936 and after some delays the dam was completed in 1942, creating what at seventy-six acres remains Dartmoor's second-largest reservoir. In 1950 the South Devon Water Board successfully sought a parliamentary order for the damming of the Avon or Aune on Brent Moor. The process lacked transparency, and the preservationists were outraged by what they regarded as underhand tactics on the eve of Dartmoor's designation as a national park. Last-ditch attempts to highlight the fate of threatened antiquities such as Gripper's Pound and associated hut circles did not prevent construction beginning in 1954.

Chronicling the process, the DPA editorialised against statutory authorities that could not 'see the matter from any point of view save that of the *enterprising* water engineer'. Demonising experts was but one part of this rhetoric, parodying technical language was another. They mocked the 'average official mind' for regarding Dartmoor's rivers as 'no more than useful water conduits provided by a thoughtful Providence solely as material for *enterprising* exploitation'.[178] That repetition of '*enterprising*' signified not only disdain for commerce but implied that the hydrologists were expansionist for its own sake. Christine McCulloch's work suggests the preservationists might have had a point, for the exponential growth in water undertakings that occurred in Britain during the post-war period cannot be wholly explained in terms of increasing demand. Large-scale reservoir building might well have vindicated Hayekian fears that governments are prone to inefficient over-expansion when they become unduly deferential to experts excited by the possibilities unleashed by technological advance.[179] Even if this was the case, rather than questioning the need for a reservoir the preservationists tended to propose it be located elsewhere. And they celebrated expertise when it suited them. They might ridicule the 'little men in raincoats with pointed town shoes' spotted inspecting an upland site, but in the post-war years they found planning much preferable to the 'present chaotic, grab-as-grab can, bags-I-this-river-and-you-take-that scramble'.[180]

Before turning to the great reservoir battles of the 1960s, the water politics of the first decade or so of the national park's history can be surveyed with the ever-vigilant DPA as guide. During this period the water boards sought to augment existing supplies by further tapping the river flows. Torquay, for instance, sought permission to divert water from the East Dart and North Teign to Fernworthy Reservoir via pipes laid in the disued Vitifer leat. They planned to sell the surplus to St Thomas (Exeter) Rural District Council. According to the DPA, during dry weather such abstractions, 'quite apart from the uglification', caused the rivers to become 'languid trickles', springs to fail and the Dartmoor water table to fall, causing fish and pond life to suffer. And although the association had not previously objected to the plan to raise the height of the Fernworthy dam, by 1958 they reckoned maintaining water levels during times of water shortage already caused unreasonable damage to river ecosystems, and raising the dam was really a ploy to get access to North Teign and East Dart water.[181] In cahoots with the National Farmers Union, the Dartmoor Commoners Association, the Dart Anglers' Association and the Teign Valley Preservation Association, the DPA lodged an opposition petition, as did Devon County Council, the Devon River Board, Totnes Corporation and Ashburton, Buckfastleigh and Paignton Urban District Councils. The environmental argument triumphed, and a Commons select committee threw out the Torquay bill. Claiming a rare victory for amenity over utility, the DPA failed to spot the significance of the ominous suggestion of the Devon River Board's consultant engineer that a reservoir be constructed instead in the heart of the moor at Swincombe. Shortly after its defeat, Torquay joined forces with the South West Devon Water Board.[182]

Preservationist opposition to the North Devon Water Board's

speculations at Taw Marsh near Belstone in the north quarter were less successful. The board was the statutory water undertaker for two thirds of the county, an area of 1,641 square miles with a population of 189,000 that stretched south from the Bristol Channel to the Exeter–Plymouth route through the national park but excluded both cities. The southern reaches of the district took water from the West Okement through an intake below Shelstone Tor, and the water was piped to a treatment works at Prewley on the north-west edge of the moor. The capacity of the treatment works was about four million gallons a day, but the board's licence required that its takings were proportional to the strength of the river flow. When the flow below the weir was less than 1,200,000 gallons a day, only 400,000 gallons could be extracted, making for water shortages during dry summers.[183] Rather than building a 'far more disfiguring' dam and reservoir, the board proposed tapping the supposedly huge underground lake beneath Taw Marsh. Permission for trial boreholes was granted quickly, followed by the passage of an enabling bill allowing construction of permanent wells and a pumping station.[184] The DPA petitioned against the bill, drawing attention to the ecological damage caused by tapping water near the river heads, quoting expert opinion on the effect taking water from deep wells had on the springs, streams and rivers of the district. As important to its case was the spoliation of the natural beauty of Taw Marsh – 'not a marsh at all, of course, but a splendid stretch of moorland surrounded by tors, perfect for walking and riding' – caused by building six-foot-diameter concrete well heads, a balancing tank, a break-pressure tank and access roads.[185]

The DPA's fuss resonated little with the general public or other amenity organisations, but this time it was proved right. In the short term, the Taw Marsh yield was only a third of that hoped and radiation levels in the water required the construction of an aeration plant that emitted radon gas. The unorthodox approach having failed, in April 1962 the North Devon Water Board's consulting engineer, Rafferty, recommended it investigate two sites for a storage reservoir on the West Okement. The river rises in the heart of the north quarter on Great Kneeset and flows north-west below Lints Tor, Steng-a-Tor – the first of Rafferty's suggested sites – and through the steep-sided valley of Black-a-Tor Beare, the site of one of the upland's three ancient oak woodlands. At Shelstone Tor it flows through another stretch of picturesque woodland before bearing north from Vellake Corner through Meldon Valley – the second proposed site – and then off the moor into Okehampton. Steng-a-Tor, upstream from Black-a-Tor Beare and several miles into the northern upland, was unacceptable to the Dartmoor Standing Committee, which steered the board towards Meldon but did not permit experimental boreholes. This forced the board to seek an order from Keith Joseph, minister of housing and local government. The preservationists, quick to condemn the proposition, reminded their readers of the board's poor record at Taw Marsh and insisted sources further downstream were investigated. New technology might provide the solution, and the DPA drew attention to the recent installation of a shallow aquifer by South Devon Water Board on the lower reaches of the Dart at Totness. When Joseph ordered a public inquiry, the DPA rightly anticipated that 'the first shots will be fired in what may later develop into . . . one of Dartmoor's major battles'.[186]

In the event, the inquiry was somewhat farcical. Held at Okehampton on 1 January 1963, appalling weather prevented witnesses from attending, and the discussion was confined to whether the boreholes would have a negative effect on amenity. No discussion of the proposed reservoir was permitted.[187] Five months passed before Joseph approved the exploratory boreholes. The real battle for Meldon began in July and August 1964, when the board applied to Joseph for the necessary order and to Devon County Council for planning consent. Following Labour's victory in the general election that October, Harold Wilson appointed Richard Crossman to the ministry. The political transition meant further delay, and it wasn't until January 1965 that Crossman announced a local public inquiry for 9 March.

A local public inquiry was required as a consequence of the Water Resources Act of 1963.[188] The act was intended to facilitate the improvement of water supply by creating regional river authorities empowered to conserve, redistribute or otherwise augment existing water supplies. Few doubted that a significant increase in capacity was needed if increased consumer demand was to be satisfied and water availability was not to stymie further industrial development. In effect, the act gave local authorities, working with the water boards, permission to acquire land, build and own reservoirs, and carry out the necessary engineering and building work. Plans had to be published, and if objections were lodged a public inquiry, conducted near to the proposed site, would solicit the views of interested local and national parties. On the basis of the inquiry report, the minister would decide whether to make the order, which could include the compulsory purchase of land. If the original opponents lodged objections within twenty-eight days of the date of the order, it was referred to a select committee representing both houses and could be made the subject of parliamentary debate.[189]

As the Meldon fight intensified, the preservationists placed great emphasis on the particular beauty of the valley. For the first time the DPA newsletter featured a photograph on its cover. It showed 'a young boy and his heritage', and the accompanying text made reference to the 'loveliness of this valley' with 'its bridleways and footpaths, bathing pools, sheltered picnic and camping places, shady spinneys of oak and ash and little waterfalls and, above all, the wildness and splendour of its great folded hills'.[190] Dr Beech of Cheshire, addressing a letter to the inquiry, encouraged them to picture the valley and imagine its effect on the visitor: 'Having passed under the Meldon Viaduct of the former LSWR,[191] the walker is confronted with as grand a scene as he could wish to see. The rocky stream and steep hillsides, granite-strewn and heather-clad, glories of russet and mauve during season, together with a deep serenity and peacefulness, make a powerful contribution to the relaxation and restoration to full health sought by many holiday makers.' Beech continued, completing the essential case against the reservoir.

> The basic factor in the present scene is the primitive, now becoming more and more difficult to find anywhere in England, and especially in the South. And man-made work in this part of the valley must, in my view, result in the destruction of the particular kind of amenity characteristic of Dartmoor, and sought by visitors, and the present proposals would seem to be diametrically opposed to the stated purposes of the National Parks and Access to the Countryside Act, 1949.[192]

A. M. Walding-White of Stoke Canon, a village near Exeter, wrote to Crossman in slightly different terms. Summarising his lifelong experience as 'walker, explorer, hill and mountain walker and climber', he compared Dartmoor to Switzerland, Austria, Italy, Germany, Scotland, the USA, Canada and Brazil, claiming 'that Dartmoor of all of these is outstandingly the most unique and fully as beautiful as all these others.'[193]

Mobilising superlatives was typical of the opposition campaign but this cannot be ascribed solely to political exigency. The course of the West Okement had been long celebrated by Dartmoor's enthusiasts, Crossing having praised its 'sylvan beauty' as one of Dartmoor's 'gems' sixty years earlier.[194] Beech rightly grasped that the amenity case against the dam was powerful because the statutory framework allowed the subjective values he extolled to outweigh the cost-benefit arguments made by the board, but no metric existed to weigh individual testimony against other criteria. Indeed, the harsh climate of the northern quarter and the gorge's distance from Plymouth and Exeter made it difficult to claim that Meldon was a popular beauty spot. Could preservationists claim that an isolated valley, exquisite but only known to a small circle of initiates, was a national asset? What was clear is that the law gave the amenity societies a right to be heard.

The Meldon inquiry was presided over by B. C. W. Wood, a water engineer appointed by the ministry, and C. Johnson, a planning assessor. The report opened with a description of the board's predicament, outlining the shortcomings of the West Okement intake and the Taw Marsh well head.[195] It described how the board sought permission to extract up to five million gallons a day from the intake subject to the prescribed minimum river flow of 400,000 gallons per day. When river flows were low, the intake waters could be supplemented with water pumped from the proposed impounding reservoir at Meldon to the Prewley treatment works. New pipes would need to be laid, roads built and a concrete dam constructed. Land would need to be purchased from the Duchy of Cornwall, the owner of Meldon Farm, the British Railways Board and seven private landowners. As Crown property, the duchy land could not be made subject to a compulsory purchase order, a privilege also claimed by the British Railways Board. A more serious obstacle, the report acknowledged, was the opposition of bodies responsible for protecting amenity, including the National Parks Commission, the Devon River Board and, most seriously, the Dartmoor committee of Devon County Council, the planning authority. The planning assessor, the report noted, had remarked 'that the rarest beauties of Dartmoor belong to its fringes rather than to the uplands and that the steep-sided valley by which the West Okement river leaves it is one of the loveliest, being all the more valuable when so much of this part of Dartmoor is in military occupation'. All parties urged the board thoroughly to investigate the proposed alternative site at Gorhuish, north-west of Okehampton and outside the national park, despite the water engineers concluding Gorhuish would be more expensive, would produce a lower yield and did not offer much potential for development as a recreation site. Although troubled by the gravity of the decision and accepting that there was no engineering case against Gorhuish, Wood allowed cost-benefit calculations to win out and recommended the Meldon order be approved, albeit with greater restrictions on motor traffic access and some modifications regarding abstraction.

The internal Whitehall discussion following the submission of the report largely reproduced Wood's line of argument. Excessive weight was given to the board's acquiescence regarding the Stenga-tor site, which was judged to be an exercise in goodwill and sensitivity rather than a political ploy, and typical of the commentary was the following note addressed to J. E. MacColl, MP for Widnes and Crossman's parliamentary secretary:

> The main issues raise the perennial conflict between the material needs of modern communities and amenity. The amenity objection here is a strong one – the planning Assessor's report makes that clear. And it is, as so often, a more attractive and comprehensible argument than the technical and financial case based on the mundane requirements of a water undertaking. There is no doubt that the Meldon scheme would materially affect the character of an attractive valley on the outskirts of Dartmoor. It clearly wouldn't involve a disastrous change – there are no large unsightly works – but it is one which some lovers of Dartmoor, at any rate, would rather not see happen.[196]

The rhetorical tropes running through this passage served to uphold the recommendation. The credibility extended to the 'amenity objection' is immediately undercut by the contrast between emotive responses and the 'material', the 'mundane', the 'technical' and the 'financial'. Note the condescension in the characterisation of the reception of the amenity argument as 'attractive and comprehensible' rather than, say, 'attractive and comprehensive'. The man at Housing and Local Government then followed these lofty judgements with the ex cathedra use of 'clearly', failing to recognise that a sheet of water in a striking river valley was judged by many to be 'unsightly'. It was somewhat disingenuous to imply that the amenity argument usually held sway. Fred Willey, minister of land and natural resources and MP for Sunderland North, was unconvinced by the cost case against Gorhuish and took the contrary view of the decision-making process. He thought there was 'a tendency in cases of this sort to give weight to quantifiable factors simply because they are quantifiable, and . . . this has been allowed to tilt the balance here'.[197]

Civil servants are prone to speculative judgements. One thought that Meldon's location, 'secluded in one corner of the national park', meant that 'in a generation people will have forgotten about the valley and enjoy the contrast of the lake in an area where other types of beauty predominate'. Crossman was wary. The difference of opinion between the engineering inspector and the planning inspector meant that he was being asked to override the local planning authority. There were problems of timing too. Manchester's request to abstract water via underground pipes from Ullswater and Windermere in the Lake District was before parliament. This already followed parliament's rejection in 1961 of a much bolder plan to draw water from Ullswater to a new reservoir in a nearby valley.[198] Further communications from the land ministry continued to play down the significance of the cost difference between Meldon and Gorhuish, and Willey observed that they could have knocked Meldon on the head earlier if the planning officer's view had been communicated to ministers more swiftly.[199]

Crossman's anxiety was evident when he met his civil servants on 25 May. Admitting they had no real answer to the planning inspector's objections, they continued to toy with the Gorhuish option. Under pressure to confirm the order, Crossman opted for delay, afraid that a badly timed recommendation could lose both Meldon and Ullswater.[200] Leonard Millis of the British Waterworks Association goaded Crossman that projects like these, answering an evident need, required a great deal of preparation. The ministry did not need reminding that insisting the board revisit Gorhuish would mean a further two-year delay, considerable expenditure and a possible recourse to Meldon. Although the preservationists suspected delay was the tactic of unsympathetic politicians who calculated that public opposition would diminish in time, in this case ministerial indecision was as much occasioned by genuine uncertainty as wider political considerations.

On 18 July 1966, following umpteen drafts, Crossman signed the decision letter. It outlined the argument, summarised the objections and explained that the decision essentially concerned choosing between Meldon and Gorhuish. Considerable emphasis was placed on time constraints. Were Meldon rejected at this stage, Gorhuish would have to be properly investigated, which would delay the opening of the reservoir on one of the two sites from 1970 to 1972. Admitting that the case for Meldon was better known than the case for Gorhuish, the decision letter revealed the degree to which the needs of the board had been allowed to outweigh other considerations. 'The Minister's conclusion is that the effect on the National Park of a reservoir in this corner of the National Park would not be so serious as to justify his requiring the Water Board to accept an alternative which would have a substantially smaller yield, would be more expensive, and could not be ready before 1972, thus prolonging by two years the prospect of restricting domestic supplies in an average summer.' As a sop to the amenity societies, the order contained three modifications. Water could only be taken from the river when the flow was above 1,500,000 rather than 400,000 gallons a day; permission was not granted for the proposed road around the eastern side of the reservoir to the intake on the West Okement; and detailed provisions regarding amenity were inserted.

Crossman did not enjoy a good press. Sayer was stinging in a BBC interview given on the day of the announcement; the *Sun* announced the flooding of a 'National asset'; the *Daily Telegraph* ran with RESERVOIR FOR BEAUTY SPOT; and, like several other newspapers, *The Times* took its lead from Sayer, heading its story RESERVOIR 'VICTORY FOR BARBARIANS'. *Private Eye* referred to 'Crossman's Moors Murder', a savage allusion to the infamous serial killers Ian Brady and Myra Hindley. The *Guardian* editorialised on

Crossman's 'impatience with the aesthetic argument for the preservation of rural beauty', noting the weakened position of the preservationists on the reconstituted National Parks Commission. Ten days later, the *Guardian* published a letter from Sayer praising its editorial and making the case for Gorhuish. Crossman suffered vilification in a string of personal letters from members of the public. One gentleman, referring to the original national parks legislation, described Crossman as 'a traitor to one of the finest pieces of legislation ever to emerge from Westminster'. He hoped the names of Crossman and Peter Wills – MP for North Devon and a leading proponent – would be 'emblazoned in large letters' across the 'concrete monstrosity'. Brian Libby, a DPA member, demanded copies of all the relevant public documents and threatened legal action. His furious letter closed with a postscript: 'In case you have contrary ideas – I am a responsible taxpaying citizen of the UK (Many people think DPA members are "barmy")'. Another voter registered his disgust in a postcard addressed to 'Mr R Crossman, Housing and Vandalism Department, London'. Several letters expressed their particular disappointment with the Labour Party, a voter from Bristol resorting to the canard: 'The only party now worth supporting are the Liberals!'[201]

The CPRE, the Commons, Open Spaces and Footpaths Preservation Society, the Ramblers' Association, the Youth Hostel Association, the Dartmoor Rambling Club, the Exeter Rambling Club, the Dartmoor Rangers and, of course, the Dartmoor Preservation Association duly lodged their objections. Parliamentary hearings that July led a nervous select committee to confirm and then immediately suspend the order pending the investigation of alternative sites. Meldon, the committee argued, 'should only be proceeded with if the Gorhuish scheme outlined to them [by the amenity societies] proved to be impracticable or significantly more expensive than anticipated'. In September the engineering firm T. & C. Hawksley was commissioned to submit an independent report on the engineering dimension of the case.

In the meantime Crossman's handling of the matter was beginning to look rather flawed. A ministerial note commented that due
consideration had not been given to the fact that thirty-nine of the
seventy-four acres were reputed to be common land, the despondent
civil servant writing that had the ministry known it would have
recommended the board lodge a private bill. A local inquiry would
probably be needed and, somewhat ironically, the power to grant
consent then would lie with Willey as minister of land and natural
resources. The application for consent was published on 15 October
1966, which meant objections could be registered until 28 January
1967. Experience suggested that the joint committee would confirm
Crossman's order, though the press speculated that in this case he
might be overruled.

In March 1968, almost two years after Crossman made the order,
T. & C. Hawksley submitted a very substantial report. 'Considered
in isolation from amenity considerations', they concluded that
Meldon 'is the natural and proper' site. Gorhuish would cost
£2,431,497 whereas Meldon would cost £1,764,000. That August,
Lord Kennet, parliamentary secretary at Housing and Local
Government and chair of the joint committee, commented to
Anthony Greenwood, Crossman's successor, 'To go for Gorhuish
would thus be a conspicuous, though not entirely unprecedented
(the Manchester scheme) swing towards paying for amenity.'[202]

In the meantime the Dartmoor Preservation Association created
a notable diversion. A vigilant member examining an old OS map
spotted that there was a disused copper mine near the confluence
of Fishcombe Water and the West Okement. Mobilising Dr Beech
and with help from an inorganic chemistry textbook, the government and the press were alerted to the danger of arsenic contam-
inating the water. Flooding Homerton mine in the valley posed a
similar danger, and Sayer helpfully supplied the government with
photographs of the 'band of arsenic bearing rock which traverses
the valley'.[203] When the DPA unearthed stories of previous chemical pollutants entering the water supply, the press obligingly printed

them. Thousands of fish were killed in 1864 by discharge from old mine workings near Plymbridge; heavy rainfall caused arsenic to be discharged into the Tavy by the Wheal Betsy mine in 1872, killing every living thing in the river. Much was also made of the growing body of research linking pollutants released by disused mines with the high incidence of cancer in the region.

With the press taking an interest, the North Devon Water Board had little choice but to commission tests. Arsenic was found in spoil heaps in the valley but not in the water, and though experts concluded that there was no significant danger if the water was treated in the normal way, flooding the valley might destabilise the land and release arsenic. Sealing off the mine workings and removing the spoil heaps was recommended as a precaution, adding £25,000 to the bill. Kennet, taking a keen interest, helpfully suggested to Greenwood that the mineshafts be filled in with earth from the dam excavations and the spoil heaps covered with butyl sheeting. The reservoir would need to be drained and the sheeting renewed in a century. The board, frustrated by the delays and the DPA's 'malice', grudgingly committed itself to removing the spoil heaps and sealing the mines.[204]

While Sayer and the board manoeuvred, a more significant development occurred. Between 20 September and 10 October the Ministry of Housing and Local Government received resolutions passed by sixteen rural, district and borough councils. All demanded a decision: most urged approval of the order; several expressed exasperation that the Townleigh site might be considered first.[205]

Townleigh?

In the mid-1960s Plymouth Corporation and South West Devon Water Board began to consider the city's water needs. In January 1967 the Water Resources Board, acutely aware of the resources being swallowed by the Meldon battle, called a meeting at Reading of the Devon river authorities, Devon County Council, the Dartmoor Standing Committee, the National Parks Commission, Plymouth Water and the Ministry of Housing and Local Government.

They agreed to establish a technical steering committee to investigate a wide range of sites and schemes, including desalination and estuarial barrages, in an effort to coordinate Devon and Cornwall's future water needs. Nine potential reservoir sites attracted their particular attention, including Huccaby and Swincombe on Dartmoor. These investigations occurred below the radar, exposed only when the 'Water for Plymouth' report of September 1968 identified Swincombe as its first choice. Lee Mill, only sufficient to Plymouth's needs, came second, Townleigh third and Tor Wood fourth.

In August 1968 H.G. Godsall, clerk to Devon County Council, wrote to the ministry on behalf of the council warning that Swincombe would prove highly controversial and suggesting that no decision be made on Meldon until Townleigh, potentially large enough to satisfy the needs of Plymouth, south-west Devon and north Devon, was fully examined. Godsall added that the North Devon Water Board, convinced it was going to get Meldon, was not engaging properly with the Townleigh discussions.[206] With two Dartmoor sites facing flooding, the Countryside Commission mustered a final intervention, asking that the Meldon question be reconsidered in the light of the Swincombe and Townleigh questions. Kennet continued to hedge, suggesting to Greenwood that the strength of public opinion was pushing them towards the least undesirable alternative. These developments frustrated the ministry. Whether out of convenience or from conviction, it had decided that the needs of the North Devon Water Board were so urgent and it had been kept waiting so long that it was entitled to a favourable decision. Townleigh threatened a three- or four-year delay. Moreover, only an act of parliament could revoke the Meldon order because the joint committee had been satisfied by the conclusions of the Gorhuish investigation. For Whitehall officials, the only outstanding question concerned when the minister would bring the order into effect. In early November officials at the ministry began to mutter that further delay might bring them before the parliamentary ombudsman on a charge of maladministration.[207]

And so in August and September 1968 the rural, urban and borough councils of north Devon coordinated their expressions of support for the Meldon solution. The Countryside Commission was tipped off as early as 1 October that the minister was about to bring the Meldon order into effect. The North Devon Water Board was told on 20 November that the order would come into operation on 2 December; H. R. Slocombe replied that the board was 'extremely pleased'. In January 1969 consent was granted for the enclosure of common land; in November detailed plans were submitted to the local planning authority. The Dartmoor Preservation Association succeeded in having the case investigated by the parliamentary omsbudman, who reported in November 1969 that he found no evidence of maladministration. Works commenced in March 1970, two months before the plans received final approval.[208]

In 1891 Sandeman had argued that Plymouth needed a storage reservoir sufficient to guarantee 120 days' supply; in 1969 Plymouth could guarantee 40 days' supply. In November 1969, following extensive investigations of several sites, Plymouth Corporation and South West Devon Water Board resolved jointly to promote a private bill that would allow them to draw water from the River Tavy and create a vast storage reservoir at Swincombe in the heart of Dartmoor. Not since London had proposed buying Dartmoor in the 1890s had such an ambitious and transformative scheme been proposed. Trenchford Reservoir had a water surface area of 30 acres and held 200,000,000 gallons; Fernworthy Reservoir had a water surface area of 76 acres and held 380,000,000 gallons; and Burrator Reservoir had a water area of 150 acres and held 668,000,000 gallons: the proposed reservoir at Swincombe would have a water surface area of 745 acres and would hold 11,000,000,000 gallons. The available water supply would be increased by almost 50 million gallons per day and would safeguard supplies for the next 50 years. If approved, 55 per cent of the resident population of Devon and 40 per cent of the resident population of Devon and Cornwall would henceforth enjoy water security.

The Water Resources Board helped promote the bill. In their view the case rested upon the relative value of the possible sites, the choice being between construction in the national park, which would cause little disruption, or elsewhere on land of agricultural value. Of the shortlisted sites, Lee Mill was sufficient only for Plymouth's needs and would mean the loss of 600 acres of good agricultural land and nineteen houses; Townleigh on the River Tamar was also good agricultural land and would mean the lost of nine houses; the Hill Bridge site on the Tavy was simply too expensive; and Huccaby was too beautiful. Among the materials submitted to the ministry was an emotive presentation showing photographs of attractive farm buildings, houses and cows at Lee Mill.[209] The Swincombe plan anticipated extracting water directly from the Dart at Totnes and the Tavy at Lopwell, with the reservoir being used to regulate the flow of both rivers. A smaller reservoir would be established at Milton Combe to hold the water extracted from the Tavy; treatment works would be built on Roborough Down for the Plymouth supply and on the left bank of the Dart at Totnes.[210]

A neutral voice helping the case for the scheme along belonged to Lady Aileen Fox, a respected archaeologist at Exeter University whose pioneering work on the Roman south-west is of exceptional significance. Her report argued that nothing of outstanding impor- tance would be lost to the flooding. Two medieval monuments would need to be removed and re-erected, as would the surmounts of a Bronze Age cairn and the cross marking the site of Childe's Tomb, which in any case was only a late nineteenth-century replace- ment of the original destroyed in 1812. Fox was dismissive of the Whiteworks tin mine, the ruins of Fox Tor farm and a furnace house because they were all post-medieval, an antediluvian attitude at odds with the new generation of industrial archaeologists.[211]

The Countryside Commission strongly opposed the scheme. It approached the case in terms of first principles, reminding the minister that under the Countryside Act of 1961 it had been mandated to conserve and enhance the 'natural beauty and amenity of the countryside' and to encourage 'the provision and the improvement, for persons resorting to the countryside, of facilities for the enjoyment of the countryside and of open-air recreation in the countryside'. A reservoir at Swincombe would be 'an artificial intrusion into one of the wildest parts of the Dartmoor moorland', and the commission was concerned by 'the continuing erosion of the purposes which justified the designation of the Dartmoor National Park'. To build a large dam and reservoir complex 'at the heart of the unbroken sweep of accessible moorland which forms the core of Dartmoor' would divide the upland into three separate and relatively small areas, seriously diminishing the value of the national park as a whole. This time opposing petitions lodged by the usual suspects were backed by the Duchy of Cornwall. The Devon River Authority and Dart Fisheries Preservation Society effectively withdrew their objections following modification of the extraction plans.[212]

Hostile preservationist screeds were duly penned, but their validity was bolstered by a remarkable report buried in a file of largely repetitive Whitehall material. P. A. Sydney, senior planning officer at the South West Regional Office of the ministry in Bristol, was asked for his impressions of the site. He visited on 6 January 1970.[213] Explaining that he did not in principle oppose the construction of reservoirs in national parks, he thought there were notably successful examples in Snowdonia and, indeed, at Burrator. He did though query the technical case, suggesting that if it would take ten years to refill the reservoir after a major drought the watershed could hardly be thought large enough for the size of reservoir. Like the Countryside Commission, Sydney argued that Swincombe was the solution to the wrong problem. He could see that there was a major water supply problem in

Devon but it could no longer be treated as a series of challenges to be addressed by separate boards. Regional planning was needed.

However this was not the essence of Sydney's case. He opened his report as follows: 'As one stands at Whiteworks overlooking the large, shallow bowl contained by the surrounding tors one is first impressed by the utter silence and the desolation of this inner wilderness. Frightening in its primitiveness yet attractive in its starkness, the moor is undoubtedly uncompromising in its rugged integrity.' The operative adjective here was 'shallow'. As Sydney explained, a large sheet of water would overwhelm the landscape's natural features, the significant landscaping and tree planting required to make the reservoir attractive creating a 'new landscape . . . entirely alien to the spirit of Dartmoor'. He was also disturbed by the visual consequences of the reservoir being emptied: 'to imagine this vast bowl drained of water with its great slope denuded, naked and baked over this length of time boggles the imagination'. Sydney concluded with a *cri de coeur*, arguing that there was 'little point' in considering the case in further detail for 'the whole concept is fundamentally wrong':

One thing is quite clear about Swincombe, that although a perfectly useless area as far as agriculture is concerned has been selected, it forms part of the wilderness, magnificent in its forbidding qualities and uncompromising in its attitude, it would reject the most ingenious attempts to integrate a reservoir in this particular area. It cannot be done, and it should not be done. Before it is even thought of, I think the Department would need to be satisfied to the hilt that none of the alternatives could not be made to work, even at considerable extra cost and sacrifice of agricultural land. It is true that our agricultural land is shrinking – but there is only one Dartmoor and this small primeval area was designated as a park for this very reason and it is unique.

In this remarkable passage an official spoke the language of the enthusiast, departing sharply from technocratic conventions and making his sense of wonder instrumental. A month later a formal iteration of Sydney's report gave more credence to the South West Devon Water Board's case, but the argument was essentially the same. Landscapes like Swincombe only could be evaluated on the basis of their 'scarcity value'. By this test, 'the wild moorland scenery of Dartmoor' was 'foremost' in the south of England.[214]

The Plymouth and the South West Water Board Bill was a private bill, so it would be presented to parliament and its proponents heard by a select committee if it passed a second reading. Opponents could petition parliament, and they too had the right to be heard at the committee stage. The pursuit of a private bill was expensive, not least because at the committee stage opponents and proponents tended to appoint counsel. A bill rather than an order was necessary because another reservoir controversy had exposed a flaw in the Water Resources Act: ludicrously, the act permitted the impounding of water but not its discharge or the river-regulation aspect of water conservation programmes. With the water companies abandoning the act, the minister could loftily present rather than actively promote controversial private bills in parliament and leave the decision to the select committee.

The *Observer* newspaper asked CAN DARTMOOR BE SAVED? It was now 1970 and European Conservation Year, already dubbed 'Desecration Year' by a *Sunday Times* insistent that Anthony Crosland, now 'Mr Environment', the secretary of state for local government and regional planning, could still stop the bulldozers going into Meldon. The *Observer* argued that the Townleigh alternative could have provided for both Meldon and Swincombe had planning been properly coordinated. Meeting on 16 February, with a general election looming, ministers considered how the government should handle the second reading of the Swincombe bill given that it was likely to be rejected. They couldn't put pressure on the promoters to withdraw the bill, for the government had helped draft it and would be accused of bad faith. Their best hope was to recommend a second reading, place the accent throughout on 'neutrality and a balanced attitude' and let the process run its course. During the debate on the second reading (14 April 1970) MPs queued up to bear witness to their love and knowledge of Dartmoor but there was deep disagreement over the particular value of Swincombe. Local MPs, under pressure from their constituents, took a particularly pragmatic view. Joan Vickers, MP for Plymouth Devonport, referred to Swincombe as 'mainly bog land', highlighting the absurdity of the *Observer* and the *Spectator* referring to the 'rape of the moor'. Robin Maxwell-Hyslop, MP for Tiverton, found Swincombe 'remarkably undistinguished by natural beauty' and saw little merit in its industrial heritage: Swincombe was not 'untouched by man' but 'abandoned by man'. By contrast, Carol Johnson, MP for Lewisham South and vice-chair of the Commons, Footpaths and Open Spaces Preservation Society, emphasised the range of opposition the bill had attracted, rightly finding the duchy's opposing petition especially remarkable. Douglas Jay, MP for Battersea North, was undecided, urging a second reading, though he too emphasised his deep feeling for Dartmoor, believing it 'the most beautiful part of the British Isles'. Overall, the tenor of debate suggested that the vote in favour of a second reading was not driven by strong support for the bill but the sense that only a select committee could give the matter the time it needed.

Over seventeen days in November and early December 1970 four MPs heard the arguments put by the water boards and their opponents. On 3 December the committee room was cleared and the MPs conferred. Ten minutes later the proponents of the bill and the petitioners were summoned back to the room. John Hunt, Conservative MP for Bromley and committee chair, spoke. 'We have come to the conclusion that the promoters have not made out the case for the bill, and in the light of this we feel it would be contrary to the public interest to allow the proceedings on the case to be continued.'[215]

★   ★   ★

To visit Meldon knowing of the controversy is to leave with mixed feelings. The reservoir is relatively well contained, for the narrowness and depth of the valley hides it from much of the surrounding country, but aesthetically it does not work. In proportion to the valley the water levels are too high and the water surface is too large for this artificial lake to be truly beautiful. And to look over the dam at the West Okement flowing north is to get a strong sense of what has been lost to the south. And yet two aspects of this 'disfigurement' are mitigating and, at least to some tastes, strangely compelling. The concrete dam, lacking the granite cladding of Fernworthy and Burrator, seems almost continuous with the industrial features of the area. The wrought-iron structure of the Meldon Viaduct hovers over the valley just downstream, the London and South Western Railway it once carried branching off just beyond to Lydford and Tavistock. Enthusiasts have recently revived a section of the railway and in the summer it carries tourists between Meldon Quarry and Sampford Courtenay. Meldon Quarry itself overlooks the river's east bank. Established in 1874, the quarry is now larger than the reservoir and a classic industrial eyesore. And yet to walk upstream beyond Vellake Corner is to quickly leave all of this behind, entering a river valley that looks just as Crossing described it over a century ago. Water churns around the great granite boulders of the riverbed; the oaks, moss and lichen of Black-a-Tor Beare still enchant.

Swincombe offers a different prospect. To retrace P. A. Sydney's journey along the single-track road from Princetown to the Whiteworks tin mine is to be confronted with a scene that has changed little in the intervening decades. Passing Tyrwhitt's Tor Royal plantation before sweeping sharply around to the left, the road reaches its terminus at the disused and ruined mine buildings and shafts. Foxtor Mires delivers its thrill, just as Sydney described it, sublime as the light begins to fail on a cold January afternoon. Standing at the Whiteworks, surrounded by the decaying remains of Dartmoor's industrial past, it is hard not to feel relieved that the Meldon order made the passage of the Swincombe bill less likely. And yet thinking of city men in pointy shoes is to be assailed by other thoughts. On what grounds can it be assumed that the Romantic gaze is superior to the engineer's gaze? When hundreds of thousands of people needed a reliable water supply, was it so narrow-minded to value the land according to hydrological and agricultural statistics and sophisticated maps describing contour lines and geological features? The politicians asked to weigh the different arguments were asked to evaluate different ways of seeing. To some it would have seemed improper for Crossman to have walked through the Meldon valley or for Greenwood to have stood at the Whiteworks mine, and yet the statutory obligation to protect amenity expressly required that this emotional response be weighed in the balance. Twenty years after the first national park designations, Whitehall still did not have an effective way of doing so. They were grappling with the political consequences of a shift in sensibility that had inverted conventional ways of valuing land. Despite the cultural authority of romanticism or the emerging politics of access, a century earlier it would have been scarcely conceivable that loud voices would have urged that urban needs be met by flooding fertile farmland when barren upland would do. Perhaps the audacity lay less with the water boards than the national parks legislation itself.

At first glance, the Whitehall correspondence concerning the Meldon decision simply exposes the ad hoc and short-termist nature of ministerial decision-making. Crossman, later reflecting on his experience at Housing and Local Government, commented that 'nearly all my technical advisers were passionately in favour of the producer and against the amenity lobby', a claim that chimed with the DPA's suspicion that on these issues ministers effectively did the work of their civil servants. 'In all these areas,' Crossman wrote, 'a Minister who actually read the documents and made his own decision was keenly resented.'[216] Crossman's hand had been forced by political considerations, though exercising power is of course a hazard of a successful ministerial career. More particularly, the process exposed the inadequate planning of water provision in the 1950s and 60s. Crossman argued that the minister should always be offered a range of choices rather than handed a single choice and the threat that he would be responsible for choosing supply or drought. That the Meldon and Swincombe proposals could be formulated in parallel, landing on the minister's desk within a couple of years of each other, reflected a system of water provision not fit for purpose.[217]

The continuing existence of multiple water boards pursuing individual solutions had become absurd. During the debate on the second reading of the Swincombe bill Jeremy Thorpe, ambitious Liberal MP for North Devon, touched a nerve when he described it as 'obscene that one or two promoters can get together and decide to come to Parliament with a view to raiding somebody else's valley or part of their national park'. He was not alone in concluding that the reservoir controversies proved the need to nationalise the water industry. Michael Heseltine, Conservative MP for Tavistock, found it ridiculous that they were considering the bill before the publication of the survey of Devon and Cornwall's water needs required by the Water Resources Act of 1967. The tendency of the Devon boards to turn to Dartmoor, thereby generating so much controversy, only drew attention to the inefficiency of the system.

Paradoxically, the failure of the Swincombe proposal partly stemmed from the fact that it was an attempt to coordinate the needs of two boards. Meldon could be sold as a discreet intervention in the north-west corner of the moor of benefit to nearby residents, whereas the scale of the Swincombe proposal offered a very different prospect, raising questions of national significance. It demonstrated that coordinating the needs of multiple boards required reservoirs on a new scale. This made it less likely that the boards could take on the Dartmoor amenity interest and expect to win, and more likely the interests of millions of consumers would outweigh the interests of relatively small numbers of farmers and residents.

When the government promulgated its new Water Resources Act in 1971, an amendment inserted by the Lords made it possible for parliament to annul any ministerial order that applied to a national park or an area of outstanding natural beauty. As was pointed out in the Commons debate, the threat of a parliamentary annulment made it unlikely that a water board would seek an order for flooding part of a national park or that it would be approved by the minister. If some MPs were disturbed by a development prejudicial to farmland, others saw the rejection of bills related to water undertakings in Calderdale, Derwent, the Dowlais valley and Swincombe as reflecting a fundamental shift in mood. Robin Turton, MP for Thirsk and Malton, argued that the fate of these bills showed that 'Parliament has at last awakened to the needs of conservation.' He believed European Conservation Year had a 'great deal to do with it'. Turton also reckoned there was 'very strong public disapproval of the river authorities' piecemeal approach to the solution of water deficiency and of the absence of a national policy for the supply and more economical use of water'. James Scott-Hopkins, MP for West Derbyshire, who was involved in the passage of the original bill, agreed that the 'whole climate of public opinion over the last seven years has changed'. Not only did he think the public was now 'much more aware of conservation and of the needs of agriculture' but it also wanted parliament to have its say.[218]

In 1973 Edward Heath's Conservative government amalgamated the numerous water boards into regional water authorities. South West Water, rarely out of the news for one reason or another, has since created large reservoirs at Roadford, north-west of Okehampton, and at Colliford on Bodmin Moor.

Christine McCulloch argues that thanks to the exponential growth of reservoir construction in the post-war period and its unanticipated industrial decline in the 1980s, Britain suffers from a net over-supply of water. More rationalised distribution systems could allow some reservoirs to be decommissioned, allowing the sites to be returned to their former state. There is not much prospect of this happening. However, the government's Restoring Sustainable Abstraction programme appears to have scored some modest successes. A poster child of the scheme nicely vindicates arguments made in the 1950s by the DPA. The preservationists rightly predicted the ecological damage caused by the Taw Marsh wells. In the 1980s English Nature and the National Park Authority became concerned by the effect on native species of reduced groundwater and lower water levels in the Taw. Particularly threatened were brown trout and plant species that thrived in blanket bog. The Water Resources Act of 1991 enabled abstraction schemes that were environmentally unsustainable to be investigated. In 1993 South West Water agreed temporarily to stop abstracting water from Taw Marsh in return for permission to tap the River Exe; further negotiations saw their licence to abstract water rescinded altogether in 2001. In 2012 the Environment Agency highlighted the restoration of the Taw Marsh ecosystem as exemplifying the success of the programme. Although the buildings have been left for fear their dismantling would cause environmental damage, bog bean and bottle sedge are thriving and a healthy brown trout population has re-established spawning and nursery areas.[219] Has Roadford made Meldon redundant? To restore the Meldon valley to its former state would be a bold act of redemption, paying not for amenity but for nature.

The Services

6. A HELICOPTER EXERCISE IN THE DARTMOOR NATIONAL PARK, WATCHED BY LADY SAYER, CHAIRMAN OF THE DARTMOOR PRESERVATION ASSOCIATION. THE RENEWAL OF MINISTRY OF DEFENCE LEASES AFFECTING A LARGE PART OF THE PARK WAS STRONGLY CRITICISED BY THE COMMISSION AND BY MANY LOCAL ORGANISATIONS.   [*Western Morning News*]

In the foreground a woman in a headscarf, knee-length coat and wellington boots, a light bag slung over her shoulder. She is taking a photograph. A helicopter, immediately above her, lifts or lowers an artillery piece. A soldier in the background has turned away,

walking off, hands in pockets, paying the scene in the foreground
no heed. The landscape is broad, featureless, swelling to the right.
The sky is big, grey, overcast. It's a damp day. The figure at the
centre compels attention. She seems calm, unperturbed, focused
on the task in hand. Hers is a familiar if increasingly archaic pose,
for now we hold our cameras at arm's length and look at a screen
rather than through a viewfinder. She could be taking a holiday
snap. Her clothes are civilian, domestic even; their ordinariness
bespeaks not lack of preparation but familiarity with the territory.
A little patch of bare flesh on her right leg is visible; she's wearing
a skirt, tweed probably. It's not just that her clothes offer little
defence against an unfriendly natural environment; it's also the
contrast they make with the military hardware hovering above her
and the fatigues worn by the soldier in the background. Above all,
her poise betrays no sense of the noise of the machine, probably
in excess of a hundred decibels, or the threatening swing of the
artillery piece. She seems oblivious to the sheer drama of the situation.

The photographer, of course, was Lady Sylvia Sayer, and the
time and location of this dramatic moment was Ringmoor Down
in 1968. But there's another photographer implied by the image
and that, of course, is the photojournalist who took this picture
for the *Western Morning News*. It's an artful composition. Sayer, the
artillery piece and the helicopter aligned on a perpendicular axis.
Perhaps Sayer wasn't taking a picture at all but merely posing for
the press, conscious of the publicity value of the image's sharp
contrasts. For this picture is evidence of newsworthiness, and an
army helicopter operating on Ringmoor Down was newsworthy
because Sayer made it so. But this photograph wasn't encountered
by leafing through back issues of the *Western Morning News*. It was
found in an archived annual report of the National Parks Commission
where it illustrated brief comments on Dartmoor. This photograph
has now been selected three times. First by an editor looking to
illustrate a news story, then by a government official looking for

an image which encapsulated the National Parks Commission's perspective on the controversies shaping Dartmoor politics in the late 1960s, and now by an historian seeking an image emblematic of what follows.

Since the late nineteenth century the services had made use of the northern quarter of the moor, establishing the Okehampton, Merrivale and Willsworthy Ranges still in use today. Okehampton (20,000 acres) and Merrivale (19,000 acres, of which 4,200 acres were non-firing) were leased from the Duchy of Cornwall and included agreements brokered with the commoners;[220] in 1900 the War Office purchased Willsworthy Manor (3,200 acres) from the Calmady-Hamlyn family for use as a rifle and field-firing range. Although over the next forty years horses were replaced by motorised vehicles, moorland tracks were developed into simple roads, and eighteen-pounders pummelled the landscape, the military did not seek to extend its territorial claims on the moor. All of this changed with the outbreak of war in 1939, as John Somers Cocks explained:

> By agreement or under the defence regulations most of the Moor north and west of the Tavistock–Two Bridges–Moretonhampstead road was used as a firing range; another rifle range was made on the flank of Rippon Tor; much of the south-eastern sector of the Moor, known loosely as Scorriton, became an area for training with rifles, machine-guns and anti-tank weapons; an airfield on Roborough Down and a hutted camp at Plaster Down were built; Penn Moor and Ringmoor Down were used for further training. Finally, an area around Haytor and Houndtor was set aside towards the end of the war to train troops for service in the Far East.[221]

Add the hundred-foot Air Ministry aerial erected on Laughter Tor–Dartmoor's contribution to the Gee radio navigation system operated by the RAF – and the full extent of Dartmoor's wartime militarisation is realised. And just as the Napoleonic and the 1812 Wars unexpectedly brought thousands of people to Dartmoor, so the same occurred during the run up to D-Day, 'when virtually the whole of the moor was used in preparation for the invasion of Europe'.[222]

There was nothing exceptional about this. Just as 'total war' led the state to mobilise its people, so too did it mobilise its natural resources, including its landscapes. Britain became honeycombed with training camps, barracks and military hospitals, its large open uplands proving particularly valuable as battle training areas. In response, Dartmoor's preservationists largely withdrew from the fray, their quietism signifying an acceptance that the national interest outweighed their particular concerns. They took comfort in the assumption that once victory was secured, the services would demilitarise Dartmoor, swiftly releasing all the land acquired since the beginning of the conflict.

This proved to be wishful thinking. The transition to peace brought large-scale demobilisation, but Britain's increased military commitments did not wholly dissolve with the end of the war. Maintaining the British presence in occupied Germany, resisting the intensifying pressures exerted by anti-colonial movements throughout the empire and fulfilling Britain's emergent obligations as a member of NATO meant the UK effectively remained on a war footing. The most visible and contentious domestic consequence of this was the continuation of conscription – National Service – but less noticed was the maintenance of a greater military training infrastructure, as Prime Minister Clement Attlee explained in a pamphlet presented to parliament in December 1947. Extending National Service and the increased range and mobility of the armed forces meant that the 252,000 acres used for training by the services in the 1930s would need to be increased to 702,000 acres. This nevertheless represented a huge reduction on the 11,547,000 acres used during the war, most of which had already been released. Attlee admitted that land of high amenity value would need to be used if agricultural land, forestry land and common land near large centres of population were to be avoided.[223]

Despite these developments and the garlands of victory, the services nonetheless found themselves in a changed and somewhat hostile world. For a few years the chiefs of staff had been at the apex of political power, the war effort predicated on meeting their needs. This unprecedented intimacy between military and civilian power had also brought the military machine under closer civilian scrutiny, and although Attlee chose not to retain the exceptional prime ministerial responsibilities accumulated by Churchill during the war, he was not prepared to allow the services to return to the relative autonomy they had exercised during the interwar period. The creation of the Ministry of Defence in 1946 marked a downgrading of the War Office, the Admiralty and the Air Ministry. Each remained a separate department of state, but henceforth the secretary of state for war, the first lord of the Admiralty and the secretary of state for air would not sit at the cabinet table. Instead, a single minister of defence, working in close cooperation with the Foreign Office, would attempt to present the government with a clear picture of the strategic needs of the services. By avoiding the unseemly and inefficient competition for resources that had characterised the politics of the services in the past, Attlee hoped the new ministry would ensure the national interest trumped the vested interests of the services. In practice, insufficient resources and a dearth of talent saw the MoD reduced to arbitrating between the services and the Treasury.[224]

These uncertainties played out in microcosm on Dartmoor. The position of the War Office in the north quarter, which included their outright ownership of the Willsworthy Range, was pretty unassailable. Meeting the needs of the Royal Marines, now an important presence in the West Country, was more contentious. In 1941 30 Commando was established and based at Stonehouse in Plymouth; 42 Commando was established in 1943 and based at Bickleigh near Plymouth; and the Commando Training Centre Royal Marines is at Lympstone in Devon. During the war the marines shared the Merrivale Range with the army and made extensive use of its other new territories; the Admiralty believed it needed to secure these training grounds if the commandos were to have a viable future in the West Country. These needs conflicted with the assumption that the military would be expelled from these areas at the end of the war. The Dower and Hobhouse Reports reinforced these expectations, ensuring that the militarisation of Dartmoor was rarely discussed in 1945–51 without reference to Dartmoor's expected future as a national park; after 1951, the orthodox position of the preservationists and the amenity societies was that the military's use of the moor was incompatible with Dartmoor's new status. However, decades would pass before the preservationist harrying of the services and lobbying of government saw the behaviour of the military demonstrate their full recognition of Dartmoor's new status.

For those apprised of the Admiralty's needs it was not surprising that in the autumn of 1946 the services announced their intention to retain use of Ringmoor Down and Penmoor, Scorriton, Rippon Tor, Plaster Down and Laughter Tor in addition to the extended Okehampton, Willsworthy and Merrivale Ranges. Scorriton (11,200 acres) and Penmoor (4,850 acres) were transferred from the War Office to the Admiralty that year, and if the initial proposals had been accepted, the public would have been completely excluded from Scorriton, which would be used for live firing, and excluded from Penmoor when used as a manoeuvre area. Added to the services' hope that training would continue on some 12,500 acres south of the Tavistock–Moretonhampstead road, this raised the prospect of an unprecedented peacetime militarisation of the southern quarter.

The response was swift. The South Devon Regional Planning Committee resolved on 3 October 1946 that 'it is vitally necessary in the interests of public recreation and amenity to preserve Dartmoor for the use of the public and that with this object in view the service departments should not be permitted to utilise any larger part of Dartmoor than they occupied in 1938'. This

quickly became the pragmatic position of those Dartmoor enthusiasts who recognised that national park status would not force the
total removal of the military from the moor. Devon County Council
took a more moderate line on 11 November 1946, resolving that
the military must not be allowed to continue its extended use of
the moor without a public inquiry. Tourism, protecting water
supplies from pollution and upholding the needs of graziers were
their stated priorities. The Dartmoor (National Park) Joint Advisory
Committee, representing Devon's various planning authorities, took
a firmer line, stating that the services needed to give up all but the
Okehampton Range, which might be extended to take account of
new long-range guns.

Hobhouse, working on his national parks report, took up the
question with the Ministry of Town and Country Planning, asking
whether acceptance of the services' proposals would mean 'the
major portion of Dartmoor . . . rendered impossible for public
enjoyment'. Members of the public and representative bodies also
made their voices heard, most significantly through a coordinated
effort in November 1946. Seven MPs, twenty-three local councils,
sixteen associations, six chambers of commerce and trade, eight
societies and fifteen other bodies supported a strongly worded
resolution passed at a public meeting held in Newton Abbot on
16 November. They demanded the prime minister direct the military to give up all areas it had not occupied prior to 1939.
Commoners, cyclists, hoteliers, farmers, naturalists, motorists,
archaeologists, ramblers, railway owners, bird watchers, pony and
sheep breeders, and members of the local hunts all desired the
restoration of the status quo ante.[225]

On 4 July 1947 the services, Devon County Council and the
government reached a provisional agreement, which then became
the basis of a public local inquiry.[226] On land south of the Tavistock–
Two Bridges–Moretonhampstead road the agreement allowed
manoeuvres using blank ammunition but forbade any use that
might interfere with the rights of the public, owners, occupiers,

commoners or statutory undertakers like the water authorities. The Laughter Tor gun site – already controversial thanks to the Forestry Commission plantation skirting the tor's northern flank – was the exception, though firing at night was forbidden and there was some suggestion that permission would be withdrawn if the services proved to be a nuisance. As for Okehampton, Merrivale and Willsworthy, the agreement reiterated the established assumption that the public entered during firing at their own risk – there was no formal exclusion – though there was to be no restriction on the public entering the eastern section of Merrivale (reserved for 'dry training'). Absent from the provisional agreement was any mention of the services' designs on Ringmoor and Roborough Downs, both sensitive locations owing to the importance of Ringmoor's antiquities and Roborough's proximity to Plymouth.

At the public inquiry, held at the Castle, Exeter on 16–17 July, Sir Charles des Forges oversaw the admission of evidence. Representatives of the services gave the impression that they needed only to explain themselves in order for their requirements to be met. Concessions were presented as voluntary and conditional. For instance, both the Royal Marines and the Royal Navy offered to relinquish Scorriton if they were given an additional portion of the Merrivale area, though the inquiry was reminded that they retained the right to conduct 'controlled hiking' – manoeuvres without arms – anywhere on the moor. Amenity organisations were concerned that the provisional agreement did not include the services' plans for Roborough, Ringmoor and Haytor, which in their view undermined the significance of the Scorriton concession. Witnesses who found the presence of the services in a national park abhorrent nevertheless generally accepted they would remain in the northern quarter. The water authorities, worried any extended use might compromise their watersheds, ensured consideration was given to where trucks and other forms of mechanised transport were permitted, but attention was mainly paid to *where* the services might train rather than *what* they might do.

Lewis Silkin received the des Forges report on 8 October.[227] It largely upheld the provisional agreement, urging that training with live ammunition be confined to the north quarter ranges and training elsewhere conducted with the utmost care. Des Forges was particularly concerned by the threat posed by the services to rights of common, the heavy burden of responsibility the removal of livestock before firing placed on sometimes irresponsible junior officers and, presciently, the inadequacy of the system of warning flags, which confused the public by being left flying long after the danger was past. Also troubling was the continuing use of Laughter Tor, the effect the gun sites north-west of Willsworthy had on the villagers of Lydford and the tendency of the services to fix boundaries according to convenient lines of private property rather than actual usage. Ringmoor, Roborough and Haytor hung heavily over the deliberations for it was hard to make firm recommendations when the scope of the services' irredentism was unknown. When the government published its Dartmoor decision later that year the pragmatists could claim a modest victory. The 72,000 acres claimed by the services in the original proposal, slashed to the 43,000 acres of the provisional agreement, was further reduced to 33,000 acres.

In practice, the services were given considerable latitude. Live firing was confined to the ranges of the northern quarter and, in addition to the small hutted camp at Plaster Down and the RAF Gee station at Sharpitor, these were the only parts of the moor over which the services could control access. But uncertainty remained about what was acceptable elsewhere, particularly on Ringmoor and Roborough Downs, both held by the War Office under a wartime directive. Ringmoor (3,000 acres) and Roborough (741 acres) were used for 'dry training' by the Royal Marines and the army, which included the use of blank ammunition and, controversially, the digging of foxholes and the use of tracked vehicles. During discussions in 1949 Devon County Council maintained that the 1947 ruling forbade both outside the northern quarter. The Admiralty, vigorously asserting its rights under the

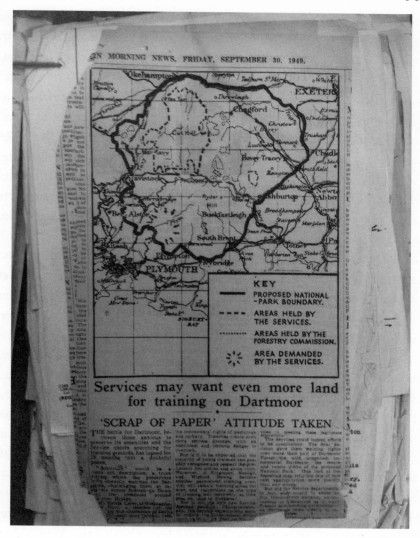

... N MORNING NEWS, FRIDAY, SEPTEMBER 30. 1949.

KEY

—— PROPOSED NATIONAL-PARK BOUNDARY.

- - - - AREAS HELD BY THE SERVICES.

.... AREAS HELD BY THE FORESTRY COMMISSION.

AREA DEMANDED BY THE SERVICES.

Services may want even more land for training on Dartmoor

'SCRAP OF PAPER' ATTITUDE TAKEN

original directive, nonetheless announced that tracked vehicles would no longer be used on Ringmoor and Roborough but only because they had recently decided that the Royal Marines would no longer be equipped with Bren gun carriers. If carriers were needed in the future, their use would be confined to the Merrivale area. But with its next breath the Admiralty threw down the gauntlet. 'The carriers are being replaced by jeeps and trailers, and as the use of light wheeled vehicles has not previously been mentioned, we are assuming that there will be no objection to their use on Ringmoor in place of tracked vehicles. We consider that the "rights of manoeuvre" might reasonably be held to cover the use of such vehicles.' As for complaints about digging trenches on Roborough, the Admiralty was prepared to forgo this 'right' and confine it to Ringmoor if the War Office did the same. The War Office suggested that training might be confined to Ringmoor altogether provided the use of light tracked vehicles was allowed. The Admiralty was less conciliatory. Their Lordships would happily share Ringmoor but were not prepared to give up Roborough. And so it went on.

In March 1950 Devon County Council informed the ministry that the Admiralty's continued use of Roborough was unacceptable and further concessions would be needed regarding Ringmoor. The response from an Admiralty official struggling to come to terms with the political economy of peacetime was that the Ministry of Housing and Local Government might 'now wish to consider whether the time has come when the adoption of a firmer attitude might prove more profitable and dispose of this haggling'.[228] Instead, new procedures dictated that the military's failure to reach an agreement with the planning authority would trigger a public inquiry. Frustrated and angry, the War Office and Admiralty privately reverted to their previous positions on the grounds that Devon County Council had not responded constructively to their attempts to compromise. They would go into the public inquiry asserting their right to train on both sites using tracked vehicles and foxholes.[229]

At the inquiry, held in Exeter on 18 October 1950, Major General J. E. Leech-Porter made a practical, emotive and conciliatory appeal. Roborough and Ringmoor were needed because they were close to military bases, and their utility had been proved by the bold raids men trained there had carried out in Korea. Denying access to either site would have repercussions in Malaya and 'any other place where aggressors are likely to stage another Korea'. Leech-Porter also explained that the informal motto of the Royal Marines was 'Dig or die' and they must be allowed to practise the techniques needed when facing enemy fire. The 'trenches' found so objectionable by the preservationists were really only holes of two by three feet, three to four feet deep and set about five yards apart. Despite the promise of a hard-line stance, Leech-Porter was ready to forgo digging and the use of light tracked vehicles on Roborough if both were allowed on Ringmoor.[230]

Although the local authorities and the amenity and commercial interests generally opposed the use of both sites, their pragmatism and differing priorities meant they did not offer a united front. Devon County Council, Plymouth City Council and Lord Roborough noted the importance of Roborough to the people of Plymouth, saying damaging activity must be confined to Ringmoor Down; the National Parks Commission and the archaeologist Aileen Fox claimed Ringmoor was a more important landscape on account of its size and importance to archaeology. Local concerns regarding livestock converging on the road were heard, as was a piqued representative of Roborough golf club who complained that under the current arrangements only the first, fifteenth and sixteenth holes were exempted.[231] Sylvia Sayer, attending as a representative of the Ramblers' Association, later complained that the chair had implied that the services would get more land for 'damaging training' and no one had been forewarned that alternative sites might be discussed. Her suggestions, brought to the attention of the ministry by the National Parks Commission, were judged 'too small and cannot be combined'.[232]

Notwithstanding some caveats, the services got their way, leading the Dartmoor Preservation Association to lament another 4,000 acres of 'unspoiled Dartmoor' doomed 'to ordeal by tracked vehicle and spade'.[233] For the prescient, however, a less high-profile but still contentious development was as significant. In the years since the US dropped atomic weapons on Hiroshima and Nagasaki in 1945, and knowing that the Soviets were developing a nuclear bomb, defence strategists had been preoccupied by the new apocalyptic threat from the air. In early 1952 the War Office was looking for somewhere to locate a heavy anti-aircraft gun for the protection of Plymouth. Needing a three-and-a-half-acre gun site and a further sixteen acres to accommodate living quarters, ammunition dump and radar control station, the middle of Roborough Down seemed ideal.

A site visit with representatives of Devon County Council left L. J. Watson of the National Parks Commisson thinking the proposed siting would be the 'last straw to those who had hoped to retain intact such unspoilt sections of Roborough Down'. As an alternative, he proposed Commonlane plantation, which abutted the western edge of Roborough and was less visible from Dartmoor. Sayer pounced on the proposals, protesting that the scheme was a War Office ruse to get Wigford Down, a site located between Roborough and Ringmoor Downs. Predictably, the War Office rejected the proposed alternative, unhelpfully proposing another site 500 yards to the east and within the park. Another alternative, according to the records of the National Parks Commission, was rejected by the War Office because Lord Roborough, the landlord, objected; it was thus somewhat dubious that in September 1952, Roborough, alongside Geoffrey Clark, Devon County Council director of planning, was deputised to negotiate with the War Office. In January 1954, a little over a year after the US first tested the hydrogen bomb and a month before its notorious Pacific tests commenced, Housing and Local Government accepted the War Office's recommendation on the grounds that Plymouth needed protection.[234] Global-political developments of terrifying import saw another small part of Dartmoor militarised.

Those same developments also effected the eventual resolution of the protracted squabble provoked by War Department plans to transform the hutted camp at Plaster Down into a permanent barracks for an infantry battalion. Used since the end of the war as a training camp for the Territorial Army, at stake was whether Plaster Down's 'permanent retention' by the services allowed the replacement of huts with brick-built buildings. Government officials maintained their replacement was the 'natural corollary to the retention of the camp' and insisted this be explained to the Dartmoor Standing Committee. Frustrated, the War Department warned Housing and Local Government that 'if the County Council proved difficult about the plans for brick buildings the War Department could probably go ahead without waiting for any clearance'. On this occasion even the National Parks Commission struggled to see Devon's case: Abrahams wrote that 'it is really unarguable that it was always intended that there should be a permanent camp here'. Godsall, in response, claimed Devon had been 'deceived' by the War Office and throughout the controversy maintained that the proposed development was in breach of the 1949 agreement. The Commons, Open Spaces and Footpaths Preservation Society intervened in March 1956 when it transpired that the Whitchurch Commoners had granted the War Office a ninety-nine-year 'lease' at fifty pounds per annum on the additional land needed. Permission was needed from the CPRE, W. H. Williams wrote, and only the Duke of Bedford, as holder of the manorial rights of this part of Whitchurch Down, could grant the War Office the lease it needed. Sayer's tireless lobbying finally awoke the National Parks Commission to its statutory responsibilities and it began to push the War Office for a fuller account of their plans. At a meeting on 24 October the War Office finally revealed what it had in mind.[235]

Plaster Down would be a permanent base for 35 officers, 54 warrant officers and 570 other ranks. It would be equipped with 100 vehicles, including 20 three-ton trucks, 30 small jeeps and 11 tracked vehicles. Five three-storey and seven two-storey dormitory blocks would be built, plus a two-storey junior ranks' club, a two- or three-storey sergeants' mess and officers' mess, each containing between 20 and 30 bedrooms, and miscellaneous other buildings, including married quarters comprising 54 houses suitable for different ranks. Referring to garrison towns in Wiltshire, Clark had not been far wrong in predicting, 'We should find a Larkhill or a Bulford being constructed in the National Park.' Illustrative photographs of barracks at Garets Hey near Loughborough were not reassuring.[236]

The National Parks Commission now began to take the matter seriously. On 31 October 1956 its chair Lord Strang wrote to W. E. Playfair, permanent under-secretary of state at the War Office, advising that the strength of opposition meant they should probably halt their negotiations to buy the site. Strang also wrote personally to Dame Evelyn Sharp, permanent secretary at Housing and Local Government. As Crossman later observed, Sharp had little time for the amenity interest and summarily dismissed Devon's objections, saying they regretted the 1949 agreement. Devon sought an alternative site, looking into whether a local firm of butchers which grazed its stock on Hurdwick Farm near Tavistock was willing to sell. Tavistock Urban District Council thought this a good option, doubtless anticipating the trade the barracks would bring to the town, but Tavistock Rural District Council objected to the transfer of the camp from 'rough grazing' to 'good agricultural' ground. Their suggested alternative was Burnford Down, a 200-acre site to the north of the town that the owner was willing to sell, but Playfair found it unsuitable just as he had found Hurdwick too small. Little more was heard of the plans until November, when the *Western Morning News* suggested the War Office had abandoned its plans and would keep Plaster Down as it was for use by the Territorial Army. On 9 December 1957 Playfair wrote to Strang explaining that following the review of the needs of the armed forces, 'we have now decided to abandon the idea of a permanent Regular Army barracks in the area. The change of plan is of course due to the reduced requirements for accommodating the Regular Army of the future in the United Kingdom.' He went on: 'I should add that we still contemplate that use of Plaster Down as a Territorial Army camp, though even on this there is at the moment no finality. However, I understand that there is no objection to such use – indeed, it has always been accepted.'[237]

Why was Plaster Down not transformed into a new garrison town? Why didn't it become the most significant new settlement on Dartmoor since the establishment of Princetown? Once again

international developments and changing national priorities determined the future of a small part of Dartmoor. The arrival of the
thermonuclear age sparked panic in Whitehall as politicians, civil
servants and the military contemplated the development of
weapons that the Soviets would surely match. Two years later the
Suez debacle confirmed that Britain was no longer a great power.
Britain's humiliation strengthened the Treasury's conviction that
the military establishment had become bloated and inefficient;
soaring costs needing to be curbed. Political pressure to end
conscription from the Labour opposition added to the sense of
crisis.[238] When Harold Macmillan succeeded Anthony Eden as prime
minister he made Duncan Sandys his minister of defence, knowing
he could be relied upon to shake things up.

The 'Sandys Era' commenced with the famous defence white
paper of March 1957, which announced a transformation of Britain's
defence policy on the basis that the country faced little threat of
a ground invasion but 'no means of providing adequate protection
for the people of this country against the consequences of an attack
with nuclear weapons'. Anti-aircraft batteries might knock out a
bomber or two, but Britain's safety required a shift away from its
reliance on conventional military forces towards nuclear deterrence
and an unabashed Atlanticism in which the US would be the dominant partner. Aspects of this analysis were already conventional
wisdom but, as Matthew Grant has argued, 'new in 1957 was the
argument that a deterrent posture allowed concomitant reduction
in conventional forces'.[239] Sandys was resolute that new circumstances made much of the present military establishment redundant. Ending conscription, slashing the British Army of the Rhine,
cutting back the surface fleet, reducing the strength of Fighter
Command and centralising decision-making at the MoD was his
prescription. The number of serving military personnel was to fall
from 690,000 to 375,000 by 1962.

Consequently, on 12 March 1957 the government announced that
the War Office and the Admiralty would release 71,000 acres of

training ground.[240] Besides halting the development of Plaster Down, Merrivale would be released if an alternative training area could be found for the marines. In the event the War Office relinquished the range, but the Royal Marines took over the Merrivale West area, allowing only Merrivale East's 4,000 acres to be released.[241] In April 1958 the War Office informed the National Parks Commission that one of the brigades being transferred to the UK from Germany would return to Plymouth. It was uncertain whether Dartmoor's use would increase.[242]

Relinquishing Merrivale East ended the first phase in the re-territorialisation of the national park. In conjunction with the collapse of the Plaster Down plans, the hold the services had over the moor was significantly diminished, establishing a status quo that would endure for some decades. Britain's long military retreat from victory, exposed by Suez and made policy by Sandys, had played an important role in this story. Dartmoor benefited from the more tolerable equilibrium achieved between military and civilian power by the early 1960s. Public inquiries and the power of the local planning authority in conjunction with Dartmoor's new status as a national park required the services to justify their use of the moorscape and could not simply presume that what they held was a permanent part of the defence estate regardless of immediate needs.

The preservationists, charged by their maximalist agenda, regarded these reductions as largely symbolic. Sayer's dismissal of the Merrivale East release as an empty gesture on the grounds that it was already little used was consistent with the DPA's focus on the behaviour of the services rather than the extent of their territorial rights. Since the early 1950s they had accepted that harrying local commanders would not alter the settlement and for a time the preservationists had taken a more conciliatory approach, choosing to advise rather than hector. Constructive relationships were helped by Captain Taylor's attendance at DPA meetings and the congenial and cooperative approach of commanding officers like Major General R. F. Cornwall and Major General C. L. Firbank. Major General E. K. G. Sixsmith, Firbank's successor in late 1954, worked to maintain cordial relations. Small changes helped. An objectionable concrete flagpole on Yes Tor was replaced with a more discreet structure; the corrugated-iron shelters on High Willhays and Yes Tors were built up with moor stone and roofs turfed or painted in order to make them less conspicuous; and servicemen were warned not to drive on the open moor except in training areas.[243] Aesthetic concerns like these were far from petty. They reflected the degree to which the services respected the cultural value the nation attached to Dartmoor's particular beauty and the rights of Dartmoor's many users.

Domesticating the services proved difficult. Large-scale multinational training events like the Sea Trout exercises of 1–4 March 1955 transgressed established territorial boundaries and indicated how easily the services could claim exceptional needs. More antagonising was the consistent failure of the services to clean up after conducting exercises, the physical damage shelling caused the surface of the moor, including tors and antiquities, and the gradual development of the military infrastructure in the northern quarter. Godsall, clerk to Devon County Council, often took up individual questions with the services and politely dealt with responses that were tardy at best. Was the army willing to repair a dangerous footbridge across Walla Brook? No, because there was no public right of way. Was the army prepared to remove a disused railway track that had been part of a wartime anti-tank training range at Holming Beacon, Merrivale? Yes, as part of the government's tidying up the countryside initiative.

Less satisfactory was the army's irresponsible approach to unexploded bombs on Blackslade Common dating back to wartime training exercises. This was raised by the Dartmoor Preservation Association in an angry letter of August 1959, prompting Godsall to remind Andrews at the War Office that this needed to be taken seriously. Three months later Godsall wrote to say that if the bombs were not being removed the decayed warning signs needed to be replaced. A month later Andrews assured Godsall the bombs would be dealt with that summer but complained that it was difficult to do anything about renewing the signs because it was private land. Wearily, Godsall expressed his surprise that they were 'talking about . . . clearing . . . unexploded missiles so long after the end of the war'. With the issue still unresolved in July, Godsall mordantly informed Andrews that the Dartmoor Standing Committee assumed the 'War Office will accept full responsibility in the event of any unwary holiday-maker being blown-up in the meantime'. In December, almost eighteen months after the original complaint, it was Godsall who let Andrews know that the chief constable had informed him that the army had finally cleared the common.[244]

Godsall's patience was surprising given the broader context. On 15 April 1958 Henry Whitfield, the son of a schoolmaster, was killed by shrapnel at Cranmere Pool when out hiking with his parents. Christopher Soames, secretary of state for war, summarised the event in the Commons: 'Firing had been in progress all morning. Visibility was good and normal safety precautions were in operation. Early in the afternoon a family of three people entered the danger area and one of them, a schoolboy, was killed by a mortar shell.' Under pressure from MPs Joan Vickers (Plymouth Devonport), Frank Hayman (Falmouth and Camborne) and Henry Studholme (Tavistock), Soames explained the warning system. Noticeboards were placed on every track leading to the danger area; during firing large red flags measuring twelve feet by six feet were flown from the highest tors and five wardens conducted patrols under the supervision of a retired colonel. These arrangements, Soames asserted, are 'as comprehensive as we can make them'. Dismissing the suggestion that the tendency to leave the flags flying meant they were not a reliable indicator, Soames resorted to sarcasm: the flags are 'very large and very red'.[245]

As is often the case in politics, media pressure ensured the initially complacent response of a minister gave way to a more comprehensive examination of the issue. Debate in Whitehall and the media, stimulated by the inquiry into Whitfield's death, raised doubts about the accuracy of the DANGER AREA marking on maps and the positioning and visibility of the warning notices and red flags, especially when the weather was bad. The old complaint that the flags were not raised and lowered with sufficient assiduity proved justified and the War Office, determined to counter the negative publicity, mounted a determined public relations initiative. Abrahams was impressed by the appearance of Captain E. L. Carter, a War Office land agent, before the National Parks Commission, chummily writing to Andrews, 'I do wish he was able to attend more often, as it does make a difference when the representatives of the War Office are prepared to come and be pilloried.'[246] Carter also attended the Dartmoor Standing Committee in Exeter on 16 September, where he agreed that the boundaries of the ranges be marked with evenly spaced red and white posts. New regulations soon followed and a list was drawn up of nearly sixty local individuals and officials who were to be informed of firing times, including farmers, local postmasters and -mistresses, the police, pub landlords, Devon County Council and the *Western Morning News*. A leaflet was drafted for wide distribution. Godsall approved the proof and requested 2,000 copies for distribution and a further one hundred for display at various locations in the county. Copies, he told Andrews, should be sent to the Ramblers' Association, the Youth Hostel Association, the Boy Scouts, the Girl Guides and sundry other organisations.[247]

P. G. Stokes of the Dartmoor Rambling Club raised a difficult question. He wanted to know the legal basis for forbidding members of the public from accessing the ranges during firing. This went to the heart of the difficulty the services had in the national park. Andrews was unable to offer a clear answer, saying that persons with a legal interest in the land had given the War Office unrestricted use when firing and that there was no established right of way. Privately, the War Office was unsure whether attempts to forbid

PROOF
WALKING ON DARTMOOR

Dartmoor National Park contains 200,000 acres, mainly open and ideal walking country. But precautions must be taken in the northern part where the Services have training areas over some 30,000 acres.

These training areas (marked on the sketch map overleaf) are:—

    A. OKEHAMPTON.

    B. WILLSWORTHY.

    C. MERRIVALE.

The boundaries of these areas are marked, on the ground, by a series of red and white notice boards and red and white posts. Red flags are flown at certain points shown on the sketch map when it is forbidden to enter the training area, but the red flags are not necessarily on the boundaries. (Red lights are shown at night).

Times of firing are advertised in "Western Morning News" and the "Express and Echo" every Friday. Notices and more detailed maps can also be seen in neighbouring police stations and post offices. The dates and times of firing can always be obtained by telephoning the nearest police station or post office.

WHEN FIRING IS TAKING PLACE IT IS FORBIDDEN, AND OF COURSE DANGEROUS, TO ENTER THAT PARTICULAR RANGE AREA.

At other times it is quite safe, **PROVIDED THE WALKER REALISES THAT ANY STRANGE OBJECT ON THE GROUND SHOULD NOT BE TOUCHED OR TAMPERED WITH.**

There is a training area near Yelverton but no live ammunition is used there and it is safe at all times. If firing is heard in the areas shown on the map when the flags are not flying it means that blank ammunition is being used and the walker may proceed.

The sketch map should be used in conjunction with whatever map the walker is using to identify the tors carrying the red flags. It is only intended as a guide.

CRANMERE POOL *is in the heart of the Okehampton range.*

PROVIDED YOU DO NOT PASS THE PERIMETER MARKED WITH RED AND WHITE NOTICE BOARDS AND POSTS WHEN THE RED FLAGS ARE FLYING OR RED LIGHTS BEING SHOWN, YOU CAN WALK ON THE MOOR IN SAFETY.

Copies of this leaflet may be obtained from the local District Council Offices.

[SEE MAP OVERLEAF.

(8911). Wt. 45998/8468 10,500 (2 sorts) 10/59 F.B.S. Gp. 999/147

access clashed with the public right to fresh air and exercise on Dartmoor's open land. Stokes pointed out that the Whitfield inquest had concluded that it was not unlawful to pass the perimeter, and it was true that the War Office had dropped the counter-charge of trespass in response to the father's compensation claim. Given that the case was sub judice, Stokes accepted that the issue was left for the time being but when new warning posters were designed in 1963 Andrews admitted that local by-laws meant they could not forbid but only strongly discourage access.[248]

DANGER

THIS IS THE BOUNDARY OF

OKEHAMPTON ARTILLERY RANGES

DO NOT TOUCH ANYTHING

It may EXPLODE   You enter at your **OWN RISK**

WHEN THE **RED FLAGS** ARE FLYING ON

ST. MICHAEL'S BUNGALOW

BLACKDOWN   WATCHETT

HALSTOCK   STEEPERTON TOR

YES TOR   HANGINGSTONE HILL

KITTY TOR   RATTLEBROOK HILL

FIRING IS TAKING PLACE AND IT IS THEN

DANGEROUS

TO PASS THIS NOTICE BOARD

The Cranmere fatality was a turning point in the development of Dartmoor National Park. Just as the reaves territorialised ancient encounters with the moor, dividing and subdividing the landscape, so encounters with the northern quarter are now shaped by the red and white boundary poles that mark the limits of military authority. Crossing into the training area can still feel like asserting a right, and the feeling of having entered somewhere 'other' never fully dissolves nor does the fascination exerted by the evidence of military activity: it is easy to imagine a latter-day William Crossing weaving this latest layer of Dartmoor activity into his palimpsest. And yet to read the red and white poles as an assertion of military authority would be to entirely misunderstand their origins. With a horrible jolt the Cranmere fatality forced the services to come to terms with the new political realities their sense of entitlement and arrogance had allowed them to disregard. It was a chastening experience which made clear they had no absolute rights in the park but were engaged in a continual process of negotiation and accommodation. Resources had to be made available if the public's safe access to all the territories under their partial control was maximised: military institutions that prided themselves on their internal discipline had to come to terms with civilian discipline and expectations.

Civilian interests were determined to seize the opportunity. Captain Carter left the Exeter meeting having not only agreed to the erection of the boundary poles but also having undertaken to see that litter was cleared away, wires removed and trenches filled in. The Disfigurements Sub Committee of the park committee became more demanding; the DPA began to catalogue vigilantly the services' sins of omission and commission; and sympathetic political allies like Frank Hayman and Lord Chorley, the Labour peer who had sat on the Hobhouse Committee, ensured the issues were discussed in parliament. Lord Jellicoe, junior minister at Housing and Local Government, not wanting to be seen to be prompted into action by Chorley, ordered 'mopping-up operations' on Dartmoor. Brigadier Acland, given five months to oversee the operation, was unofficially and begrudgingly guided in his task by a list of DPA complaints. Smashed-up vehicles, derelict army trucks, abandoned tanks and unexploded or buried bombs and missiles were dealt with. Acland's effectiveness allowed Jellicoe's public relations push to culminate in a speech at Moretonhampstead in May 1962, in which

he insisted that the services would stay on Dartmoor but recognised the problem of 'military muck' which 'of all dereliction . . .
is just about the dreariest'. Acland's claim that much of the litter
was civilian elicited from Abrahams another chummy response,
this time making light of the services' most determined antag-
onist – 'dear Sylvia . . . never does worry much about her facts!'
'She tells me that she has just become a grandmother and that
her grand-daughter is very attractive,' he wrote. 'I wonder if my
successor will have to deal with her in 30 years time.' This was a
little duplicitous, for when V. G. F. Bovenizer of the War Office
wrote to justify the continuing presence of the services on
Dartmoor in response to the latest complaints, Abrahams
forwarded the letter to Sayer for a response. She found Bovenizer's
claim preposterous that 'nature rapidly assists in recovery and
permanent damage to the natural beauty is not caused by military
activities', lamenting the disfiguring 'craters, trenches and ruts
caused by military use' and the 'damage and fracturing' done to
tors and prehistoric menhirs used as targets.[249]

Much to the irritation of the less energetic Dartmoor Standing
Committee, the DPA's influence was evident in other ways too. It
provided the template for a memo of May 1962 calling for the
services to be out of the park by 1970 jointly submitted by the
Standing Committee on National Parks, the CPRE, the Commons,
Open Spaces and Footpaths Society, the Ramblers' Association, the
YHA, the Cooperative Holidays Association, the Dartmoor
Rambling Club and the Dartmoor Rangers. Two years later Sayer
and Somer Cocks submitted to the National Parks Commission a
new statement on Dartmoor's use by the services. They argued
that Dartmoor was excellent for 'adventure-training for the young
soldier, in reasonable numbers and under enlightened supervision',
but 'the military roads and installations, the random military
vehicle-driving, the bombing, firing, digging and helicopter flying
would go, for these have no part in, and are inimical to, adventure
training of the right kind in wild country'. Though voices at the

commission thought 'Lady Sayer's latest effort' showed a naive view of military training, her broader points would expect the commission's 'approval and backing'.[250]

This was part of a much larger campaign aimed at exposing the superficial nature of the improvements prompted by the Cranmere death. The DPA's list of complaints was extensive. Use of Burrator Reservoir by the Marines was a 'new incursion', raising particular questions about hygiene (could the public facilities cope?); plans to construct a hardened track at Baggator Gate on the ancient Lich Way were deemed illegal; Conies Down had been used for target practice, causing damage to natural features, a stone row and hut circles ('inexcusable vandalism', 'outright lawlessness'); pits had been dug in White Barrow, which was about to be scheduled an ancient monument; throughout the northern quarter tracks were being metalled with crushed stone and surfaced with lighter stone, making semi-permanent roads; motorcycle training – long highly contentious – was churning up great stretches of the Okehampton Common; and damage was caused by heavy bulldozers being used to fill in craters.[251]

The DPA did not rely on words alone.

Particularly offensive to the preservationists was the failure of the services to prevent civilian motorists from using their new roads. Much of this concerned whether civilians had the right to use roads branching off the Okehampton loop that cut a little into the moor. A photograph published in the *Western Morning News* in December 1959 of a car parked near the top of Yes Tor attracted much attention. The DPA objected to the effect improved roads had on the surface of the moor and how they changed the experience of the moor. 'What thrill,' the association asked in 1959, 'in climbing to the top of Yes Tor, when around the corner one may come upon a car with its radio blaring and its occupants immersed in *The News of the World*?'

If the constant distraction of the transistor radio and the vulgarities of populist newspapers signified to the DPA all that was wrong with modern life, this question was not simply motivated by snobbishness. In a written answer Soames explained that the military had only improved a track that had always been in existence. Making Yes Tor easily accessible by Land Rover meant the warning flag could be lowered as soon as possible after firing. Using the Cranmere death to justify development of Dartmoor's military infrastructure was perhaps a low blow, but more significant was how the old question of whether the agreements of 1947 and 1950 allowed the services to improve existing infrastructure in ways that made it more permanent was related to new questions. Were the services responsible for preventing civilians from using roads that facilitated access to the ranges? And if they were responsible, on what legal basis could access be prevented? If the military could drive its Land Rovers along the improved roads, why couldn't civilians drive their cars? Thus the new generation of car-owning tourists raised difficult and paradoxical questions about governance and access in the north quarter of the moor that would go unresolved for a decade.[252]

On 29 July 1966 representatives of several ministries met to consider 'the increasing volume of Ministerial correspondence on the topic of military training within National Parks and especially

Dartmoor'. Granted that responsible officials needed to be more au fait with local issues if they were to deal efficiently with correspondence, the surviving notes from the meeting nonetheless suggested ministers needed to respond more decisively to issues raised by the National Parks Commission. In particular, the legislative framework guiding the management of the parks allowed a more proactive response to issues highlighted in the commission's annual reports. 'Recreational management plans', for instance, might draw attention away from the military's activities by making non-militarised parts of the parks more attractive to visitors. Perhaps the management of the park might be rethought in 1970 when the duchy's War Office and Admiralty leases came up for renewal?[253]

If nothing very concrete came of this suggestion, the renewal process nonetheless brought the MoD under unprecedented pressure to justify its use of Dartmoor. The opening salvo in the campaign was a thirty-three-point petition presented to the duchy by the CPRE, the Commons, Open Spaces and Footpaths Preservation Society, the Ramblers' Association, the YHA, the DPA and the Dartmoor Livestock Protection Society. Asking with one voice that any request for a renewal be refused, they could hardly be ignored. If the gist of the case against renewal was familiar enough, recent developments in training regimes gave the case new force. 'Widely damaging and disruptive use of the land,' the petition argued, was caused by 'the introduction of new weapons, the mechanisation of the services with the widespread use of heavy vehicles, the requirement for massive military road-making, the use of low-flying aircraft and particularly of helicopters, the dropping of one-ton loads by parachute, etc.' Particularly effective was point seventeen, which held up the National Trust as an exemplary landowner. As the owner since 1964 of part of Ringmoor Down – designated by Nature Conservancy as a site of special scientific interest – the Trust had refused the military training licences permitting hole-digging and vehicle-driving.

Acutely aware of the delicacy of its position, the duchy immediately sought advice from the MoD regarding its future needs. Everything, it seemed, was contingent on whether the Royal Marines remained in Devon. Recent investment in barracks and training centres meant this was likely but there would be no certainty until 1970. While the duchy was advised to offer the petitioners a holding response, the MoD made preparations for a meeting at Taunton on 23 October 1968 when the minister and the military would make their case for the continuing use of Dartmoor. Much Whitehall effort was invested in preparing for this meeting. Who should be invited? How should they respond to hostile questioning? A long list of interested parties was drawn up, including several government departments, the licensees, the county authorities, the local authorities, and interested civic associations, including all who had signed the duchy petition plus the National Farmers Union, the Country Landowners' Association, the National Trust, the Parish Councils Association, the Devonshire Association and the Dartmoor Commoners Association. These latter groups could be dealt with through the CPRE, but it seemed advisable to invite them.[254]

Although the MoD's confidence was boosted when the Treasury said it supported the renewal because relocation costs would be very high, the pressure to offer concessions was still great. Anthony Greenwood, minister of housing and local government, accepted the MoD's case in principle but looked for a reduction in the intensity of firing, the use of helicopters and the number of acres held, and asked whether they really needed to retain the Rippon Tor Rifle Range and if Roborough or Ringmoor could be given up. Initially, the MoD saw little room for manoeuvre on any of these matters, but a confidential internal communication from Southern Command highlighted a very significant development. From 1 April 1969, Dartmoor would be downgraded from a 'major training area' to a 'local training area' and thereafter principally used only by units based in south-west England. The military's knowledge and experience of the moor, bolstered by the classes already given by Lady Fox on Dartmoor's antiquities, should mean more sensitive use in the future. Although this proposed change in use was pitched in terms of evolving strategic needs rather than concession, there is little doubt it was partly a response to the minister's expectation that Dartmoor would no longer be seen as a free for all, becoming a more carefully controlled environment. Managing concession, of course, was highly political. Another internal MoD minute urged that if the rifle range at Rippon Tor could be given up, it should nevertheless be retained for the time being as a bargaining chip.[255]

A meeting on 26 February 1969 left the MoD in little doubt that the Duchy of Cornwal wanted the services off the moor and was not prepared to sign another twenty-one-year lease. Much the same was being said by the Countryside Commission and the National Trust: they accepted the land couldn't be released immediately but a twenty-one-year lease was unacceptable because it would create the impression of permanence when the intention must be an early exodus. MoD sweeteners, like the offer of tighter controls on Ringmoor and Roborough and clearer marking of antiquities on Merrivale, made it clear that uncertainty over the renewal of the leases meant the MoD was finally taking its responsibilities seriously. When in March these promises of good behaviour were presented to the Dartmoor Standing Committee, Devon County Council and local MPs it simply became another opportunity for further complaint. Lord Foot of Buckland Monachorum, a Liberal life peer and influential Dartmoor voice, insisted that nothing could change the fundamental incongruity of the military presence in the national park; Turnball, Devon's planning officer, regretted that the proposals had been made without consulting his office; and few were impressed by undertakings the amenity societies had been demanding for decades.[256]

The duchy remained nervous. It recognised that the application had the backing of the government but felt it needed to know what conclusions the Countryside Commission had reached and whether it was satisfied that sufficient consultation had occurred. As the duchy reminded the MoD, the Countryside Commission had 'a statutory responsibility for National Parks and is therefore the governmental body to which all the other bodies concerned with amenities naturally turn'. If satisfied that a full consultation had occurred, the duchy would renew the lease for five years. Despite MoD concerns that a short lease would inhibit investment in training facilities and encourage their opponents by creating the impression of impermanence, an emollient response was required. They admitted the commission opposed the renewal but said they were prepared to publicise this when the decision was announced. Expressing regret that it continued to need Dartmoor, the MoD pledged to work closely with the Countryside Commission, the county council and its Dartmoor committee. The *Western Morning News* got wind of these exchanges and reported that the MoD's push for a twenty-one-year lease with possible breaks at seven or fourteen years was running into trouble.[257]

In the meantime, Devon County Council, the Dartmoor Standing Committee and the Countryside Commission met to consider their position. Their conclusions were uncompromising. There should be phased withdrawal by 1975 and an immediate scaling back of activity. Clearly rattled, E. H. Palmer at the MoD took up these demands with Southern Command. He wanted to know the implications of restricting Dartmoor's use to units based in Devon and Cornwall and whether the Rippon Tor Rifle Range and Plaster Down could be released. Though unconvinced that much could be done about the use of helicopters or the new roads, Palmer did ask whether heavy lorries or digging on the open moor really was a big problem. Above all, he hoped Southern Command could agree to no firing at weekends between Easter and September and none at all in August. The response was somewhat encouraging. Rippon could not be released, but Plaster Down would probably not be needed after 1972, though the implications of giving it up would need to be considered then. Limiting Dartmoor's use to local units could mean no firing at Willsworthy in August and most weekends and, Southern Command noted, having bowed to earlier pressures, there was already little weekend firing on Okehampton and Merrivale between Whitsun and the end of September. Agreeing to no firing at all in August should be possible, though helicopter use could not be reduced, for this was essential to the training of marines based at RNAS Culdrose in Cornwall and RNAS Yeovilton in Somerset; digging and the use of heavy lorries was equally necessary.[258]

Accepting that this was the best offer the MoD could make, and given uncertainty about future deployments, Palmer proposed they push for a seven-year lease with the promise of an inquiry after five years, albeit on the understanding that they were not anticipating a full withdrawal. With the full force of government now behind the MoD, the Countryside Commission, the Dartmoor Standing Committee and the National Trust accepted a fourteen-year renewal, with a full inquiry at five years and a break clause at seven. The duchy, as convention demanded, followed suit. Local press comment registered a defeat. The *Western Morning News* reported that operations would be curtailed during the holiday season and over most weekends but it played down the review and break clauses and criticised the duchy for 'giving way'. The DPA reckoned the MoD, with the backing of the cabinet, had 'bulldozed' the duchy into acceptance, though it loyally insisted that Prince Charles and Prince Philip could not have been happy with the outcome.[259]

Hindsight suggests that with the renewal of the leases in 1970 the services overcame the greatest challenge they have faced on Dartmoor since the creation of the national park. Despite the Dartmoor Preservation Association's insistence that it would not give up the struggle, the process had comprehensively demonstrated that the government considered the training ranges essential to the national and regional interest. The services, and the Royal Marines in particular, were embedded in the West Country, and only a seismic shift in Britain's strategic military position was likely to change this. This sense of permanence was reinforced in 1973 with the publication of the Nugent Report of the Defence Lands Committee. Nugent recommended that 2,400 acres of Dartmoor upland be freed from firing and that Plaster Down and Rippon Tor be released, but apart from these long-anticipated easements, he affirmed the status quo. John Cripps, chair of the Countryside Commission, publicly criticised Nugent's failure to investigate alternatives sites and demanded a full public inquiry. The principle of a national park, he maintained, should have been upheld and the consequences grappled with. Instead, the government adopted Nugent's recommendations in the defence White Paper of August 1974.

The amenity societies got a non-statutory public inquiry. Presided over by Lady Evelyn Sharp, it considered whether the training needs of the army and marines could be met elsewhere 'without unacceptable loss of efficiency' and what changes, if any, should be made to the extent and use of the Dartmoor holdings. Its particular remit owed something to the National Trust, which had informed both the MoD and the Department of the Environment that it would not renew its licences until it was satisfied that alternative sites outside the national park had been fully explored. When Sharp sought clarification as to whether she could recommend the transfer of units to other parts of the UK, John Silkin, minister of environment, made it clear that the government 'did not intend that the need for the army and the Royal Marines units in the south-west to retain their present bases and to having training facilities in the area should be open to discussion'. With that clarified, hearings were held at County Hall in December 1975 and alternative training sites were considered the following May.[260]

Sharp's report concluded that there could be 'no doubt that, on Dartmoor, military training is exceedingly damaging to the national park'.[261] This statement is not as simple as it might seem. Sharp did not claim that military training was exceedingly damaging to Dartmoor but that it was incompatible with its status as a national park. As Sharp explained, the military presence affected access, caused noise pollution and damage, spoiled scenery with its warning signs, flags, notices and observation posts, and 'perhaps worst of all the heart of the northern moor is invaded by cars, both military and civilian, using the military road and tracks'.[262] In particular, she followed Sayer in criticising access restrictions to the 'incomparable upper Tavy valley and Tavy Cleave' in the north quarter and was evidently uneasy that the services occupied the most dramatic and challenging parts of the moorscape. By recommending that firing might be transferred from the Willsworthy Range to Davidstow on Bodmin, it was clear that to her mind the rights of the MoD as freeholder were outweighed by the national park principle.[263] Sharp's second major recommendation, that training on Ringmoor should be reduced, owed much to recent archaeological advances, particularly those made by Andrew Fleming, then a lecturer at the University of Sheffield. As Fleming explained with respect to Ringmoor:

Dartmoor offers us the opportunity to visit and study the best preserved prehistoric settlement pattern in Europe. Because the Bronze Age peoples of Dartmoor built their foundations of stone, and because the Moor was sparsely occupied in later periods, there is now an archaeological landscape rare in Britain and even rare in Europe as a whole. To ceremonial sites like barrows, cists and stonerows may be added settlement sites varying from isolated huts through enclosed farmsteads to sizeable villages. To this richness we have recently been able to add reaves – low, tumbled walls which have been shown to be equally of the Bronze Age date. Sometimes there are sets of single reaves, perhaps demarcating grazing areas; sometimes we find groups of parallel reaves, representing a remarkable development of land surveying techniques, and presumably land apportionment, in the second millennium BC. The existence of substantial traces of dense settlements and contemporary land boundaries is of extreme rarity. When a site is destroyed or damaged on the Moor, we are not simply seeing the loss of just one of hundreds of similar examples; we are watching the gradual erosion of a rare landscape of incalculable use both for scientific studies and for the recreation of a public which is becoming more and more aware of our archaeological heritage.[264]

Evidence presented by Nature Conservancy accounts for the emphasis Sharp placed on the damage caused to Dartmoor as a national park. Explaining that its concern was the conservation of flora and fauna, geological and physiographical features, Nature Conservancy insisted that it was not concerned with landscape or visual amenities, or with ancient monuments and other human artefacts. Dartmoor's ecological importance was incontestable. It hosted national nature reserves and twenty-one sites of special scientific interest, the largest of which was the North Dartmoor SSSI, covering much of the Okehampton and Willsworthy Ranges. However, by claiming that eighty years of military use had 'not significantly damaged the importance of the land from a nature conservation point of view' Nature Conservancy turned conventional thinking on its head. In general, it argued, damaged or depleted natural flora and fauna quickly re-establishes itself, and on Dartmoor there was no evidence that animal or plant species had materially suffered as a consequence of military training. Indeed, military training brought 'indirect beneficial consequences'. Thanks to the compensation payments offered to commoners, training helped 'maintain the traditional grazing regime on Dartmoor . . . generally favourable to nature conservation'; tracked vehicles had not, as claimed, caused bog regression – which was happening anyway; and by inhibiting the recreational use of the moor, training might 'have prevented excessive disturbance or wear and tear' from pony trekking and other activities. The old DPA complaint that the land was damaged by shelling was downplayed:

no more than 0.1 per cent of the whole range was cratered, and Nature Conservancy was satisfied by the obligation in the standing orders to rehabilitate the landscape after use. In essence, Nature Conservancy opted for the status quo, believing intensive afforestation or agricultural development rather than the military posed a greater threat to the ecosystem.[265]

Over the course of her long and brilliant career in Whitehall Sharp had been a passionate advocate of local government and planning. Confronted at Exeter by the bitterness and division that animated local politics when land use was in dispute, Sharp chose not to emphasise local needs in her conclusions. She showed little regard for the many local authorities and commercial interests appreciative of the economic benefits the services brought to the region or the preference for the military over tourists expressed by the Dartmoor Commoners Association.[266] Nor did she make much of the Nature Conservancy argument. Instead, although concluding that her terms of inquiry meant that she could recommend 'no change in the defence land holdings on Dartmoor', she argued that the MoD's future training needs should be evaluated in the light of the powerful case made for their termination on Dartmoor.[267] Highlighting the national park's scarcity value, she seems to have come to share Lord Foot's view that at stake was 'a basic conflict of *national* interests'.[268] Dartmoor was small, access was already restricted, and the 'wild and rugged scenery' of the northern quarter could not be found 'on this scale' elsewhere in England south of the Peak District.[269] Almost thirty years after the creation of Dartmoor National Park, Sharp's report found the case for its designation more than vindicated.

Sharp followed up the submission of her report with a brief note highlighting the importance of her Willsworthy and Ringmoor recommendations. The lord chancellor, uneasy about the proposed transfer from Willsworthy to Davidstow, asked what would be gained given probable local opposition. Peter Shore and Fred Mulley said they could not drop Sharp's 'most important amelioration' for they had to give the Countryside Commission something and wanted to avoid any criticism that the whole inquiry was 'a meaningless charade'. Sharp's suggestion that the loop road be closed to the public and policed by the army was not accepted, though her other recommendations were upheld by the government, including the need for regular consultations between the armed forces and the Dartmoor National Park Authority and the possible transfer of artillery and mortar firing from Dartmoor to Salisbury. A note dated 21 October 1977, addressed to the secretary of state for defence from the parliamentary under-secretary of state, threw these deliberations into sharp relief: 'The conclusions and recommendations of Baroness Sharp's inquiry into training on Dartmoor are at E34. I do not believe that it will cause much difficulty for the MoD. Its suggestions, which are put forward in a rather rambling way, seem fairly minor, favourable for the most part to MoD, and couched in terms which give us considerable flexibility to choose our response.'[270]

Conserving a Resource

To some sifting Whitehall evidence is intrinsically interesting; to others it appeals little, resembling an elaborate soap opera, but can provide valuable lessons in how government works. Tracing the genealogy of a decision tells much about how representatives of local needs, sectional desires and the national interest interact and compete for primacy in the political process. Planning decisions do not as such require comprehensive knowledge of Dartmoor but access to a wide range of representative and expert opinion and a good understanding of the statutory framework. The civil servant's role has been to assess the varying desires of interest groups in the light of the legal powers and obligations at the disposal of the minister and the government.

Dartmoor planning decisions were subject to the statutory framework provided by the National Parks Act of 1949 and later amending legislation; however high-minded the ideals that shaped the original legislation, the advocacy role of the National Parks Commission and the subordinate role of the Dartmoor committee of Devon County Council reflected the limited powers conferred by the act. The commission could take the initiative, but in practice it had limited resources, was distantly located from the parks at a grand address in London, and tended to exercise its responsibilities only when harried by the local pressure groups it was obliged to represent to government. In essence, the 1949 act asked ministers to decide if a reservoir could be built, a forest planted or a military licence renewed without damaging Dartmoor's amenity value. If the development would undermine amenity, the minister had to decide whether the development was sufficiently important, the damage sufficiently light, and the alternatives sufficiently explored and found wanting before making a decision. Recognising the ad hoc and subjective aspect of these processes is relatively straightforward; more difficult to recover fully is the vague but palpable power of political pressure.

To make sense of these highly fraught and protracted processes, it is necessary to recognise that the national park designations subjected some of the least valuable rural landscapes in England and Wales to closest control. This diminished the resources available to local authorities and challenged the operative assumptions of the water undertakers, the foresters and the military. Conflicts over Dartmoor's use were intensified by the tendency of any proposal to raise questions about land use throughout the region. In particular, the blasé attitude of the amenity societies to sites outside the park offended people who were economically and emotionally invested in other Devonshire and Cornish landscapes. Wide swathes of public opinion instinctively opposed using agriculturally productive land to

meet rising consumer and industrial demand when moorland was
suitable, the uplands seemingly plentiful and good farmland scarce.
Car bumper stickers in the 1970s demanded a return to Swincombe.
For those whose interests were largely – and legitimately – practical
the old notion that the moor was a barren waste proved resilient.

Part of the difficulty arose from the fact that the national parks
were designated on the grounds that they were part of the national
patrimony, but the ad hoc constraints imposed on their development relied more on the effectiveness of local activists than efficient
government oversight. Land within and outside the national parks
had no intrinsic or absolute value beyond that assigned to it by the
cultural politics of any given moment in time. In terms of the
practical political calculations underpinning any decision, this made
the amenity value of Dartmoor or any other landscape proportional
to the capacity of its champions to mobilise support and make
their case in terms of the statutory framework. Despite the national
interests at stake, this meant that conflicts over land use often
became localised and personalised. No single individual was more
influential than Sylvia Sayer, but her prominence, notwithstanding
her undoubted personal qualities and extraordinary determination,
reflected the weakness of the National Parks Commission.

Reservoirs were built, conifer plantations established and the
army maintained extensive rights on the moor, but the record of
Sayer and the preservationists was better than she thought. Much
she found objectionable, like the creation of new car parks, public
toilets and extensive signage was near inevitable. Not only did the
1949 act encourage the development of the park's tourist infrastructure, but pressure from increasing visitor numbers was a predictable
consequence of post-war prosperity. What the archival traces left
by the many campaigns Sayer orchestrated reveal is the effect
lobbying had on the decisions reached. Delaying tactics, including
pushing for a local public inquiry, ensured proposals endured much
closer scrutiny, and if Swincombe was the only outright preservationist victory, every other decision that looked like a defeat for the

amenity interest was conditioned by their lobbying. The Forestry Commission's expansionism was checked and more sensitive planting introduced; after the Meldon fight another reservoir on Dartmoor became almost unthinkable; and the military, having lost its right to treat Dartmoor as its private demesne, gave up large parts of its wartime estate, curtailed its firing and learned to clean up after itself. Perhaps the North Hessary television mast does blight the Dartmoor panorama, but its exact positioning, which minimised damage to the fabric of the tor and has left its restoration more plausible, owed much to DPA lobbying.

When Sayer retired in 1973, having chaired the DPA for twenty-five years, nobody thought the management of the national parks was satisfactory. The legislators of 1949 had not anticipated government becoming so burdened by questions arising from developments within the parks, while experience had taught the preservationists that the amenity principle did not provide an adequate defence against the development and territorialisation of the upland. The indifference of many visitors to the concerns of the preservationists did not help. Forestry Commission plantations, private forests and reservoirs managed so they became beauty spots in their own right, were regarded with affection; restricting firing significantly reduced the effect military training had on access to the most forbidding part of the upland, weakening the case against the renewal of training licences. More significant, however, was the challenge coming from the new politics of nature conservation. Ecological thinking created new ways of thinking about the landscape that exposed the shortcomings of the preservationist agenda, rendering emotive arguments calibrated to contest the 'rational' and 'quantifiable' claims of their traditional opponents increasingly null. As Marianna Dudley has shown, the defence estates proved agile, quickly responding to new concerns by publicising in their magazine *Sanctuary* the environmental benefits that stem from their occupation and management of land. If much of this was fortuitous and some of it 'greenwashing', *Sanctuary* still cannily

reflected the shift in emphasis away from preservation to nature
conservation.[271]

Two legislative changes reflecting the new thinking affected the
national parks. Harold Wilson's Labour government passed the
Countryside Act in 1968. A classic example of well intentioned
social democratic legislation, the act sought to improve the working
of the 1949 act by enhancing the capacity of local authorities to
protect the countryside, agree new public access agreements and
improve facilities for 'enjoying' the countryside. Conservation here
denoted conserving natural beauty and amenity rather than ecosystems, but by replacing the National Parks Commission with the
Countryside Commission, Wilson's government questioned the
segregationist attitude of preservationists whose anywhere-but-the-
park outlook accommodated neither ecological realities nor the
amenity needs of millions of people whose first priorities were
local parks and what might be called suburban countryside. Of
greater consequence for the national parks was the Local
Government Act of 1972 passed by Edward Heath's Conservative
government. It created for each park a new national park authority
(NPA), largely funded by central government, which would be the
planning authority for the whole park and a 'senior' committee of
the county council – as the preservationists had demanded back in
1951. In 1976 Wilson's third Labour administration adopted the
recommendation of the Sandford Report (1974) that when there
was an irreconcilable conflict in the parks between the conservation
of the landscape and its enjoyment by the public, the former must
take precedence. This was not enshrined in law until 1995. Parliament
was thus relieved of much unwanted business, and the new NPAs
were granted an unprecedented power to manage the parks
according to conservationist principles.

In 1977 each national park authority was to produce a ten-year
plan. The Dartmoor National Park Plan, published by the National
Park Department of Devon County Council, constituted the most
complex and coherent overview of the park's state and predicted

needs published since the designation in 1951. That the plan was credited to a Dartmoor NPA staff that included an ecologist, a park planning officer, a park land agent, an interpretation officer, a national park officer, a head warden, and a plans officer testified to the revolution in park governance wrought by the 1972 act. Nearly forty years later, the plan's confidence and optimism, comprehensiveness and clarity shine through. That confidence is evident in numerous ways. First, the importance of the landscape was not in doubt, and the authors of the plan saw no need to make grandiose claims on Dartmoor's behalf. When listing, for example, the factors to be taken into account when seeking to educate the visitor about the value of the park, the plan stated: 'Dartmoor is not noted for its dramatic scenery – except for its indefinable quality of isolation – but it is much more remarkable for such qualities as small scale beauty and hints of very long occupancy by man. Fascination is a question of walking and seeing, and not gasping at a dramatic view as you pass by on a quick car ride'.[272]

Dartmoor had nothing to prove.

Confidence was also evident in the faith placed in managerialism and good planning. Pages could be spent summarising the 'management objectives' and 'management policies' and systems of 'implementation' outlined in the plan, but what they amount to is nicely summarised in a paragraph surveying the NPA's landscape management objectives:

Achievement of the National Park purpose involves the maintenance of the typicality of the overall landscape and its component parts. Such conservation – a positive relationship between preservation and change – must be based on an understanding of the processes and practices which have created the living landscape and which must continue to evolve. While conservation of the landscape is the prime concern, all of the management objectives must reconcile this with the social and economic well being of communities, individuals and visitors.[273]

A further gloss on this 'positive relationship' was offered in the distinction the plan made between conservation, mandated by government as the priority of the NPAs, and preservation. '"Preservation" may be happily applied to buildings, monuments and other inorganic fixtures, but the living landscape cannot be frozen. Conservation is an attitude to life, activities and resources on which the NPA will base all its actions.'[274]

By distinguishing preservation from conservation, the NPA represented Dartmoor not as an ecology whose characteristics had survived in spite of human activity, but as a landscape whose unique value was produced by centuries of human activity. By this reading, Dartmoor was not a pristine wilderness needing protection from humanity but a resource that had to be managed and developed so its essential qualities could be freely experienced by increasing numbers of people. Licensing coach tour routes and constricting the movement of certain kinds of traffic might have a conservationist effect, but as important was the influence new information and interpretation centres could have on how visitors behaved in the moorscape. Providing visitors with the right kind of knowledge of the moor was part of the NPA's remit.

This was not all. The plan's foreword closed by arguing that 'the total resource which is Dartmoor goes well beyond the National Park purpose', and the NPA's responsibilities extended to all who lived within the park boundaries. Quoting government policy (Circular 4/76), the NPA declared that good park management, which included seeking government finance for new projects, would enhance the '"social and economic well-being of the Park"'. Planning decisions must reflect not only the needs of visitors, but also the housing, transport, educational, recreational and occupational needs of residents, including those who lived in the park but commuted for work elsewhere. People resident in the park, often from families whose Dartmoor roots long predated the designation, should not have their expectation of a decent life curtailed by conservation. In particular, young working people living within the park needed affordable new homes and, as a socially just planning authority, the NPA would allow new building.[275] In much the same way, 'environmental protection' needed to be 'realistically reconciled with an efficient land economy' in which farming and forestry could thrive. The NPA's task was 'to co-ordinate the activities of all users, in order to ensure the continuing optimum yield from the Dartmoor resource for all those who can benefit from it'.[276]

Confidence that accurate data was at the heart of properly informed decision-making was powerfully present in the plan. Representing this data through maps allowed distinctions to be made about the significance of various landscapes within the park and the particular pressures they faced. Maps plotted Dartmoor's natural and historic characteristics, mapped contemporary usage, tracing patterns of work, residence and movement through the park. Each map showed a different way of territorialising the moor, whether as a complex ecology, a natural resource, a tourist destination, a place of work or a transport network. According to the NPA's pluralist and developmental understanding of conservation, Dartmoor was the totality of these maps.

If the arcadian note sounded softly in the technocratic prose of the Elizabethan park official, the moor of the 1970s was still characterised as an anthropic landscape important for its beauty and recreational value. Managerialism, faith in planning and the strengthening of state agencies were pitched as responses to a conception of the moor that chimed with Crossing's late-Victorian realism and the national parks legislation of the 1940s. New, however, was the institutionalisation of nature conservation, and as this increasingly dominated the decisions made by the NPA, so the politics of the national parks edged away from the conflict that dominated their first thirty years, agitators and activists gradually finding themselves co-opted into a more consensual system of governance. Despite this, just as the 1977 plan should be read as aspirational rather descriptive, so did NPAs not prove beyond reproach. As unelected bodies exercising considerable power over extensive territories, they were sometimes criticised as suffering from a significant democratic deficit, while the claim that privileging national over local needs was authoritarian did not go away.[277]

That said, among the most significant achievements of the Dartmoor NPA in the 1980s was the new system of governance it established for the common and the accent this placed on self-help. Like so much else in the history of modern Dartmoor, Victorian enthusiasts provided the connective tissue between the deep past and the most pressing concerns of the present. And to tell the story of the formation of the Dartmoor Commoners Council brings into our narrative not only Dartmoor's sheep and cattle, but also the Dartmoor pony, symbol of the national park and the freedoms the designation seeks to protect.

4

Commoners and Conservation

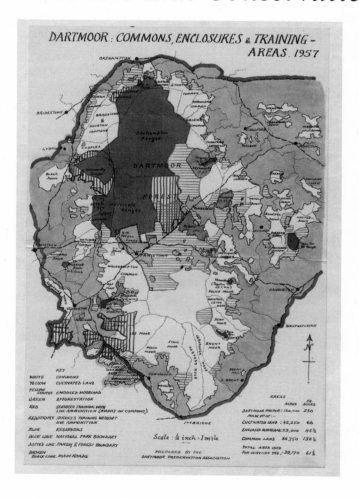

Perambulation

Of all the Dartmoor villages that edge the common, Belstone feels most like a frontier town. Close to the most northerly point of the common, it lies a few miles south of the A30, the dual carriageway that connects Exeter to Okehampton and Cornwall to the rest of the country. It is the smallest of the three villages overlooked by the great mound of Cosdon Hill and the least likely to be chanced upon by a motorist taking the scenic route from Widdon Down to Okehampton. South Zeal and Sticklepath are charming in their way, but the rustic setting below Ramsley Hill of the former and the National Trust's domestication of the latter do not make for Belstone's misty atmospherics. Belstone was given Phoenician or Viking origins by early Dartmoor writers, investing the village with an ancient significance, but Bellestam's appearance in Domesday is probably a more sensible starting point for thinking about the village's origins. The chapel-like Telegraph Office is one of a number of moodily evocative buildings. Clad in weathered granite with contrasting red telephone box and letterbox, its current state tells of successive but now archaic forms of communication and the limited reach of gentrification in these parts. Also telling of how we used to live is the hand-painted notice explaining that the little enclosure nearby was once used to impound stray animals but 'is now a garden to be enjoyed by all'.

The road leading into Belstone branches off to the south for about half a mile before terminating at a gateway to the moor. Deteriorating as it advances, the road passes through a small stretch of common that falls away to the River Taw and then becomes a

steeply winding passage through dilapidated housing decorated by an old tractor and other agricultural detritus. Another gateway to the left guards the driveway of the large house and grounds of Tawcroft; that to the right opens directly onto Belstone Common, the genteel wealth of Tawcroft separated from the open moor by a long granite wall. The path follows a short narrow strip of newtakes before reaching a sheepfold, a steel and concrete assemblage that like the blackface sheep grazing on the common tells of commoners that still extract a living from the moorscape.

Scrabbling along the Taw is fun, though the view of Belstone Tor and its riotous clitter is obscured down in the river's shallow but steep-sided valley. It would be easy to miss the Irishman's Wall, which stretches a mile or so west of the Taw right over Belstone Tor. So named, the legend has it, on account of the Irish labourers who attempted to enclose a large part of Belstone Common in the nineteenth century. What they achieved before the Belstone commoners drove them off was too extensive to be worth dismantling and stands today as a peculiar memorial to the moorscape's unruly past.

As the path passes through the grassy valley formed by Belstone Tor and Cosdon Hill, it opens out into the expansive heathland flats of Taw Marsh, and here a sense of the vastness of the northern quarter starts to be felt. Closed at the southern end by Steeperton Gorge and overlooked by Steeperton Tor, passing through the gorge means leaving the reassuring ambit of Belstone and committing to the long hike down through blanket bog to Two Bridges. Heavy-hanging skies threaten rain, delivering the thrill of danger that draws the hiker further into the isolated parts of the northern quarter.

The Taw flows through the gorge, dissected first by the boundary posts of the Okehampton Range and then the military road connecting Hangingstone Hill, one of the moor's highest points, with the long tracks leading north to Okehampton Camp. At a junction a military road leads west, and the slopes of Lints Tor are suddenly lit as a break in the cloud flashes yellow across the landscape. The purist, expert with map and compass and properly equipped, abhors the road as a violent intrusion into the loneliest landscape of southern England; the novice concedes that the roads ease access to the most dramatic and bleakest parts of the moorscape, just as the disused railway lines open up the southern quarter. To the east, beyond sodden blanket bog, the conifer stands of Fernworthy sharply delineate domesticated moorscape and old enclosure from the open moorland of the northern plateau.

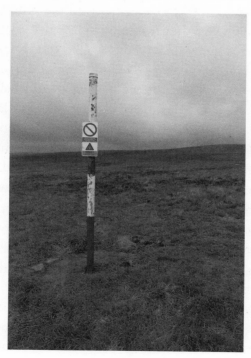

South of Hangingstone Hill the military roads come to an end and the going becomes less certain, though metal military huts punctuate the way and boundary posts threatening death aid orientation. As the boundary posts take a sharp right and then left, we pass from the Okehampton Range into the Merrivale Range. The great tors are to the north, and here, a little closer to Two Bridges than to Belstone, the driving rain gives the northern epic its first longueur: the sward of the northern plateau rests on quickly saturated blanket bog. It's a trudge through surface water.

Rough Tor – decorated with light military infrastructure – completes the second third of the hike, and knowing that here begins the gradual descent to Two Bridges delivers a pulse of relief. The rapidly flowing West Dart promises river valleys and shelter. Freshly shorn and marked sheep add an incongruous luminosity to the scene.

On the east bank of the Dart Wistman's Wood heaves into view, but today the right bank is more appealing, for here the path running

alongside the leat leads into the small conifer plantation established by Mr Bray at Beardown Farm. It comes out just above Two Bridges. After a day of frequent rain it is comforting to follow the man-made leat through the muffled soundscape of carefully managed conifer stands, anticipating the open fire in the Two Bridges Hotel.

The military's long presence on Dartmoor provides the most palpable human influence in the northern quarter. Boundary posts, military roads, metal huts, radio aerials and flagpoles intrude on the sward in a manner unmatched by ancient boundary walls, the peat passes cut through the most treacherous parts of the blanket bog and the remains of tinners' huts. Partly this is simply a function of current use, for the military roads and other structures are maintained where the redundant features of earlier Dartmoor industry are gradually going 'back to moor'. But it is also because the materials used by the military do seem 'alien', as the preservationists always insisted: even the granite cladding on the military huts on Hangingstone Hill looks suburban rather than of the moor.

All of this, however, is a diversion, for this section of *Quartz and Feldspar* is really about something else. Another politics etched onto the surface of the northern plateau outweighs its importance as a military training ground. This politics is so fundamentally constitutive of the moorscape's materiality that it can almost go unnoticed. Tracked vehicles, rifles or artillery do not make the modern moorscape; sheep, ponies and cows do. The route from Belstone to Two Bridges does not just pass through the Okehampton and

Merrivale Ranges but through land designations far more ancient, first Belstone Common and then into the Forest of Dartmoor. Had we hiked clockwise around the edge of the unenclosed upland from Belstone Common to Postbridge, we would have passed through South Tawton Common, Throwleigh Common, Gidleigh Common, Chagford Common and then into the forest. No indication on the ground signals the passage from one common to another, and the boundaries are only lightly marked on the OS. The uninitiated might assume these inscriptions are archaic designations, no more relevant to Dartmoor's present use than cartographic representations of the peat passes or the cairns. In fact, the distinction between the individual commons and the much greater central tract known as the Forest of Dartmoor is more important to the management of the moorscape in the early twenty-first century than it was for much of the preceding century. To explain this, the controversies associated with Dartmoor's modern history as a pastoral upland need to be understood. What follows considers the consequences that flowed from the decay of the manorial system, the corruption of the ancient principles of levancy and couchancy, the effect introducing new breeds had on how the common was grazed, and the influence agricultural subsidies had on stocking levels. Sense needs to be made of the tensions that arose between relatively large-scale agribusiness and the interests of small breeders and small stockholders, how animal welfare concerns drew local and national attention to the common's management, the reasons why reviving old rights and obligations could not provide a satisfactory solution, and the difficulties that beset the attempt led by the national park authority to bring the commons under a new management system in the 1970s and 80s. This issue has been as important to the development of modern Dartmoor as the politics of wood, water and the military combined.

If the detail of this history might fascinate anyone with an interest in the modern politics of the moor and its leading personalities, the significance of the issues raised extends beyond the boundaries of the national park or, indeed, beyond the politics of common land itself. The modern debate concerning how access to 'common-pool resources' should be managed began with Garrett Hardin's seminal essay 'The Tragedy of the Commons', published in 1968.[1] A common-pool resource, according to an authoritative recent definition, 'is a valued natural or human-made resource or facility that is available to more than one person and subject to degradation as a result of overuse'.[2] Speaking specifically of graziers, Hardin argued that the 'tragedy' of the common is that if all users restrain themselves, then the resource can be sustained, but a user limiting his use to his share when others do not risks losing out on the short-term gains possible before resource collapse. Rational self-interest, Hardin insisted, leads to the collapse of the common, for 'man is locked into a system that compels him to increase his herd without limit – in a world that is limited'. History demonstrates that sustainable use of common land could not rely on the exercise of customary practices because the temptation to abuse them was too great. Modern farming techniques make such abuses all the easier. According to Hardin, 'mutually agreed coercion' works, but only when the state – rather than a lower, less authoritative body – is empowered to enforce the rules. As a liberal who viewed the strengthening of the state as inhibiting the exercise of individual freedom, Hardin argued the only way to prevent the overuse of the common was to privatise it, dividing it into privately owned units. Privatising nature, Hardin's argument ran, was morally preferable to increasing state power.

Elinor Ostrom has led the challenge to this view, arguing that collective action is at least as rational as individual action when 'multiple appropriators' are dependent on the same common-pool resource and 'are jointly affected by almost everything they do'. Rejecting Hardin's abstract reasoning in favour of a close focus on particular case studies, Ostrom found users to be 'fallible norm-adopting individuals' in which informal and formal negotiation ensured the sustainable exploitation of the common resource.

Recognising that the 'key challenge all commoners face is main-taining commitment to their rules', Ostrom and her collaborators argue that evidence from Africa, Latin America, Asia and the United States suggests that top-down state regulation is rarely the best solution. The evidence suggests that when resources governed as common property by local communities were brought under state governance, rules were not enforced and overuse quickly became a problem because sufficient numbers of trained personnel were rarely deployed.[3] Alongside overuse, the other problem associated with the use of a common-pool resource is the 'free-rider', a user who does not contribute to the costs of developing and maintaining the resource but cannot be easily prevented from using it.[4]

On Dartmoor preventing overuse and minimising the effect of the free rider gave rise to a solution that embodies the pessimism of Hardin and the optimism of Ostrom. The role eventually prescribed for the state largely consisted of authorising a self-regulatory approach to the management of the common under the supervision of the national park authority. That it took the best part of a century to establish a new system of local control backed by state authority is not the least remarkable part of the story. As important has been the subsequent transformation of the commoners, not without a great deal of soul searching, from production-orientated stock raisers to custodians of a fragile upland ecosystem. Thanks to the further empowering of the national park authority in 1995 and the introduc-tion of new forms of agricultural subsidy, the graziers have been co-opted into a broadly conservationist agenda predicated on bio-diversity principles. Through this process, a new transformation in subjectivity has occurred, generating in the commoners and their state-sanctified collaborators a new 'environmentality'.[5]

Ancient Rights and Modern Usage

From the middle of the nineteenth century, successive attempts by the British government to revive enclosure allowed enclosure commissioners to consider schemes to bring about the better management or improvement of commons *as commons*. Permissive clauses in the General Inclosure Act of 1845 and the Commons Act of 1876 encouraged commoners and landowners to work with the commissioners towards the introduction of improved systems of self-regulation under elected conservators and to agree new rights of access for the general public. Conservators could oversee improvements to commons that included drainage, manuring and levelling schemes as well as planting trees for shelter and introducing changes that enhanced the natural beauty of the land. Conservators were also empowered to determine how much stock could be grazed on each common and the consequent stint of each individual commoner. Central oversight and local consent characterised the new approach. New by-laws needed the approval of the Home Office, the rates conservators could levy to finance their activities had to be approved by the Land Commissioners, and new schemes of regulation required the consent of two thirds of the commoners and the approval of the lord of the manor. By allowing district councils to initiate drainage or levelling schemes intended to increase the productivity of common land, the Commons Act of 1899 widened the remit of the legislation, though again the agreement of the landowner and the commoners was needed.[6]

By encouraging new systems of self-regulation, these acts signalled how the critical gaze had shifted away from the common and onto the commoner. If in the early nineteenth century the existence of commons was seen as incompatible with a thrifty, self-disciplined, productive life, by the late nineteenth century influential voices no longer demanded the dissolution of common rights and the enclosure of upland commons, but they often identified the commoners and their landlords as living in a state of degradation. As the amenity agenda identified commons as a valuable national resource, so the critique became less structural (the commons were the problem) and more moral (the commoners were the problem). Dovetailing with the peculiarly triumphal preservationist narrative of the Dartmoor upland's resistance to improvement was a narrative dismayed by the decay of the customary practices that had once kept order on commons and upheld the dignity of commoners. If commons were to be newly integrated into the national patrimony, something had to be done to reconfigure the relationship between commoners, their landlords and other users.

To the preservationists the solution seemed self-evident. Ancient rights and responsibilities needed to be re-established and dutifully exercised. This would restore good governance to the moorscape and raise a legal barrier against further encroachment. Percival Birkett, speaking for the Dartmoor Preservation Association, took up the charge in a lecture to the Devonshire Association at the Athenaeum in Plymouth in October 1885. Birkett's preservationist polemic took equal billing with his sophisticated exploration of the historical origins of the Dartmoor common and an outline of the various classes of rights holders.

As Birkett explained, the Dartmoor common was not a single phenomenon but comprised the Forest of Dartmoor and the Commons of Devon. Largely contiguous with the parish of Lydford, the forest was the great central mass of the Dartmoor upland, a territory that stretched far to the north and south of Two Bridges. Known as a forest because it had once been subject to forest law, Birkett maintained that the title was now an anachronism. As he pedantically observed, the forest became a chase in 1239 when Henry III packaged it up as part of the Duchy of Cornwall and granted it to his brother, thereby transferring it from the Crown, when it had been a forest, to a subject of the Crown, when it became a

chase. Reverting to the Crown in 1300, in 1336 Edward III made the
Duchy of Cornwall part of the patrimony of the Principality of
Wales, and it passed into the hands of the Black Prince.[7] It has
remained the possession of the Prince of Wales ever since. The
Commons of Devon were the smaller upland commons that encircled the forest and occupied the territory between it and the lower-
lying enclosed land or in-country. As late as the 1950s the ownership
of the Commons of Devon was uncertain.[8] Most newtakes – the
small enclosures opposed by the preservationists as 'encroachments'
– took bites out of the Commons of Devon, although the largest
of the newtakes, those of Dartmoor Prison, were in the forest.
Little on the ground marked the boundary between the Commons
of Devon and the Forest of Dartmoor: then as now it was possible
to cross from common to common or into the forest without being
aware of changes in ownership or jurisdiction. This invisibility on
the ground reflected significant uncertainty in law: some thought
Dartmoor was a single common over which all commoners had
rights; others insisted that each common belonged to a separate
parish, and commoners could exercise rights only on the common
that formed a part of their parish.

The rights holders fell into two principal groups. The first were
the holders of ancient tenements, those farms that lay within the
forest boundary and were clustered near the East and West Dart
rivers in the east and south quarters. First documented between
1260 and 1560, they occupied sites whose evocative names are
familiar to any Dartmoor resident or enthusiast: Warner, Runnage,
Hartland, Merripit, Pizwell, Dury, Lakehead, Bellever, Riddon,
Babeny, Brimpts, Brownberry, Huccaby, Hexworthy, Sherberton,
Dunnabridge and Prince Hall. Historically, there were 34 holdings
on these 17 sites, but by the late eighteenth century they had been
consolidated into 14 separate farms. A further 12 farms were created
with duchy consent after 1780, including several at Postbridge, Two
Bridges and on the East Okemont, locations suggested by the new
turnpikes.[9] The holders of the ancient tenements paid nothing to

graze their beasts in the forest provided they could overwinter them in their tenements. This was known as levancy and couchancy. The holders of the ancient tenements could also act as middlemen, taking in the livestock of down-countrymen from throughout Devon for summer grazing on the payment of a penny-halfpenny headage to the duchy. This fee was known as agistment, and the broader system of summer grazing whereby farmers drove their herds from farms throughout Devon to the Dartmoor commons is known as transhumance. The most peculiar aspect of this right to summer grazing was the exclusion of the men of Barnstaple and Totnes. Most commentators fail to explain the anomaly, though the late Harold Fox speculated that the exclusion has pre-Norman origins, dating back to when both towns were fortresses in which men were provisioned and had to be always available to fight off Viking raids.[10]

The second major group of rights holders were the venville tenants. They occupied moor-side farms in parishes bordering the forest – the commons of these parishes comprised the Commons of Devon. On payment of a small rent, venville men had the right to graze in the forest what stock they could overwinter on their farms and hold during the day and sometimes at night the rest of the year (levancy and couchancy once again). They could not act as middlemen, but they could graze additional stock on the common on payment of agistment.[11]

Birkett lamented the decay of the old system of forest law, insisting that although its courts had meted out harsh justice, the duchy's officials had ensured that customary rights were protected and complaints expeditiously addressed. Now the common was farmed out by the duchy to moor men who squeezed what they could from the venville tenants and other commoners wishing to graze stock on the moor. Quite what Birkett objected to is a little obscure, but it seems the old customary payments had given way to commercialised grazing in which the duchy's quartermen charged stock owners as much headage as they could get for summer grazing and maximised the volume of stock put on the moor. Though the level of summer grazing through agistment had always been at the discretion of the moor men, Birkett felt current practices demonstrated that the 'promiscuous exercise of the Rights of Common tends very much to obscure the ancient customs'.[12] Garret Hardin might well agree. In essence Birkett doubted whether traditional rights and customary restraints could be simultaneously revived without the introduction of a new regulatory body. He imagined the provisions of the 1876 Commons Act bringing Dartmoor under the management of 'a body of Conservators elected to represent all the various interests'. Birkett attracted the sympathy of Frederick Pollock, an eminent jurist, who hoped 'that in the course of time some arrangement may be come to whereby friendly and concerted action may be taken, with the Duchy of Cornwall, as is fitting, at its head, for the better government of Dartmoor'.[13]

W. F. Collier, a Quaker radical, agreed, though in two addresses to the Devonshire Association in the late 1880s he showed less deference towards the Duchy of Cornwall. Each time an enclosure was permitted, Collier insisted, the duchy illegally violated common rights while the role now played by its quartermen as gatekeepers to the moor had led to 'disputes and riotous behaviour' and the introduction of 'vulgar notions about money' where custom had once presided.[14] Nonetheless, as he explained at Plympton in July 1887, the proper governance of the common required the restoration of the customary rights of all commoners *and* the duchy. He urged that the duchy 'propound a scheme for the management of the Forest of Dartmoor and the Commons of Devonshire, by which the rights of the Duchy and the rights of venville tenants and commoners would be greatly protected.'[15] By August 1889, when he addressed the association at Tavistock, Collier's thinking had advanced considerably:

Schemes have been established for the management for the public advantage of several royal forests – the New Forest, Epping Forest,

Dean Forest, and others – to preserve these ancient, beautiful, and most interesting tracts of land for national pleasure-grounds, where the people of a closely populated country may see trees, flowers, birds, and animals in their wild and native beauty. These forests are in nearly every respect similar in their character to Dartmoor, and such a scheme might be established for Dartmoor, by friendly agreement with the Duchy, which would be satisfactory to the public, who delight in Dartmoor; satisfactory to the commoners, whose ponies would increase and multiply exceedingly; satisfactory to the Duchy, whose conscience would be relieved of a heavy burden, and whose property would be secured on a solid and equitable basis; and satisfactory to the Devonshire Association, who would no more be worried by papers like this, and would have the stone circles, stone avenues, and the rest of it, for their members to indulge their speculative fancies.[16]

Although Collier's teasing prescription foreshadowed much that would become orthodox in the late twentieth century, his remarkably prescient thinking failed to recognise the absence of any urgency in Dartmoor's case. The New Forest, Epping Forest and Dean Forest were not inhospitable uplands but relatively accessible lowland commons facing intensifying pressures from tourism and commercial exploitation. The reconstituted Verderers' Court and the access guarantees that formed the centrepiece of the New Forest Act of 1877 could indeed be adapted to Dartmoor's needs, but what had mobilised support for the act was the massive threat posed to this metropolitan playground by the navy's hunger for timber in the 1870s. Threats to Dartmoor, incremental rather than comprehensive, did not generate an equivalent sense of urgency. And when strident demands for reform were heard after 1945, they took the form of an attack on – rather than a defence of – the commoners.

Had that urgency been felt, reformers would have faced difficulties that could not be solved by reviving customary practices. Birkett and Collier both failed to notice the historic transformation in how the common was grazed occurring in the last decades of the nineteenth century. In the late eighteenth century the cows pastured on the moor were mainly North Devons (Red Rubies) and South Devons, a breed similar to Limousins and possibly a cross enabled by cross-Channel trade. Dartmoor's sheep were a distinct breed, generally Whiteface Dartmoors possibly crossed with Dorsets and new Leicesters.[17] During the late nineteenth century Scotch Blackface sheep and Scottish black cattle (Galloways) were introduced to the moor. Tough and resilient, these animals found favour with farmers because they can be wintered on the moor and do not need to be brought in for lambing and calving. Existing rights were rendered redundant by this shift in stocking: levancy and couchancy no longer provided an effective means to determine levels of stocking or fair stints. Year-around grazing also reduced the scope for down-countrymen to rent additional grazing in the summer and gave the moor less opportunity to recover, which reduced its summer grazing capacity and changed the ecology of the moor by inhibiting the growth of the 'sweet' grasses downland stock could digest. Consequently, much of the common became suited only to the most hardy breeds whose digestive systems and metabolism were adapted to rough grazing. It is hard to exaggerate the significance this change in stocking brought to Dartmoor's future as a common. No customary practices or regulations existed to control this unprecedented exploitation of the land. And there is no more telling symbol of the continuing isolation of Dartmoor's upland and its marginality as an agricultural landscape than the unruly way blackface sheep marginalised their whiteface cousins.

'The tenants in venville,' Collier explained, echoing Tristram Risdon, the seventeenth-century topographer, 'have the right to take anything off Dartmoor that may do them good except green oak and venison, or more properly vert and venison.' Right of turbery (to cut turf for fuel), right of estovers (to cut and take wood, reeds, heather and bracken) and right of piscary (to take fish from ponds and streams) was limited to the household's needs and could not, in contrast to animal products derived from right of pasture, be exercised commercially: nothing the commoner took from the moor could be sold to a third party. Discussing these rights, Collier's lecture made its radical turn. The injunction against taking deer from the common, he reminded his readers, came with the forest law of the Norman yoke. Before then deer was the 'food of man', but as 'industrious man cultivated land and bred sheep and oxen, so idle man – a very superior person – clung to hunting and eating venison; and it came to pass that laws and penalties were required to protect the deer'. Hunted to extinction in the late eighteenth century, the ascendancy of the deer, 'representing the lords', came to an end, and the Dartmoor pony, representative of the commoners, symbolised the new social order on the commons. '[H]e is the great wanderer, defies all boundaries, lives out on the moors winter and summer, night and day, and the law of levant and couchant, if there is such a thing, is totally inapplicable to him. For hundreds of years he has ranged Dartmoor at will, asking for no care, no shelter, and no winter quarters.' Invoking the free-ranging, free-spirited pony as symbol of the liberties of the common did more than provide a little romanticism to a lecture on common rights. Collier believed the pony's liberty to wander was proof that for all practical purposes the Commons of Devon and the forest had for time immemorial been a single common: the undisputed freedom of movement across the commons enjoyed by the pony reflected the historic rights of the venville men.[18]

This principle, clung to by the preservationists, proved highly contentious in the post-war period, and it is somewhat ironic that it was the feral promiscuity of the 'scrub' stallions that first focused attention on the common *as* common. In January 1949 Sylvia Calmady-Hamlyn, sometime justice of the peace for Devon, made herself known to officials at the Ministry of Agriculture, Fisheries and Food. A Dartmoor pony breeder with holdings at Hatherleigh on the south-east edge of the park, Calmady-Hamlyn played a prominent role in the Dartmoor Pony Society and had good contacts with the National Pony Society and the British Horse Society. She was prompted to action by the introduction of new legislation requiring the licensing of stallions, and her farm fell just outside the Dartmoor area where licensing stallions was not required. H. F. Lovell, an official at MAFF, explained that the new licensing system did not extend to large parts of Dartmoor because where ponies ran free, licensing was thought impractical, a condition that did not exist at Calmady-Hamlyn's Hatherleigh holdings. Lovell thought registration would benefit breeders, adding that the veterinary inspection included in the one-guinea registration fee gave value for money. Calmady-Hamlyn thought it was absurd that pure-bred horses faced closer scrutiny than free-range ponies running unrestricted on the common. They threatened the purity of the Dartmoor line, particularly since free-range Shetlands had been introduced to the moor during the war: 'a commercialised pitter', she explained in a pamphlet, 'is being bred by skewbald Shetland stallions on certain quarters of the moor to the great detriment of the breed'.[19]

Internal MAFF correspondence indicates that Shetlands were exempted from licensing thanks to concerted pressure from the Shetland Pony Society. The society feared licensing would undermine the trade in pit ponies and insisted on there being little evidence of interbreeding, ignoring those Shetland stallions on Dartmoor inclined to 'serve' Dartmoor mares. And though Scottish breeders could accept licensing Shetlands south of the border, this would not address the problem on Dartmoor given the common's exemption from licensing. A Scottish official's suggestion that it was the responsibility of the owners of mares to ensure they weren't served by Shetlands indicated, at best, little understanding of the common and, at worst, crass stupidity. Either way, the Dartmoor Pony Society wanted the Shetlands off the moor but it had no legal means to bring this about. When Calmady-Hamlyn demanded the act authorising Shetland stallions and colts to run on Dartmoor be amended, Lovell patiently explained that no such act existed. It was not the government's job to decide what stock farmers kept.[20]

Calmady-Hamlyn is easily stereotyped as another well connected upper-middle-class figure whose social connections got a hearing for her special interest. Not everyone could present to the annual meeting of the Dartmoor Pony Society a personal message from Princess Margaret, as she did in August 1950. 'I am dismayed. Will you go back and give breeders a message from me. Ask them, with all my heart, to do everything you can to keep your beautiful breed alive and not to lose for Dartmoor and England so charming a heritage.' Similarly, in January 1952 Lady Elizabeth Pease wrote on Calmady-Hamlyn's behalf to Sir Thomas Dugdale, the new minister of agriculture. Pease explained that although Calmady-Hamlyn had devoted her life to the preservation of the Dartmoor pony 'in its own native habitat', the Dartmoor Pony Society had struggled for years to get a fair hearing because officials loathed admitting their mistakes. Incidentally, Pease added, Calmady-Hamlyn, a Cleveland woman, was a niece of her late husband Sir Alfred and had 'spent all her childhood at Hutton', the family seat in Guisborough, North Yorkshire. Shortly after, Dugdale received a letter on the same question from Joseph Pease, second Baron Gainford and hereditary Liberal peer. Calmady-Hamlyn, Gainford explained, was his cousin. Reason enough, it seemed, for the minister to give the matter his personal attention. 'Tom' promised 'Joe' that he would look into it. In a handwritten note thanking Dugdale for his eventual reply, Gainford promised to forward the minister's letter to his cousin.[21]

Further down the pecking order, Calmady-Hamlyn could not expect such gentle handling. When Austin Jenkins at MAFF asked a livestock officer about the problem, E. A. Farey sniped that the Dartmoor Pony Society was practically a 'one-woman show' and before any 'drastic step could be contemplated . . . the men who really get their living from Dartmoor should be consulted'. This response was characteristic of the tendency of MAFF officials to dismiss the concerns of special interest groups whose activities were considered marginal to the proper use of common as rough grazing. Shetlands had been brought to Dartmoor, Farey explained, because breeders wanted to take advantage of the continuing demand for pit ponies. Pure Dartmoors were too big for the pits, too small for draught work and often too wide and heavy-shouldered to make good children's ponies. Interbreeding made Dartmoor ponies a marketable commodity. Moreover, Farey reckoned Shetland ponies were less of a problem than the scrub stallions left running free before castration at two years old. Lack of effective management saw some pretty arbitrary responses to the problem. A big horse dealer said to exercise 'dictatorial control' over the western side of the moor was known to castrate any pony stallion he considered inferior regardless of who owned it. When Jenkins, 'pestered' by the pony interest a year later, sought further insight into the problem, Farey admitted that although there was no evidence that the Shetlands were inferior or mongrel, Shetlands and some crossbred colts were covering mares against the wishes of their owners. Discreet inquiries made at Tavistock market that July again indicated that Shetland breeders thought the main culprits were uncastrated colts of mixed provenance which ran freely until rounded up for slaughter.[22]

MAFF's purpose was to regulate British agriculture and help maximise its productivity and profitability. When facing a tricky public relations question, it tended to defend the agricultural interest, its regulatory purpose seguing into an advocacy role that reflected the close relations between local officers and livestock owners whose principal purpose was food production. These views were reported as representing good sense, and as they were communicated up through the Whitehall hierarchy en route to the minister's desk relatively nuanced views could accrue greater certainty. As such, it was characteristic that the change brought to the equestrian ecology of the common by the recent introduction of the Shetlands was not seen as a problem because the efficient use of rough grazing trumped the marginal interests of the pure breeders.

In a frank minute of 10 September 1951 J. H Locke admitted that Dartmoors might die out, observing: 'This would, no doubt, be a pity from some points of view, but would not seem to be any loss from the agricultural point of view or any reason for action by the Ministry. Since pure-bred Dartmoor ponies are little use economically whereas Dartmoor–Shetland crosses are useful, we should have to go very carefully in supporting any proposal to restrict the Dartmoor National Park to Dartmoor ponies.'[23]

Locke was not entirely happy with this prognosis and elicited the views of other parties. The National Parks Commission vaguely acknowledged the problem, proposed a meeting and advised Locke to consult with Nature Conservancy. When the commission met in January, it was told that Shetlands were more pugnacious than Dartmoors, inclined to oust heads of family, and when kept off the common to avoid contamination tended to suffer a dilution of their characteristic 'hardiness'. Add to this how the particularity of Dartmoor's ponies had formed part of the case for designating Dartmoor a national park, and the commission concluded that the Shetlands should probably be kept off the moor. Nature Conservancy took a different view, reporting that the Dartmoor pony question was not their affair because it did not strictly concern 'wild life'.[24] Indeed, the emerging consensus was that it was not really a problem for government at all. Farey told Jenkins who told Maher who told Dugdale who told Lady Pease, Gainford and Sir Henry Studholme, MP, that, quoting Farey, 'the only answer to these complaints is the old one that the Commoners should help themselves, in the first instance, by forming a proper Association to deal with these troubles under the Commons Act'.[25]

Somewhat against expectations, MAFF was moved to act. With backing from the National Farmers Union, the ministry sponsored a meeting in early 1952 as a preliminary to the establishment of a Dartmoor Commoners Association. At Exeter on 7 January 1952 MAFF officials, representatives of the commoners and the Duchy of Cornwall considered the purpose of the new organisation. The general tenor of the meeting was that if commoners could speak with a single voice they could respond more effectively to criticism – at that time only the breeders' associations had any lobbying capacity – and if they overcame their mutual antagonisms and acted as a corporate body they could improve the common in order to raise the productivity of all and respond more effectively to sudden problems. The much-admired Cornwall Commoners Association offered a humbling contrast to the rudimentary organisation evident on Dartmoor, where active commoners' associations existed only at Peter Tavy, Roborough, and Whitchurch and Sampford Spiney, all on the western side of the moor.

A lead was needed but where it should come from? The ministry and the NFU could advise but no more and L. S. Mutton, the perfectly named duchy representative, said he intended to act only as an observer because a duchy initiative would likely rouse suspicion among commoners fearful its purpose must be the diminution of their rights. An effective association could develop only if the commoners themselves willed it. The meeting concluded by planning to launch the initiative at a high-profile gathering that March. Representatives of the commoners, encouraged to organise on a parish basis, would be invited, alongside a ministry official to explain the Commons Act, a representative of the Animal Health Department of the War Office, and Russell Perry, who had played an instrumental role in establishing the Cornwall Commoners Association.[26]

Held at Princetown on 19 March 1952, the meeting boasted a tranche of Dartmoor grandees and MAFF officials as a well as a good representation of the commoners. Austin Jenkins gave a hard-hitting speech on national meat needs and the failure of Dartmoor farmers to maximise production. He highlighted the effective management and superior marketing regimes established by the hill farmers of the Scottish and Welsh Borders, and he pronounced lamb production on Dartmoor poor and claimed that the Galloways only calved every two years because they were malnourished. This stinging attack set the tone, and the delegates gradually came around to recognising the need for an association. Reflecting on the meeting, an official ruefully concluded that if the association got off the ground it would prove effective in negotiations with outside parties like the War Office and the National Parks Commission, but it was unlikely the commoners would prove sufficiently united to empower the association to make majority decisions that could be imposed on all.[27]

For the next thirty years, the Dartmoor Commoners Association functioned as a voluntary body dependent on the efforts of a small number of committed individuals. It issued guidelines, gave advice and purported to represent the interests of all commoners. Enjoying the confidence of MAFF and the NFU, it achieved a semi-official status and state authorities often found it convenient to regard its views as representative and its spokespersons as authoritative. What the DCA did not have was statutory power. The clearest expression of its views in these early years is found in the evidence it presented to the Royal Commission on Common Land on 30 April 1957.

The commission, appointed on 1 December 1955 under Sir Ivor Jennings, grew out of a felt need to consider the use of common land in the interest of national efficiency and in the context of post-war reconstruction, particularly given the house-building programmes pursued by successive governments. The commission was asked 'to recommend what changes, if any, are desirable in the law relating to common land in order to promote the benefit of those holding manorial and common rights, the enjoyment of the public, or, where at present little or no use is made of such land, its use for some other desirable purpose'. Over the long term the findings and recommendations of the commission had an important effect on Dartmoor, but more immediately the hearings revealed the conflicting ways the commoners and the amenity societies regarded the great upland common.

The DCA's submission to the commission contrasted the ancient rights described by Percival Birkett with modern usage. It explained that the right to graze on the moor was no longer attached to land lying outside the forest and the surrounding parishes, thereby denying 'foreigners' their old right to summer pasture. Although the forest remained a single common, modern usage differentiated the Commons of Devon, treating them as separate parish and manorial commons over which reciprocal rights could be exercised. Outwintering had more or less destroyed the principle of levancy and couchancy, and although the DCA reckoned five acres could produce the additional winter fodder needed between 1 January and 1 April for eight cattle or ponies or forty sheep, there was no way of enforcing this as a new means of establishing a commoner's stint, not least because it left unresolved how to account for the fodder a commoner could buy. In essence, the DCA's submission attempted to redefine ancient rights according to modern usages, while at the same time accepting that those usages were not subject to sufficient regulation. The DCA left the commission in no doubt that it wanted the established practices of the commoners treated as rights to be identified and regulated by statute.[28]

The DCA evidence was thrown into sharp relief by the Dartmoor Preservation Association's submission. This predictably wide-ranging and coherent account of the amenity interest was predicated on two basic propositions. First, the moorscape was a single common and any notion that the Commons of Devon should be treated individually must be rejected. To dilute the 'universality and homogeneity' of the common would assist a landowner seeking to buy out those rights by requiring the agreement of a relatively small number of commoners. If, as a consequence, tenant rights superseded common rights, open land would be fenced and the public's right of access to the common would come into question. According to the DPA, upholding ancient rights of venville protected customary rights of access: the feral pony now symbolised the liberties of the hiker rather than the rights of the commoner. Second, the DPA argued that the agricultural purpose of Dartmoor no longer constituted its 'main value'. Any new management body would have to include representatives of the amenity interest, the defenders of the customary rights of visitors: 'health and happiness', the commission was told, 'are Dartmoor's best exports'. Under examination, Sylvia Sayer proffered a classic DPA formulation: 'It is important that Dartmoor should produce better cattle, ponies and sheep; it is more important that it should help to produce better human beings.'[29]

Animal Welfare

The health and happiness of whom or what? By the early 1950s animal welfare was beginning to shape public attitudes towards the commoners and each spring there was considerable commentary in the press on the condition of livestock outwintered on the moor. In 1954 horse enthusiasts and tourists alike were shocked by the emaciated condition of stock encountered that spring. A petition jointly organised by the RSPCA and the International League for the Protection of Horses demanded MAFF issue a compulsory order that ponies and cows be taken off the common during winter. Although 350 signatures were gathered, MAFF explained to its initiators – Brigadier General Sir George Cockerill was the recipient of one letter – that the ministry had no power to control grazing on common land.

Something of the bad reputation of the commoners was captured in an observation of R. V. Vernede, spokesperson for the Commons, Open Spaces and Footpaths Preservation Society. Driving the ponies off the moor over winter and into small fields or yards 'out of the public gaze', he wrote, meant the ponies 'would starve more effectively than on the Moor, where at least they can forage freely for what there is, and where their condition can be seen by everyone'. Vernede repeated the old panacea. Needed was 'a complete overhaul of the commoners' own customs and duties, once so willingly and effectively carried out by themselves, in the days when they were really proud of their ponies, and husbanded and farmed their holdings in the old traditional hard-working way'. Commoners who no longer put out hay when there was snow on the ground or during an icy frost were found especially wanting. Moreover, echoing demands made by Birkett and Collier in the 1880s, Vernede wanted to see the traditional autumn drifts re-established. This practice was once central to the duchy's management of the common and a pivotal moment in the Dartmoor year. The drift saw the ponies rounded up and claimed by the commoners, new foals branded, fees paid, unclaimed stock impounded and eventually sold, and weaker stock taken off the moor.[30]

The Dartmoor Commoners Association was acutely aware of these criticisms and its responses tended to be at once defensive and constructive. Its Nine Point Plan, agreed on 15 July 1954, was its most significant early achievement and had been presented as its calling card before the Commission on Common Land. In terms of good husbandry, which the DCA considered indistinguishable from the maintenance of acceptable standards of animal welfare, there was nothing exceptional about the plan. It called for old or 'unthrifty' ewes or cows to be culled; sheep to be dipped in November (in addition to statutory summer dipping); calves to be weaned at an age that allowed cows to be wintered free; additional feed to be provided as conditions warranted; keepers of cattle to press on with voluntary attestation (for it was to the commercial advantage of all that Dartmoor was declared an attested area); and ponies to be kept as suited the commoners provided they were sufficiently hardy to withstand severe climatic conditions and were properly fed and watered. Stallions kept on the moor should be of 'good type', gelded as necessary, and every effort made to prevent inbreeding. Unthrifty or 'broken-jawed' animals were those whose teeth had been worn down through grazing the rough pasture of the uplands and were no longer able to extract adequate nutrition from the moorscape over winter; attestation referred to the official certification of herds as free from tuberculosis, a major priority of post-war agricultural policy.

At a DCA meeting in July 1955 prompted by the latest wave of criticism, H. H. Whitley, the dogged chair of the association, plainly articulated the challenge faced by the commoners. Demanding that common law be re-codified in line with current usage was not enough; they must adhere to the DCA plan if the threat of a ministerial order to clear Dartmoor for the winter was to be staved off. 'If you want to protect your rights you must acknowledge your responsibilities,' Whitley thundered. 'If you don't do that, you'll have someone else telling you what to do.'[31]

By the time the DCA appeared before the Common Land Commission it represented twenty-three commoners' associations and fifteen lords of the manor, and it evidently saw its appearance as a further step in its campaign for statutory powers. Its plan was posited as a constructive response to the animal welfare criticisms of 1954 and evidence of the DCA's capacity to manage the Dartmoor common if empowered to do so. At the same time the DCA insisted recent criticisms were largely unfounded and it challenged the RSPCA to prosecute commoners not properly looking after their stock, saying such legal steps would have its support. Attempted prosecutions needed, however, to take account of how visitors feeding ponies at the roadside lured them away from their grazing grounds in the inner moor. Another problem that could be tackled more effectively through coordinated action was increased bracken growth and the controversy caused by apparently indiscriminate 'swaling,' the local term for its annual burning. Given that commoners no longer cut bracken as winter bedding for their animals (the right of estover), increased swaling was needed to maintain the common's grazing capacity. The old problem of kids setting fire to bracken for fun was thought as bad as ever, but if

the commoners agreed on what was to be burned and when, the
local constabulary could distinguish legitimate from illegitimate
burning.[32]

The royal commission published its recommendations on 18 July
1958. Recognising that common land ranged from village greens
and metropolitan commons to great upland areas, it judged that
'as the last reserve of uncommitted land in England and Wales,
common land ought to be preserved in the public interest'.
Improving facilities for public access and increasing the productivity
of the land was judged equally necessary, and schemes for the
improvement of commons were thus encouraged, whether by local
authorities, the Highway Agency or, possibly, new commoners'
associations. The commission was keen to see neglected commons
revived as public amenities, preferably by local authorities, but if
necessary the commissioners could take the lead. Most importantly,
the commission recommended that rights pertaining on common
land be registered, an immensely complex undertaking they
predicted would take more than twelve years. An eight-year registration period was needed, followed by an additional four years for
the registration of objections and then further time for the commissioners to resolve outstanding difficulties.

MAFF did not have any fundamental difficulties with the commission's recommendations, but the Dartmoor Commoners Association
was worried. Writing to Christopher Soames, the new minister of
agriculture, T. J. Brown made the case for the separate treatment
of Dartmoor once the commons registration process began.
Highlighting the 'divergence between Ancient Rights and Modern
Usage', Brown argued for the statutory redefinition of common
rights before the register was opened. Without this prerequisite
contested claims could only be decided on the basis of ancient
rights, and Brown feared the amenity interest would encourage
people to register dormant rights which, if upheld, would undermine the capacity of the upland graziers to retain control over the
source of their livelihood. Brown's fears were not unfounded.

During the pony controversy R. Hansford Worth made a classic preservationist move. Finding an act of Henry VIII requiring undersized ponies to be taken off the moor, he asserted that Shetlands were small and the legislation still stood.[33] This was the just the kind of evidence Sayer might triumphantly present to the commissioners as incontrovertible. Such shenanigans, Brown implied, could be avoided if the government passed an act before the registration process began which re-codified the rights of common on Dartmoor and established a 'Council of the Venville' to determine stints, as the Verderers' Court did in the New Forest.

MAFF was sympathetic and in November 1961 invited the DCA and the duchy to discuss the matter at Westminster. Confident of duchy support and facing a friendly audience, the DCA made bold. It admitted that year-round stocking was technically unlawful and could not be justified in terms of rights of common, but the practice was backed by MAFF and firmly established on the moor. For these practices to be challenged during the registration process and possibly declared illegal would be a fundamental assault on modern upland farming and risked undermining national productivity. Adapting the archaic principles of levancy and couchancy to the new circumstances did not make sense either, and the DCA reiterated its case for a Council of the Venville empowered to determine stints.

MAFF's response was weighed down by practical political calculations. Registration according to existing laws was the government's priority, and separate provisions for Dartmoor could not be attached to the forthcoming bill. The possibility that the DCA might seek to establish a council according to the provisions of the Commons Acts of 1876 and 1899 was also dismissed. Those acts were now thought unworkable, and lack of parliamentary time gave little prospect of MAFF promoting a separate bill. A private bill would be received sympathetically, but again pressures on parliamentary time meant it stood little chance of success. A month later an official confirmed that there was no prospect of a ministerial bill and advised the DCA to place its trust in the capacity of the registration process to settle contested claims. Rattled, the DCA successfully sought another meeting in January, but this brought no change in the ministry's position. It acknowledged that defining admissible rights according to current practice would be helpful, but primary legislation remained impossible and the DCA would have to trust the appeals process.[34]

As these discussions haltingly progressed, the crisis began to unfold that would prove the most testing time in the history of the commoners between 1945 and the foot-and-mouth disease epidemic of 2001. On 22 December 1962 a high-pressure system thrust bitterly cold winds over the north-east of the British Isles. Glasgow had its first white Christmas since 1938 and the Big Freeze began. The first of four major blizzards hit Wales and south-west England on 29 December. Others followed on 3 January, 18 January and 6 February. High winds caused huge snowdrifts. Roads and railways were blocked, telephone wires brought down, villages cut off, schools closed and Britain was bequeathed its most iconic images of a modern winter. Mild south-westerly winds did not begin to blow until 4 March. It was the coldest winter on record since 1740.[35]

Dartmoor and Exmoor were snow-covered for 60 days, some 15 days longer than was typical over winter. Around Princetown, 20 inches of snow fell by the end of January. Snowdrifts 25 feet deep were reported. Farms were cut off. Access to the moorland roads was exceptionally difficult. The military and the RSPCA airlifted supplies, including animal feed. Sheep and ponies stuck in snowdrifts, sometimes for days, were dug out. Many died. The apparent unpreparedness of the commoners for the worst unleashed a blizzard of criticism. Stories appeared in the press describing sheep buried in drifts being eaten alive by foxes; MPs condemned 'a most deplorable state of affairs'; local activists established the Dartmoor Livestock Protection Society; the DCA responded with a characteristic mix of defence, offence and attempts to reform;

and MAFF tried to head off the criticism while investigating whether it was justified.

As the crisis unfolded, fatalistic agriculturalists who regarded mortality among outwintered stock as inevitable met an impassioned animal welfare lobby outraged by the appalling suffering of stock exposed to blizzards and caught in snowdrifts. Public discussion tended to reduce the controversy to a crude problematic. Was high mortality in early 1963 the unfortunate consequence of exceptional circumstances or did it reveal disregard by commoners for the welfare of their stock? Agricultural subsidies came under close scrutiny, critics convinced they led to overstocking, though some national commentators picked up on local hostility to outwintering on the moor, particularly the opposition of traditionalists and preservationists. Newspapers ran articles under headlines like IS DARTMOOR TOO TOUGH FOR ANIMALS?[36] Animal welfare concerns quickly became the basis of a more wide-ranging assault on modern upland farming.

In the heat of the moment the focus initially fell on the aircraft drops of animal fodder and other supplies. Marcus Lipton, Labour MP for Brixton, demanded to know what the operation had cost the taxpayer. Keenly baiting the agricultural interest, he referred to 'the callous and mercenary farmers in Dartmoor . . . exploiting the humane instincts of the public in efforts to minimise the appalling suffering caused by the selfishness and negligence of these farmers'.[37] When Soames put the cost at £26,000, Lipton provokingly claimed during the adjournment debate of 26 February that this made the taxpayer 'a shareholder in the livestock which still survives on Dartmoor' – that neatly placed 'still' a scathing rhetorical flourish.[38] The airlifts, organised from Okehampton, Tavistock and Yelverton, relied on close cooperation between the army and the DCA, the latter identifying need and taking responsibility for ensuring farmers paid for the fodder. By 6 February, of the 178 farms helped in Devon, Cornwall, Somerset and Dorset, 52 were on Dartmoor. Mostly this help was supplied by road, sometimes using army lorries, though of 264 airlifts made throughout the region 38 directly supplied Dartmoor farms and 57 dropped fodder on the moor. According to MAFF, this was a small proportion of the total need, for which most Dartmoor farmers made their own provision.[39]

In parliament Lipton cited angry locals, an editorialist at the *Daily Telegraph* and the agricultural correspondent of the *Western Morning News* to the effect that the crisis had exposed overstocking and outwintering as problems on Dartmoor. Picking up on arguments made by the preservationist R. E. St Leger Gordon, Lipton demanded that the principles of levancy and couchancy be reinstated on the moor. Lipton's criticisms got short shrift at MAFF. P. Holmes, author of the most substantial MAFF brief on the crisis, rejected much of the criticism, arguing that taking stock off the moor during the winter would jeopardise the livelihoods of hundreds of farmers and weaken the stock, leaving it less suited to rough grazing. Maintaining Scotch Blackface sheep all year round on the common maximised productivity. And though Holmes did not think the introduction of headage payments in 1949 had led to an increase in the volume of stock on the moor, he accepted that on other British commons effective regulation ensured stockholders received payments limited to their stints, whereas the anarchic system on Dartmoor made it difficult to calculate either the moor's overall capacity or the fair stint of each commoner. Much had to be taken on trust, but Holmes reckoned informal systems of self-regulation inhibited overstocking, and although subsidy regulations did not prevent additional unsubsidised stock being grazed, the shared interests of all commoners ensured few abused the system.

The public debate was not helped by a tendency to confuse the gradual take-up of subsidy with absolute stock figures. MAFF figures stated that in 1949/50 195 applications were received for the subsidy of 26,703 ewes, of which 26,345 were paid, whereas in 1961/2 201 applications were made for 42,133 ewes, of which 37,729 were paid; in 1953 227 applications were made for 3,028 cows, of which

1,934 were paid, whereas in 1962 182 applications were made for 5,876 cows, of which 4,559 were paid. According to MAFF, these figures did not show a significant increase in stock levels on the moor but were evidence that by 1962 a much higher proportion of stock on the common was subsidised. The discrepancy between the level of subsidy applied for and the level of subsidy paid suggests judgements were being made about appropriate stock levels, limited evidence that the subsidy system had not become a free-for-all. Holmes did concede, however, that the 'accurate stock carrying capacity of Dartmoor will only be known when registration, examination of individual claims and assessment of commons usage has been carried out'. Either way, he was contemptuous of the idea that Dartmoor farmers kept sheep purely for the subsidy: 'A few shillings against the loss of £4–£6 for a sheep would be economic nonsense.' Lipton's view that the farmers had plenty of warning of the blizzard was not thought credible either. As a senior local official commented on a cover note to Holmes' brief, 'No prudent farmer with exposed lands could be expected to have anticipated and planned to meet the weight of trouble that he had to encounter.'[40]

Back in parliament Lipton found his accusations derided as 'extravagant' in a long and often sarcastic retort delivered by Sir Henry Studholme, MP for Tavistock. Not for the first time this dogged defender of the agricultural interest, expressing a ruralist prejudice no less predictable than its urbanite equivalent, wondered what a Brixton MP could know of conditions on Dartmoor. Similarly, the draft speech prepared for James Scott-Hopkins (joint parliamentary secretary to MAFF) referred to Lipton's 'cosy urban seclusion in Brixton', though at the despatch box Scott-Hopkins, a Cornish MP, was more civil.[41] F. H. Hayman, MP for Falmouth and Camborne, sympathised with the plight of the commoners and their livestock but did not disregard the complaints about animal husbandry he had followed in the correspondence columns of the *Western Morning News*. Accepting that hardy stock should be outwintered in the uplands, Hayman did not share the fatalistic attitude of the agriculturalists and insisted that alleviating the suffering of stock when conditions were exceptionally harsh must be the responsibility of its owner. Commons legislation might be some years off, as Scott-Hopkins had admitted, but still Hayman wanted to know what provisions had been made should the severe weather recur.[42]

Despite the bad-tempered debate, few doubted that the organisation of relief efforts could be improved. In an attempt to out-manoeuvre his critics, Studholme had already convened a closed meeting at Westminster on 14 February that brought together the duchy, the National Farmers Union, the RSPCA, the Dartmoor Commoners Association, the War Office, the Air Ministry and the Horses and Ponies Protection Society. Lipton's contemptuous description of it as 'snug and smug' was not too wide of the mark. All agreed that overstocking – and by implication agricultural subsidies – was not the problem on the common, although the RSPCA dissented from the view that outwintering was in the best interests of stock. Most importantly, cordial relations were restored between the commoners and the RSPCA. These had been stretched to breaking point by the RSPCA's decision to act unilaterally during the crisis, consulting neither MAFF nor the DCA before airlifting fodder. When a flock of sheep frightened by a helicopter scattered, causing some to drown in a river, resentment of the RSPCA's 'interference' became anger. During the adjournment debate Lipton was scathing about the DCA's criticism of the RSPCA, lambasting the ingratitude of commoners; more emollient, Scott-Hopkins reasoned that well intentioned interventions could have negative consequences. MAFF officials privately criticised the RSPCA for its refusal to cooperate, its decision to scatter fodder rather than supply individual farmers and its apparent desire to compete with the state for public acclaim. To the commoners the RSPCA's actions constituted trespass and illegal interference in private property. Prosecutions were threatened should it recur.[43]

As the immediate crisis passed, the DCA considered what provisions could be made to prevent it recurring. Central to its approach was the provision of strategically located fodder banks and closer liaison with the armed services and local authorities. All discussions sympathetic to the commoners, whether local or national, placed the emphasis on self-help. Thanks to undertakings made by Scott-Hopkins in parliament, MAFF kept an eye on the DCA's preparations, providing advice and chivvying them along, but it was careful not to offend local sensitivity to outside interference or to make commitments that exceeded its obligations or limited resources. It was telling that when the RSPCA sought a meeting with MAFF to discuss how to coordinate assistance in the future, officials accepted that the charity had an essential role to play because the ponies fell outside its agricultural remit but advised them to work directly with the DCA. A top-down attempt to integrate the RSPCA into DCA plans would encounter fierce resistance. At the same time internal discussions established that the state was under no obligation to help farmers feed or care for their stock; airlifts could be only authorised by the Treasury and then purely on humanitarian grounds. Fruitful talks between MAFF and the local authorities established that intervention by the services was only possible when there was a clear threat to life and property, and the police were unwilling to assume a general coordinating role during peacetime agricultural emergencies. Recommendations already made to the DCA had been taken up. Old ewes should be culled, ewe hoggs (young ewes) should not be wintered on unenclosed land, and plans should be laid for stockpiling emergency fodder. Holmes thought most commoners would adopt this advice, but new legislation for the management of the common must remain the long-term goal. Nonetheless, officials had 'to realise that as a result of extensive sheep stealing, personal feuds and fodder stealing, many of these farmers did not trust each other and, in his opinion, had no intention of co-operating. It seemed that they only co-operated to any extent when they were faced with a really serious emergency.'[44]

At the end of September the *Western Morning News* reported that agricultural merchants had agreed to stockpile winter feed in case of a fresh emergency, and new fodder banks were being established on Dartmoor: 26 in the north, where the climate was fiercest, 7 in the east, 12 in the south and 19 in the west. In an accompanying letter puffing this development T. J. Brown of the DCA applied a characteristic blend of astringent and emollient. Insisting the 'disastrous interference with stock that took place last winter' must not recur, Brown reminded readers that all stock on the moor, including the ponies, was privately owned, and access to the common was a privilege rather than a right, though the commoners saw no incompatibility between the common and the national park and did not want that privilege withdrawn. Brown also stated that the DCA welcomed specialist advice and was willing to work with the RSPCA on the basis of a new agreement – MAFF officials discreetly observed contacts between the two organisations over the coming months – but again he warned that legal action would be taken against anyone interfering with commonable stock. The healthy stock seen on the common that summer, Brown pointedly observed, had spent the winter on the moor, and it was absurd to suggest it should be otherwise.[45]

Not all were satisfied. On 21 October 1963 the inaugural meeting of the Dartmoor Livestock Protection Society was conducted in an atmosphere of righteous indignation. The new society declared its intention to alleviate livestock suffering on moors and common land, assist any official body working for the improvement of moorland and common management, and ensure that any new measures or laws were adhered to. The society opposed outwintering, believed Dartmoor was overstocked, and planned to press for changes to the subsidy payments it believed incentivised irresponsible stocking and neglect. Protest meetings were to be called, campaigning postcards printed and distributed, and demands made to widen the powers of the RSPCA and police and place greater restrictions on swaling. J. C. Clotworthy suggested patrols of the moor during the winter and bringing prosecutions when they encountered incidents of animal suffering; Barbara MacDonald, the most energetic member of the society, said they should oppose the new fodder banks on the grounds that they were a stop-gap solution that would prolong year-round grazing; and Sylvia Sayer, introducing the methods of the Dartmoor Preservation Association to the new society, suggested they compile evidence of neglect and hand it to the authorities. Unsurprisingly, the Dartmoor Commoners Association cast a cold eye on this sudden eruption onto the Dartmoor scene, which, if anything, drove them closer to the RSPCA.[46]

Some significance was attached to the new society's decision to name itself the Dartmoor Livestock Protection Society rather than the Dartmoor Animal Protection Society. The minutes of its inaugural meeting noted it feared coming into conflict with the 'hunting fraternity', but that alone is not an adequate explanation. Focusing on livestock welfare certainly gave the society moral purpose, but it was equally evident that members wishing to see changes in how the moorscape was utilised were only partly motivated by animal welfare concerns. Members wanted to see traditional Dartmoor pastoralism revived. According to the DPA, 'experienced Dartmoor farmers' did not outwinter their sheep, and those taking advantage of subsidies 'may not be farmers at all'. The DPA wanted the common re-dedicated to summer grazing, 'traditional' breeds re-established as the dominant flocks or herds and land use to accord with traditional common rights. Deep offence was caused by the absence of external regulation, the alleged incentive to overstock provided by state subsidies, which mirrored the tax breaks available to forestry, and the apparent unilateralism with which the commoners disregarded what the preservationists imagined was the proper use of the common. The perceived pre-eminence of the commoners in the moorscape challenged the notion that a landscape of Dartmoor's national importance should be managed according to amenity principles. 'Any new Committee of

Management,' the DPA insisted, 'must *effectively* represent all those who have rights on the Dartmoor commons – and this should include walkers and riders, whose *de facto* rights of air and exercise should be made *de jure*.' The commoners needed to be brought under control, their independence curbed, their behaviour subject to strict discipline. In particular, the DPA fingered the DCA as an 'association of graziers' representative of powerful agricultural interests seeking to marginalise holders of venville rights.[47]

It rankled that the commoners and the military had successfully cooperated for decades, that the commoners did not always show due regard for antiquities, and that the government treated their needs as broadly contiguous with the national interest. Whether or not the commoners were practitioners of good or bad husbandry, preservationists and animal welfarists found something intolerable about their sense of entitlement, apparently effortless defiance and infuriating obliviousness to other concerns. Despite the state's increasing presence in the upland, whether in the form of the army, the water authorities or the Forestry Commission, the agricultural *habitus*, with its intensely felt loyalties, rivalries and suspicions – as cooperative as it was competitive – remained barely penetrable to outsiders.

Noting the absence of sympathy for the commoners during the severe winter of 1962/3 is not to deny the neglectful practices of some stockholders, but it is to see as significant the shrill readiness with which outsiders unhesitatingly condemned a class of workers facing exceptionally arduous conditions. But there was something else at work here, barely acknowledged but palpable once discerned. The phlegmatic acceptance by the agriculturalists that increased animal mortality, consequent on outwintering on the moor, was inescapably natural needs to be challenged. Sheep – whether Dartmoor Whiteface or Scotch Blackface – are grazed on Dartmoor because they are of agricultural value. There is no natural sheep presence on Dartmoor. They are there for human purposes and human purposes alone. If they suffer, this is not, first and foremost, a natural phenomenon but a consequence of human actions. Sheep suffer during a severe winter because humans have put them at risk; stock dying of hunger or thirst do so because human beings have exposed them to danger. Writing of stock instead of animals, like writing of dehydration and malnutrition rather than thirst and hunger, might be emotionally distancing but it also reminds us that these animals are emphatically the responsibility of the farmers who raise them for the slaughterhouse – and the consumers that buy their butchered carcasses.

Taking hardy breeds off uplands over winter and feeding them on the sweet grasses of lowlands changes their digestive systems and metabolism, which makes them less able to cope with rough grazing once they return. This scientific truth is quite distinct from the assumption that uplands should be used for grazing because this sustains an ancient way of life, makes the land productive (though unprofitable without subsidies) and helps to sustain an upland ecology and a type of scenery that enthusiasts are accustomed to. By being alert to the assumptions underpinning agricultural orthodoxies, it is possible to develop a more historically nuanced understanding of farming practices, whether on Dartmoor or elsewhere. Some in the 1960s found it plausible to argue that because the skinny, undernourished ponies of March become the healthy, glossy ponies of September their capacity to survive these fluctuations provides evidence of their resilience and evidence that this annual cycle is natural. However, once it was accepted that the lives of the ponies or the sheep were not natural, so their emaciated state come early spring required justification for which their health in September seems hardly adequate.

Debate about animal welfare on Dartmoor was revived in spring 1968. The previous November the *Animal World* (official magazine of the RSPCA) led with the threat to moorland animals posed by a harsh winter, drawing particular attention to their plight in the West Country and the New Forest. In April Barbara MacDonald wrote to John Reid, chief veterinary officer at MAFF, about the daily reports she was receiving describing the 'parlous state' of Blackface sheep outwintered on the moor. 'As taxpayers,' she complained, 'we are utterly sick of donating money to enable these so-called hill-farmers to litter the moor with the carcasses of their stock.' Her bête noire remained agricultural subsidies paid on the basis of capital (the number of animals owned) rather than productivity (the number of lambs or calves successfully reared). '"Subsidy farmers"' overstocked Dartmoor with animals they had 'not the faintest intention of caring for'. In the *Western Morning News* Henry Forward, chief inspector of the RSPCA, voiced his concerns about the number of sheep found dead on Dartmoor that spring. A spokesperson for the DCA rejected these claims, saying he had not received many reports of dead sheep, though conditions were difficult because fresh grass growth had been late thanks to a bad winter and drying winds that spring. Editorial comment in the newspaper voiced some sympathy for the view that subsidies incentivised neglect. 'Will we have to wait for several years until there is new legislation on common rights before this situation is satisfactorily dealt with?'[48]

On 6 and 7 May these allegations were investigated by a team of officials under the direction of H. D. Abercrombie, MAFF divisional field officer at Exeter. Although the team did not speak with one voice, its results did not justify the usual MAFF apologia. A. E. Ridgeway investigated reports that feral mink were feeding on dead sheep carcasses near the River Mardle on Holne Moor and Buckfastleigh Moor. He followed the river for three miles, found three dead ewes and a dead lamb but no evidence of mink. Mortality did not seem higher than usual, though stocking levels were low. The problem, as Ridgeway saw it, was that the sheep tended to 'die by moor gates or on the lowest parts of the moor by streams where hikers and trekkers cannot fail to see them which repulsive sights outrage their unhardened sentiments'. Other reports drew back from this classic expression of agriculturalist realism.

J. Pritchard, finding thirteen dead ewes and five dead lambs on

Bridestowe and Sourton Common, helpfully supplied the ministry with the grid reference of the carcasses. J. C. Cotter and U. R. Perry, both MAFF field officers, offered contrasting views of the situation. Cotter found no evidence of dead sheep and said the farmers were constantly riding the common and were doing their best, as was their usual practice, to take the wool and bury the carcasses. 'The sheep appear to be in good condition and the lambs are going ahead.' His concern seemed to be less the level of mortality than whether the farmers were responsibly dealing with the dead. Burials at least removed the bodies from the public gaze. Perry, by contrast, wrote vaguely of the 'ghastly situation', reporting hearsay suggesting Whiddon Down in the northern quarter was the worst area, though Widecombe-Blackaton and Trendlebere Down near Lustleigh were reportedly bad. P. D. Sullivan, Totnes area officer, was specific in his report on the situation he found in the Burrator area. In four hours he located twenty-seven dead sheep. 'Carcasses were found inside the wood, on the open moor, under the shelter of walls and by, or in, water.' Sheep in poor condition suffer loss of wool, and Sullivan said he had never seen such extreme examples as those in the woods near the reservoir. A forester, a former farmer who still sheared 4,500 sheep a year, described the state of the Blackfaces as 'a scandal'. He and his 'mates were forever succouring weak sheep from brambles, etc. but many eventually died and then they had to move rotting carcasses when they were extracting timber. He estimated that there were about 30 carcasses in the woods at the time and took me to three within 200 yards of where we were standing. The carcasses are not readily observable in the woods. This man, too, stressed he did not wish to be "involved".'

Maps preserved in the ministry files show the location of carcasses near the River Erme. There were some seventeen near the Plymouth Water Works, twenty-one near Ivybridge at the small Harford Reservoir, eight further south near the A38, seven just north of Two Bridges and twelve in the Peter Tavy area near Horndon. O'Sullivan, who worked for the ministry and farmed three miles south of Dartmoor, admitted that until then he had not realised the extent of the problem. 'On more than one occasion he had come across a ewe, in trouble with lambing, with her eye pecked out, her lips and part of her tongue pecked away and her back end equally ravaged – and yet still alive.' He believed paying the subsidy on the basis of capital rather than production had incentivised neglect and, more unorthodox still, he questioned whether outwintering Blackfaces really was the most productive use of the moor. O'Sullivan urged the restoration of traditional common rights, the introduction of more productive breeds and the revitalisation of the moor as summer grazing. 'Many times during the checking of flocks for the new Basic subsidy this autumn,' he wrote, 'I encountered owners of other breed flocks who had Common right but said they were quite unable to exercise them because of the dominating presence of the Black Face on the open moor.' His informants, of course, did not want to be involved personally 'for reasons of neighbour relations and business'.[49]

Abercrombie's memorandum did not offer Whitehall officials a reassuring picture. His personal observations reinforced those of his officers. On 2 May 1968, in a half square mile near Ivybridge, he encountered seventeen carcasses. They had 'been degutted by foxes, ravens and crows and such flesh as there was had gone'. A freshly dead ewe, not yet eaten, displayed signs of malnutrition. A dead lamb, just a few weeks old, had definitely died of starvation.

As a boy it was common practice to leave a sheep until it was rotten in order to pluck off the wool, but I have never experienced a sight to compare with the veritable graveyard of sheep carcasses on Harford Moor; there cannot have been any concern shown whatsoever by the farmer for those sheep. They had been given no additional food – the grass – such as there was – was brown and arid and on 2nd May barely growing and lacking in nutrition. On top of the difficulties these sheep had in keeping alive more than

76 Galloway bullocks had been put on the area with the last few days just to aggravate the position.

Abercrombie's conclusion was uncompromising. 'You will note that the dead sheep have been observed in similar circumstances far and wide and the officers' reports are sufficiently graphic to leave no doubt regarding the callous and disgraceful lack of proper management.'[50]

Whitehall officials continued to fall back on established shibboleths. When a cattle farmer wrote outlining problems associated with lambing on the moor, highlighting the disputed ownership of carcasses and the failure of farmers to retrieve animals that strayed off the common into enclosed land, surviving drafts of the response suggest he was fobbed off with MAFF's pat realism: animal mortality was a fact of life for farmers, clearing carcasses from enclosed land was easier than clearing them from the common, previous investigations into paying subsidies on the basis of yield could not be made to work (lambing varied from year to year) and, despite some uncertainty in the law, the onus on keeping animals out of enclosed land lay with the landholder rather than the commoner. The high mortality that year was not attributable to individual farmers, 'the great majority of whom are fully conscious of their responsibilities and do their utmost to discharge them', but to exceptional circumstances, this time the drying winds that had damaged grazing. Animal cruelty, complainants were reminded, was a criminal offence and a matter for the police rather than the ministry. T. J. Brown of the DCA, they were told, could identify miscreant owners on the basis of sheep markings. More revealing than this reiteration of MAFF's standard apologia was an instruction marked on a later draft. A junior official was told to omit the claim that there was 'no evidence to suggest' that the high mortality rate was 'due to negligent owners exploiting these hardy breeds of sheep beyond their limit'.[51]

The controversy simmered on over the summer. At the annual

meeting of the Dartmoor Commoners Association in June the failure of some farmers to follow the 'technical advice given to them' and the reputational damage this inflicted on hill farmers was discussed. The DCA reiterated its intention to support the RSPCA in bringing any cases of animal cruelty against members. Meanwhile, Barbara MacDonald voiced first her disappointment that the Dartmoor Livestock Protection Society had not been invited to a meeting between the DCA and MAFF and then her astonishment when MAFF explained that the meeting was convened by the DCA – a non-statutory body! – and it was for them to issue invitations. In July, Anthony Greenwood received a letter from the CPRE, co-signed by the Dartmoor Preservation Association, the Ramblers Association, the YHA, the Dartmoor Rangers Club and the DLPS. Demanding a public inquiry, the letter expostulated: 'This is no longer a local disaster, but a national disgrace, as a National Park is being degraded and defiled by animal-suffering and its inevitable aftermath. Cruelty on this scale – and it is cruelty – can only demean those who condone it and lower the whole concept of National Parks as areas of special amenity value.'

When Housing passed the letter to Agriculture, MAFF responded that it was really about amenity rather than animal cruelty and was not their domain.[52]

Agriculture and Housing both thought a public inquiry would do little good and MAFF continued to insist that the subsidy system did not run 'counter to the farmer's normal commercial inducement to keep his animals alive and productive'. The old tendency to dismiss the lobbyists as unrepresentative re-surfaced, and MAFF was unenthusiastic when Housing revived the dormant notion that the solution was 'a statutorily constituted committee of management' empowered to 'adjust . . . individual grazing rights as established by registration, so that they might accord with the stocking capacity of the land'. Ideologically MAFF was committed to voluntary solutions and offered in counterpoint the code of practice promulgated by the DCA in October, essentially a repeat of its 1963 guidance. MAFF's reluctance to address the problem was clear in the response it furnished a query by Kenneth Robinson, minister for planning and land. 'Basically,' an official wrote, 'I am sure that this is problem of individual management.' This was not as offhand as it seems, for filed alongside Whitehall papers on the plight of Dartmoor livestock was a similar correspondence regarding the ponies of the New Forest. Problems occasioned there by the weather had engendered a similar public outcry, which somewhat undermined the argument that the solution to the problems on Dartmoor was its own version of the Verderers' Court. Still, opinion in Devon looked to statutory solutions. In September 1968 the Dartmoor National Park's head warden told the *Western Morning News* that it was well known that the DCA thinks 'animal care on the Moor leaves a lot to be desired' but 'there is no action that the Commoners Association can take against any of its members who are guilty of negligence'.[53]

The DCA's opponents were on the warpath again in early 1969, this time taking their complaints to Devon County Council as well as MAFF. MacDonald provided a list of sixty dead sheep that had been brought to her attention, made her usual points about subsidies and observed that the national park had been described to her as a 'national cemetery'. 'The sad account of sheep losses, such as you have catalogued,' came the condescending response from Whitehall, 'is unhappily the not unusual sequel of a sudden severe snow blizzard.' The official then airily waved away MacDonald's notion that a properly nourished sheep could survive more than forty-eight hours in a snowdrift: sheep were dying because of the weather, not because of neglect. County Hall's concerns were similarly dismissed, as again was MacDonald when she met an official at Exeter that June, though K. Harrison Jones admitted in his report that a close examination should be made of the quality of hill farming on Dartmoor and suggested the ministry seek the opinion of the NFU, the police and the RSPCA.[54]

★ ★ ★

There can be little doubt that in the springs of 1963 and 1968 the unwary hiker might have encountered some grisly sights on the common. Figures suggest that in 1963 the mortality rate among sheep was about 10 per cent, twice the usual average, which in absolute terms amounted to as many as 5,000 of some 50,000 sheep outwintered on common.[55] The carcasses that came to public attention or were reported to the ministry thus represented a small proportion of the total, suggesting most farmers dealt fairly efficiently with their dead. This would have provided scant comfort to those who regarded the problem as less a question of agricultural hygiene and more whether 10 or indeed 5 per cent mortality was an acceptable price to pay for the outwintering of sheep. Concerns about the welfare of pony and cattle notwithstanding, this was most pressingly a problem with ewes, and there is little doubt that the relatively low value of sheep helped account for the risks many farmers were prepared to take. Sheep farmers accepted that they would lose stock over winter in a way that cattle farmers did not.

The Ministry of Agriculture's performance during these crises was dissembling. They perceived a problem they were not prepared to admit publicly and for which they made little effort to conceive a solution. Institutional bias and a 'realist' mindset were certainly at work here, but so too were pressures on parliamentary time and the simple fact that Dartmoor was low on the political agenda. Wilson's governments confronted rather more pressing issues in the late 1960s. None of this was helped by the commons registration process, which by November 1968 had seen just 720 of an estimated 1,250 Dartmoor commoners register their interests, and it seemed likely the process would not be completed before 1974. Consequently, the ministry tended to take refuge in voluntary solutions and to trumpet the DCA's guidelines, comforted by the organisation's condemnation of the recalcitrant minority who failed to live up to their responsibilities.

Some experts conceived of the problem differently. Holmes appended a series of 'personal observations' to the long brief he wrote on the 1963 crisis. He supported the initiatives announced by the DCA and ventured no criticism of outwintering, but believed things needed to go further, adumbrating a new agenda for Dartmoor's improvement. Providing additional feed was of negligible importance but erecting fencing throughout the common would be transformative, allowing the even distribution of stock and grazing. Fencing would also facilitate the application of lime and phosphate dressings to lower the acidity of Dartmoor's soil and produce sweeter and more nutritious grazing. Planting shelter belts of trees on the exposed open hills would dissuade sheep from congregating in the valleys and on lower-lying lands, again minimising overgrazing in some places and undergrazing elsewhere. If fixed ideas about the common were overcome, Holmes believed that many of the problems that so offended public opinion could be resolved through some elementary measures of improvement. However, only with a statutory body of commoners could they be made to work. 'It is felt that there is a genuine desire on behalf of the Commoners, the Duchy of Cornwall, and the landowners', Holmes wrote, 'to further improve the contribution made by Dartmoor to the "National Larder".'

C. D. Spencer, another official, made a similar case during the 1968 controversy. He thought the situation on Dartmoor was typical of difficulties evident throughout the country that year. High animal mortality was attributable less to the weather or even food shortages in the crudest sense and more to deficiencies in trace elements (vitamins and minerals) in diets. Lime dressings and other treatments were the solution. A hundred and fifty years earlier Sir Thomas Tyrwhitt wanted to apply lime so that flax could be cultivated on the moor, modern agriculturalists thought the same methods could be used to improve the moor for grazing. Spencer identified two key impediments. First, implementing improvements on a large open common was difficult because it was hard to guarantee that the stock of an individual farmer would be the sole beneficiary of his investment – the free-rider problem – and second, such measures were likely to be most vigorously opposed by those complaining loudest about animal welfare.[56]

Improving the effectiveness of grazing by inhibiting free movement across the common or altering its look was unacceptable to the preservationists. During the hearings of the Commission on Common Lands Dudley Stamp had challenged Sayer's view that grazing on the common was no longer Dartmoor's primary purpose. Given that Dartmoor was an 'artificial creation' formed by grazing, a hybrid mix of human and non-human agency, surely it was in the amenity interest that the common be better managed through the introduction of additional fencing?[57] Stamp's provocation, as much a playful response to Sayer in Joan of Arc mode than anything else, raised significant questions.[58] In the decades that followed Sayer not only had to deal with a revitalised National Park Authority but also a political consensus that sought to achieve 'balance' in how common land was used, including that of the national parks. The Nuffield Report into Common Land (1966) questioned agriculture's pre-eminence on common land and affirmed the importance of amenity, but it also raised the status of activities the preservationists found objectionable. Agriculture, the report argued, had 'no more than an equal claim along with public forestry, conservation of natural resources, preservation of scenic beauty and the personal interests of owners of the soil and of the common rights upon the use of common land'.[59] As balance was elevated to a political principle, its meaning on Dartmoor would be played out in the long struggle to constitute the Dartmoor Commoners Council.

Achieving Balance

In the immediate post-war years the preservationists hoped the beneficent exercise of state power would secure the Dartmoor status quo, and they considered the moorscape's designation as a

national park as largely a defensive measure. Over the decades that
followed experience taught the preservationists to think otherwise,
and by the 1970s the state was seen less as a prophylactic against
change and more its facilitator, the reservoirs, the plantations and
the military training areas making a mockery of Dartmoor's
national park status. A consequence of this deeply ingrained
mistrust of the state was the sceptical welcome given the new
National Park Authority in the 1970s with its interventionist talk
of planning and management, its access to professional and expert
opinion, and the evident determination of its leading officers to
deploy their new powers.

When Ian Mercer addressed the Dartmoor Preservation
Association in 1975, his wide-ranging and ruminative talk was
begrudgingly written up in the DPA newsletter as a 'philosophic
sermon', an unfriendly phrase that reflected resentment at an
outsider – Mercer was a conservation professional born in the Black
Country – telling them what Dartmoor was. Highlighted in the
write-up was Mercer's observation that 'the whole satisfaction and
pleasure perceived by modern man as an attribute and value of the
countryside was a spin-off from a primarily food-producing
process'.[60] If this was no more than an expression of the assumptions that had underpinned attitudes towards the park since its
beginnings, those alert to the nuances of Dartmoor politics would
have noticed the coming shift in emphasis it signalled. For Mercer
there was no essential tension between the authority's duty to
conserve the park's natural characteristics, preserve its beauty and
historical features, facilitate access for visitors, and manage the park
in accordance with the socioeconomic interests of its inhabitants.
A harmonious Dartmoor in which different interests were balanced
was only possible if the commoners were brought in from the cold.
Mercer invested a great deal of energy in building a close relationship with the Dartmoor Commoners Association, probably the
most instrumental of his long career as the national park officer.

According to the first Dartmoor Plan, published in 1977, 'The

NPA will assist the Commoners to achieve a sound management system, incorporating safeguards for the maintenance of vegetation appearance. The first step is legislation giving statutory backing to such a system, the second will be establishing the system, under the legislation. The aim should be to achieve this within the first five years of this Plan.'[61]

The authority was as good as its word. Working closely with Devon County Council, consultations regarding a draft bill were in train by June 1977. The bill had a dual purpose. It would constitute a commoners' council with statutory powers to manage Dartmoor common lands in the interests of the commoners, and it would grant the public right of access on foot to all the common and empower the county council to regulate this access. The commoners' council would be empowered to ensure that all commoners practised good animal husbandry, maintaining the health of all animals on the common, fulfilling demands that had been heard for decades: the number of animals depastured on the common would be fixed; uncastrated stallions and rams ('entire animals') would be closely controlled; unthrifty animals or animals in such condition 'that it would be likely to cause unnecessary suffering or offence to members of the public' would be removed from the moor; swaling would be more closely regulated, and the council would be empowered to order that the moor be cleared for dipping or for treatment during an outbreak of disease. A register of all commoners would be kept, and Devon County Council would appoint reeves to enforce the new council's rulings. Concern about inexperienced or negligent farmers placing stock on the common was addressed by a clause in the bill preventing any commoner from selling rights attached to land: with the bill's passage, it would become impossible to exercise rights independently of a landholding.

If little of this was controversial, trouble was soon found. Clause 9 contained the dread word 'improvement', allowing each parish to improve twenty-five acres a year, which one calculation suggested

could total 750 acres per annum. This was broadly consistent with
the 1976 White Paper *Food from our Own Resources*, which looked
to improvement as a means of sustaining a growth rate of 2.5 per
cent per annum in agriculture – ministers had already brushed aside
fears that improvement would undermine the traditional appearance of the countryside.[62] When the Department of the Environment
sought Nature Conservancy's view of the Dartmoor bill, alarm
bells rang. In the section of the draft bill that follows the italicised
passages are those Nature Conservancy highlighted in pencil and
the square brackets contain its marginalia.

> Subject to this Act, it shall be the duty of the Council to take all
> steps appearing to them reasonably practicable for the maintenance,
> management, improvement and regulation of the commons, with
> a view to ensuring that they are put to the *best* [optimum!] use for
> the purposes of agriculture; and in discharging that duty the Council
> *shall have regard* [taken into account] to the *preservation and enhance-
> ment of the natural beauty of the commons* [conservation? does natural
> beauty mean conservation flora, fauna etc etc] and its use as place
> of resort and recreation for the enjoyment by the public.[63]

Nature Conservancy's fears reflected wider conservationist
concerns about the effect intensive cultivation and new systems
of grassland management were already having on the refuges wild
plant and animal life needed. Was the purpose of the Dartmoor
legislation to permit the moor to be improved in order to maximise
agricultural yields? Would the protections offered preserve not
only natural beauty but also the ecosystem? Would provisions
allowing the prevention of access in order to protect the 'physiological' character of the moor extend to the conservation of the
natural environment? R. B. Nicholson, Nature Conservancy
regional officer, feared that the commoners' council might quickly
become a self-appointing club, and he insisted that the protection
of 'natural beauty' must be explicitly extended to include flora,

fauna, physiographic and geological interests. Clauses allowing 'very drastic operations on the natural vegetation that would materially damage its wildlife interest' must be altered. Although the bill forbade ploughing, it contained no safeguards against either the clearance of flora using harvesters or tractors with swipes or spraying large areas of the common with pesticides or fertilisers that would favour certain grasses and gradually see the destruction of heathers and other naturally occurring plants. A chemical like dalapon, Nicholson explained, could completely destroy natural vegetation in order to allow reseeding with more agriculturally productive species. Perhaps the technology to allow this without ploughing did not yet exist, 'but agricultural situations change with astonishing rapidity and it might easily become practicable in a few years' time – given new chemicals and new tested strains of grasses and other crops to carry out agricultural "improvements" on this scale'.[64]

Nicholson reported that Mercer was privately unhappy with the improvement and conservation aspects of the bill and sought outside support for its amendment, evidence perhaps of the complexity of his relationship with the commoners. Recognising that the bill needed more thought, Devon County Council abandoned it in September and invited Nature Conservancy to help Mercer draft an interpretation of 'natural beauty' that could be inserted in a new bill the following year. In a carefully prepared letter Nicholson explained to the county council that Nature Conservancy considered the whole of Dartmoor a Grade I site and that in drafting a new bill it needed to understand that the exceptional value of the common was the 'result of an interaction of climatic and soil factors, coupled with the traditional land uses'. These conditions had survived because the land had not been subject to intensive agricultural use. Herbicides, fertilisers and reseeding would endanger the ancient ecosystem of the common while the danger of run-off into Dartmoor's rivers threatened to pollute adjacent areas.[65]

The new draft of the bill went some way towards satisfying these concerns. The preliminaries explained that references in the act 'to the conservation of natural beauty . . . shall be construed as including references to the conservation of its flora, fauna and geological and physiological features'. The improvement clause now included an obligation to notify Nature Conservancy 'before undertaking, or giving permission to others to undertake, any operations which might be detrimental to the flora, fauna or geological or physiographical features' of Dartmoor's national nature reserves or sites of special scientific interest. Notification did not grant Nature Conservancy any significant new powers, and Nicholson still found the bill 'fundamentally unacceptable': it was too permissive and the inclusion of 'improvement', a term whose Victorian implications were at odds with modern nature conservation, was unconscionable. The bill's promoters attempted to provide reassurance, arguing that there was a symbiotic relationship between amenity and good management of the common. 'Both unfettered access and lack of good management lead to problems for the farmer and the visitor. Poor grazing and tatty vegetation go hand in hand. The yield of store sheep and cattle, and the yield of enjoyment, are both diminished by the lack of discipline.' This could not satisfy Nature Conservancy, whose interest was ecological rather than aesthetic or productivist. In an internal confidential note their fears were spelt out. Potential developments in pesticides, herbicides and fertilisers meant the prohibition against ploughing or breaking the surface of the soil did not provide an 'insuperable obstacle to reclamation and major impoverishment of the wildlife interest'. The bill's failure to protect the common against the future became the foundation of their opposition. By their reckoning, inadequate protection in the draft bill imperilled up to 20 per cent of the most ecologically valuable landscape in southern England.

Nature Conservancy had a choice. It could either persuade the Department of the Environment to report against the bill, effectively killing it, petition against it or appear as an expert witness

for another petitioner at the committee stage. Further correspondence confirmed that the promoters would not yield on improvement but would consider strengthening existing controls over the
use of chemical herbicides and might allow a Nature Conservancy
representative to sit on the commoners' council as one of the two
national park authority nominees. As the promoters explained, the
bill established no new rights, and to introduce new controls over
the use of chemical fertilisers would in fact create new restrictions.
Nature Conservancy, though unable to accept the injunction to
improve, decided to keep its powder dry. It would decide its
approach once the bill was deposited, and in the meantime hoped
the Department of the Environment would push for its representation on the council.[66]

Despite these early-warning signs, Devon County Council and
Dartmoor National Park Authority pressed ahead. The procedure
for a private member's bill is relatively straightforward. It is initially
presented to the Commons or the Lords depending on where the
sponsor sits. At the first reading the bill's headings are read and
then it is printed. At the second reading there is opportunity for
debate and a first vote. If the bill passes the second reading it passes
to the select committee stage. Time is given for petitioners opposing
the bill to register their objections, and they can expect to be called
to testify before the committee. The committee then reports to
parliament on the revised bill, explaining any amendments that
have been made. If the bill passes the third reading, it transfers to
the other house, where the same process occurs. When both houses
have completed this process, amendments need to be approved by
both houses before the bill can acquire royal assent and pass into
law. If the bill fails to pass a reading in the Commons it fails, whereas
if a bill passed by the Commons does not pass a reading in the
Lords, the Commons can return it to the Lords up to three times
before the Commons can pass it.

Lord Sandford presented the bill to the Lords in 1978, and it
quickly progressed to its second reading before being scheduled

for examination by a Lords select committee in June 1979. That February opponents of the bill began to lodge petitions. Collectively, they constituted a pretty comprehensive critique of its provisions. A list of one hundred commoners, mainly from the western side of the moor, argued that the bill was premature, for the commons registration process was still incomplete, and, more fundamentally, its implementation would constitute unwarranted interference in a system of local management that already functioned well. They also imagined the new council would require a large staff, creating an unacceptable financial burden on the commoners, and the new rules on access were found too unspecific.

Petitioner Alfred Johnson, of Jurston near Chagford, thought the proposals with respect to good husbandry excessive and found the bill gave the commoners too much power at the expense of the national park principle. Barbara MacDonald and the Dartmoor Livestock Protection Society objected to the improvement clauses and opposed the bill because it did not require the winter clearing of stock and gave too much power to the larger graziers. Rosemary Hooley and David Moore of Skaigh Stables Farm, Belstone also thought the council would allow large graziers to consolidate their control over the common. In separate petitions the Dartmoor Preservation Association, Sylvia Sayer and Kate Ashbrook, one of a new generation of DPA activists, while not opposing the bill in principle, opposed the improvement clauses and demanded that the council be required to consult more widely. Nature Conservancy, the NPA, the Countryside Commission and, where relevant, the Ancient Monuments Department of the Department of the Environment were among those the preservationists thought were entitled to veto any act of improvement. Sayer was particularly disturbed by the bill's failure to expressly uphold venville rights. The Royal Society for the Protection of Birds insisted that nature conservation and amenity interests should be given equal billing with agriculture on Dartmoor, and they too joined the chorus arguing for much greater protection against possible damage from improvement. The Camping and Caravan Clubs of Great Britain and Ireland, the Ramblers' Association and the British Horse Society, all organisations with considerable political clout, objected to new restrictions that would affect their members.[67]

Although careful amendment might satisfy some of the petitioners, there could be little doubt that Ian Mercer, speaking for the bill's promoters, was going to have a tough time before the select committee. As counsel for the promoters, Roy Vandermeer opened proceedings, explaining the objects of the bill to Lords Listowel, Craigavon, Gridley, Halsbury and Morris. Central to his case was the fact that although 1,511 people claimed rights of common on Dartmoor, 1,279 of whom claimed grazing rights, only 400 commoners currently put stock on the common. Defending the decision to open membership of the commoners' council to commoners grazing 50 or more sheep (according to the standard calculation this was the equivalent to 10 cows or horses), Vandermeer explained that only those with 'a significant interest' in the common should be entitled to decide how it was managed as an agricultural resource. As for the maximalist reading of the 25-acre clause, whereby 800 acres might be subject to improvement each year, this was 'a quite artificial way to look at the matter'.[68]

As Mercer explained under questioning, the kind of improvement they anticipated was limited to remedying mineral deficiencies or restoring overgrazed parts of the edge of the common. Adding cobalt or lime dressings was already permissible and unlikely on a large scale unless there was significant facilitating action from the government. 'My experience over the last twenty-five years,' Mercer joked, 'suggests that very few farmers do anything unless they get a grant to do it!' And should the commoners prove sufficiently cooperative to attempt large-scale dressing, fencing any part of the common would still need the permission of the secretary of state, who would be bound to consult other interest groups.[69] Facing hostile questioning by George Laurence, counsel for the DPA and the Ramblers, Mercer argued that conservationist fears were based on a misunderstanding of the commoners. Not only should the preservationists accept that the commoners wanted to maintain the rough grazing because it suited their stock, but they should learn to recognise the pride and expertise of the farmers and not demonise them as intent on doing everything possible to avoid restrictions based on environmental principles. Mercer insisted that good husbandry maintained the 'mosaic of vegetable patterns' that was so valued on the common.[70]

During close questioning by Robin Purchas, counsel for the British Horse Society and the camping and caravan interest, the redundancy of aspects of the bill's improvement provisions was exposed. Purchas' insistence that the bill did not provide sufficient protection against the commoners acting against Mercer's expectations was met with Mercer's insistence that the national park authority was already powerless to prevent the commoners from improving the common once they had reached agreement with the landowner.[71] Mercer also conceded that drainage schemes might mean breaking the surface of the moor, though Herbert Whitley, chair of the Dartmoor Commoners Association, played this down, saying commoners only wanted to drain eyes, small waterlogged areas that contained good grazing but were a danger to animals. He admitted these small operations involved using mechanical diggers.[72] Stewart Houndsdon, representing the RSPB, pursued the same line of questioning, forcing Mercer to admit that although the commoners had to seek Nature Conservancy's advice regarding actions that might affect Dartmoor's extensive sites of special scientific interest, they did not have to adhere to this advice. Public opinion, Mercer believed, would keep the commoners in line.[73] Whitley's testimony did not entirely bear this out. He provided a good account of the need for swaling and the essential role the council would play in ensuring negligent commoners properly hefted their flocks. On this point, he educated the Lords:

It is a question of getting animals that you wish to keep together, you take them to a certain place, and you keep driving them back to this place until such time as they have got an understanding that they have got to be there. Then the flock remained hefted for all time, because you keep ewe lambs, they remain with their mothers, and you get your return from the old ewes and the weather lambs. But once a flock is hefted, it is hefted for ever.[74]

He was less diplomatic when it came to the workings of public opinion, expressing what was often only implicit in Mercer's testimony:

On the Council, we want to confine our business to looking after the interest of the people who put the £7 million of capital on the moor, and we think that it is perfectly fair that the agricultural people should run the agricultural interests. It is a very unfortunate thing at Dartmoor . . . that it is more or less accepted that everybody in the whole of England knows more about how to run cattle, sheep and ponies on Dartmoor, than the farmers responsible themselves.[75]

Whitley represented all that the preservationists feared about the powers the bill would extend to selected commoners, and this was just the kind of comment liable to confirm their worst suspicions. Whereas the typical stock level of an individual commoner might have been 50 breeding cows and 100 breeding ewes, Whitley farmed in a partnership, keeping 250 breeding cows and 700 breeding ewes on enclosed land and 850 breeding ewes on the common. After lambing Whitley and his partner had something like 1,500 head of stock on the common. When given the opportunity to question Whitley, the preservationists quickly got their teeth in. Sayer pointedly informed the Lords that hefting was called lairing on Dartmoor, implying that Whitley's use of the more commonly understood term revealed his inadequate appreciation of the moor and its ways; she then followed a line of questioning intended to undermine Whitley's integrity regarding his respect for antiquities and nature. MacDonald gave vent to her usual preoccupations, asking questions about overstocking, outwintering and the pernicious effect of subsidies; Ashbrook focused on the twenty-five-acre limit, arguing that it was not possible to restrict the effect of spraying to such a small area; and Dr Beech voiced the DPA's concern that although it broadly approved of the bill, it was concerned that it gave the proposed council power to improve without being responsible to an external supervisory body.[76]

On the access question the bill received a similarly rough handling. Few objected to the principle that pedestrians should have full access to the common, but when questioning turned to the national park authority's desire to exert greater control over caravanning, camping and horse riding the depth of opposition from special interest groups became clear. To Vandermeer the position seemed straightforward enough. Common rights for unshod horses and ponies existed on the common, but there was no existing right to ride shod horses on the common except on designated bridleways. The bill confirmed what was already in law and gave the park authority the right to restrict the use of bridleways where overuse was causing damage. In particular, Mercer explained, organised parties riding in single file caused much damage, which of course increased when the same itinerary was routinely followed. Under questioning it was established that it was not an offence to ride on the common when notices were not displayed to this effect.

Articulate and in command of his brief, Mercer put in a creditable performance during those gruelling few days, but as the questioning gradually highlighted the flaws in the bill so his answers were increasingly undergirded with a weary irony. A permissive bill that relied on faith in the goodwill of the protagonists it sought to empower rather than formal limitations, it did not stand up to scrutiny. Too much was left to chance. But more than this, the bill failed to recognise that public opinion no longer accepted that the agricultural interest should by right be the ascendant interest on the common. When the bill made it to the Commons and the second reading debate on 24 April 1980, it stood little chance of succeeding.

Contentious procedural questions created a fractious mood in the House, and Ray Mawby's (Totnes) introductory speech faced repeated interruptions. Denis Howell (Birmingham, Small Heath), former minister in the Department of the Environment and the bill's most trenchant critic, was particularly disturbed that the promoters hoped the bill would be voted through and its difficulties dealt with at committee stage. He objected to the assumption that these questions could be settled between the promoters and the petitioners in committee because the issues at stake embodied 'extremely important principles concerning the role of national parks, how they should be controlled, rights of access to them and, most important of all, the relationship between national parks in a national setting and local interests'.[77] Early in the debate Anthony Steen (Liverpool, Wavertree), a centrist Tory, provocatively raised the problem of outwintering, claiming that 'the people who are abusing the commons of Dartmoor include the very people who will serve on the commoners' council'. Those convinced that outwintering was wrong regarded almost all active commoners as abusing the common, and it was thus hard to see the bill as doing anything but strengthening the perpetrators of bad practice. As such, Mawby's insistence that the commoners would not elect representatives guilty of 'malpractice' missed the point.[78]

Mawby's explanation of the access clauses, the new powers to create by-laws relating to particular forms of recreation and the appointment of wardens to assist the public and enforce the by-laws was met with further objections. To Mawby's point that under existing law access to the common was only allowed with the permission of the landowner, Steen rhetorically asked why it was 'important to be bureaucratic and establish a legal right which has not been needed for 1,000 years'. James Wellbeloved (Erith and

Crayford) took up the point, arguing that establishing a statutory right of access provided the government with the power to limit access to Dartmoor, which 'British subjects' have enjoyed 'almost from time immemorial'. Howell upped the ante, suggesting the notion 'that the millions of people who visit our most important national park every year are there on sufferance' was 'the most extraordinary and breathtaking proposition that has been advanced in the House since the war'. Howell surely grasped that establishing a statutory right of access undermined the legal rights of the land-owners, so it seems that what underlay his display of outrage was the way the bill reified property rights thought to be redundant. These rights might not have been tested in the courts for many years, but the bill's access clauses renewed their credence in a way that Howell found unacceptable. If Steen's intervention reflected classic anti-state Toryism, Howell offered a leftist reading of common land and the national parks. Why 'agricultural interests', important thought they were, 'should predominate over the recre-ational interests of the entire nation' was beyond him.[79]

Mawby struggled to deal with these objections. Explaining that the park authority needed greater powers to manage the behaviour of ever-increasing visitor numbers, not of all of whom could be expected to behave 'reasonably', he unwisely cited the example of the pollution caused when a caravan was parked next to a stream. Wellbeloved was on his feet in a moment, requesting that the remark be withdrawn. As the national vice-president of the Camping Club of Great Britain and Ireland, Wellbeloved insisted his members adhered to its code of practice, which ensured they did not pollute the countryside or leave litter. Mawby wearily took the point but went on to argue that by-laws were needed to prevent 'caravans, Dormobilies and high coloured frame tents' from pitching wherever they chose, which 'spoils the enjoyment of the open landscape for others and gives rise to the risk of disease among commoners' animals and to the risk of injury from certain types of litter'.[80]

Mawby came to the end of his speech exasperated that members were offended that regulation, long seen as necessary to a properly ordered society, should be thought so offensive when extended to Dartmoor. The debate that followed offered him little comfort. In a long response Harry Greenway (Ealing North) described the attitudes he had encountered on the moor as 'almost antediluvian' and questioned how eight million visitors to such a large area could be considered a threat. In the bill's attempts to regulate access to the moor Greenway detected rural resistance to its obligation to provide space for recreation to an ever more urban society, but above all he opposed the restrictions on horse riding. As a member of the British Horse Society and foundation chair of the London Schools Horse Society, he spoke emotively of how extraordinarily beneficial horse riding was to disadvantaged and mentally or physically disabled children and lamented the prospect of a situation when riders would need a 'chit' to ride, losing the spontaneity that made riding so exciting. It was a little fanciful to think that children who were 'blind, deaf, physically handicapped, mentally handicapped, educationally subnormal, autistic, maladjusted and delicate' could spontaneously pitch tents on the common but such is the nature of parliamentary debate.[81]

Howell spoke again from a similar perspective, declaring his interest as the chair of a trust in Birmingham for deprived children. He found it extraordinary that the powers of regulation would lie with significant stockholders who could become members of the Dartmoor Commoners Council. Steen called this the 'worst form of discrimination' he had 'ever heard of', prompting Howell to say it 'knocks into a cocked hat all the racist and sexist [sic] discrimination legislation that the House has ever proceeded to enact'. All of this was getting a bit silly, and Mawby corrected the misconception, explaining that under the legislation the national park authority rather than the council would be authorised to regulate access. Howell blithely passed over the point of detail before restating his fundamental objection that questions of such national significance should not be settled with a local bill.[82]

The lukewarm support in the House for the bill was no match for Steen's long closing speech. It included a 'poetic and descriptive view of the moor', condemnation of urban attempts to 'defile' it, a populist swipe at planners, an exceptionally pedantic intervention from Douglas Jay ('Yes Tor is not 2,029 ft but 2,028 ft'), a brief paean to Sayer ('remarkable, august lady'), and a strongly Tory peroration exulting small government in which the bill was identified as 'a brave attempt to control something which, I am happy to say, is out of control'.[83] MPs filed into the division lobbies, nineteen for, sixty-three against. There were enough opposition MPs to ensure the bill's defeat, but still the noes were evenly divided between Labour and Conservative members.

At the 1983 general election Steen changed constituency, moving from gritty Liverpool Wavertree to bucolic South Hams, where he quickly went native, buying the country mansion that proved his undoing during the parliamentary expenses scandal of 2010. In June 1984 Steen reprised his rhapsodic account of the moor, but this time he was introducing the second reading debate of a redrafted Dartmoor commons bill. Explaining that the 1980 bill fell 'largely because it got the balance wrong between agricultural interests and environmental and conservation interests', Steen championed the new bill, accepting the symbiotic logic pushed by Mercer and the national park authority: 'The conservationists need the commoners to maintain their livestock and so keep the commons accessible. The commoners need the ramblers, the Open Spaces Society and other conservation groups to ensure that the land is safeguarded.'[84]

Peter Mills (Torridge and West Devon) had approved the original bill, so he had no difficulty approving its reincarnation. 'The question is one of balance. No one interested party can have it all its own way. The farmers cannot, nor can the preservationists or tourists. There must be balance. The Bill achieves that balance.'[85] And as Steen explained, no Conservative need be wary of the bill for the commoners' council 'will be self-regulatory and self-financing',

fulfilling 'the old adage and Tory philosophy that self-help is probably best'.[86] Out went his celebration of Dartmoor's 'out of control'
character and in came his concern that better management of the
common was needed:

> Although common rights are in theory enforceable, in practice no
> one will take the necessary action. Stories of cowboys unloading
> cattle trucks on the commons in the middle of the night are rife.
> But nothing is done to follow up those stories or to pursue the
> cattle that have been put out on the moor. Some farmers do not
> seem over-concerned about the sickly state of their animals, and
> that is not being challenged either. Without power to deal with
> overgrazing, the vegetation of many commons, especially those
> near the road, have been eroded so badly that besides looking
> unpleasant there is little nutrition left for the increasing numbers
> of stock. The less meticulous farmers are driving out the more
> conscientious grazers who will not risk their animals being affected
> by sheep scab or brucellosis.[87]

During the drafting of the new bill petitions lodged by a wide
range of local interests were accommodated by the promoters, and
the Dartmoor committee of the county council agreed to establish
a standing subcommittee on which the amenity societies would be
chiefly represented to advise on the use of the common.[88] The bill's
final form could have hardly been more responsive to the criticisms
levelled at the first. The preamble acknowledged venville rights;
places were reserved on the council for four members who grazed
less than ten livestock units on the common; the council was to
'have regard to the conservation and enhancement of the natural
beauty of the commons and its use as a place of resort and recreation'; and planting 'clumps' rather than 'belts' of trees was
permitted providing they were 'of a broad-leaved species naturally
growing on Dartmoor'. As regards the council's obligation to maintain the common, notice had to be given to Nature Conservancy

of any operation that might be 'detrimental to the flora, fauna or geological or physiological features' of any part of the common designated an area of special scientific interest, which by 1985 included most of the upland. Reeves would be appointed to ensure the observation of new regulations relating to good animal husbandry, including stocking levels, hefting, the control of male entire commonable animals, the efficient removal of dead animals, the exclusion from grazing of shod animals and bulls exceeding the age of six months, and swaling. The preservationists were pleased that the 'dangerous' improvement clause had been dropped, though they remained concerned outwintering was not be prohibited and thought the power to plant and fence clumps of trees was 'outside the realm of mere livestock husbandry'.[89] The animal welfare interest was also satisfied: Conservative MP Janet Fookes (Plymouth Drake) recalled her 'sickening sense of disappointment' when the first bill failed.[90] Much she said reflected the claims of the Dartmoor Livestock Protection Society, which, notwithstanding its opposition to outwintering, were now accepted on all sides of the House. On this issue at least Labour MPs could no longer bait the shire Tories.

A major concession to the equestrian interest was made on access. Clause 10(1) simply stated that 'the public shall have a right of access to the commons on foot and on horseback for the purpose of open-air recreation', though the park authority reserved the right to restrict horses let to members of the public 'for hire or reward' to particular tracks. No one sought to contest Steen's justification that commercial riding stables with thirty or forty horses going over the same track several times a day 'badly erodes the land'. There was nothing specific for the caravan, camping or hang-gliding interest to object to, though the county council was not short of power to make by-laws in that respect. And though Steen acknowledged that a minority of landowners objected to statutory right of access, he did not detect any fundamental desire on their part to restrict customary rights of access.[91] Few objections were raised against the park authority's new right to temporarily deny access to parts of the common when necessary to protect the general public against man-made dangers such as quarries, pits, ponds or mineshafts, or to protect any of the common's natural, agricultural or historical characteristics, which included freshly planted trees.

The government's support for the bill was unambiguous. William Waldegrave, junior minister at the Department of the Environment, was in Munich at a pollution conference so Ian Gow took a break from privatising the nation's social housing stock to express the government's support for the bill.[92] After almost two hours of debate during which few dissenting voices were heard, MPs approved the bill. A little over a year later the Dartmoor Commons Bill was read for the third time in the House of Lords and returned to the Commons. On 30 October 1985 it received royal assent.

The Dartmoor Commons Act of 1985 was the most significant legislation of its sort passed by parliament since the passage of the New Forest Act in 1877. Its passage was possible thanks to the cooperation of the commoners, the national park authority, Devon County Council, the government, including its independent agencies, and numerous amenity societies. It is hard to think of another moment when most Dartmoor interests were in agreement, and as an exercise in consensus building it was a personal triumph for Ian Mercer. Anthony Steen was a useful ally, and the political lessons of 1980 were quickly learned, the new bill recognising that the first failed because it was too subservient to traditional agricultural interests, naively relied on unverifiable assumptions rather than statutory limitations, unnecessarily antagonised special interests, and failed to take proper account of the new environmentalism. Notwithstanding the Labour Opposition's ritualistic march into the noes lobby, the accent placed on local control, rights of access and environmental protection ensured the act satisfied sensibilities across the political spectrum. By establishing a legal right of access on foot or horseback to the common, the Conservative majority

in the House of Commons fulfilled the broadly left-of-centre aspirations of the original national parks legislation and the recommendations of the Common Land Commission. Formalising in law
what already existed in practice represented more than a little legal
tidying-up. In conjunction with the commons registration process,
the act was a decisive moment in the gradual transition from
customary practice to legal right, and from productivist agriculture
to conservationist land management, both parts of the broader
paradigm shift that brought unruly Dartmoor under control,
making it modern.

To return to the debate about how common-pool resources can
be managed, Garrett Hardin maintained that the pursuit of individual self-interest is liable to undermine the sustainability of the
resource. This tragic flaw only ceases to be operative when the
state imposes discipline on commoners, a claim that seems to be
upheld by the state's role in bringing about an effective management system for the Dartmoor common. However, Elinor Ostrom's
scepticism about the capacity of the state to effectively manage a
common-pool resource in the interests of sustainability appears
vindicated by the overstocking incentivised by agricultural subsidies.
MAFF's reluctance to publicly concede that there were systemic
problems with animal husbandry on Dartmoor reflected a broader
refusal to admit that post-war agricultural policy was deeply flawed.
Genuine fears about food scarcity shaped agricultural policy in the
post-war decades, but there is still little doubt that the farmers'
lobby and MPs who represented rural constituencies happily
exploited memories of wartime scarcity and Cold War anxieties.
Negative publicity surrounding heavily subsidised food overproduc-
tion during the 1980s gradually silenced those voices, and
campaigners were right to insist that wealth transfer by the state
from the taxpayer to farmers through agricultural subsidies incentivised the overuse of land, particularly through the use of artificial
pesticides and fertilisers. However, common rights on Dartmoor
meant that improvement, and with it a different kind of overuse

of the land, was inhibited because the problem of the free-rider undermined moves by the commoners towards closer cooperation. By the time the political will existed to provide a body with the statutory power to tackle the problem, particularly through the obligation to heft flocks, changes in the political climate had decisively moved against developments that would alter the ecological condition of the moorscape. The tragedy of the Dartmoor common was its saving.

New Thinking

As early as 1972 the Dartmoor Preservation Association had made the case for the commoners in terms of amenity: 'No one who understands the value and challenge of wild country in an increasingly over-populated island would for one moment contend that hill-farming should be allowed to die and Nature left to take its course. Matted, ungrazed vegetation two feet thick is no pleasure to walk in or ride in and it can obliterate green trackways, ancient monuments and natural features.' The DPA envisaged a new role for the commoner based on the revival of old practices made sustainable through modern means: 'And we do not want him to become a tame park-keeper or preserved rustic exhibit. We want him simply to mind his own traditional business, which safeguarded and conserved the land so well in the past, and could continue to do so in the future – and we are willing as taxpayers to underwrite his effort, and to regard him as a respected partner in a great enterprise, geared all the time to the national benefit.'[93]

This was strikingly percipient. The much-criticised headage payments of the post-war period have been phased out and British hill farmers are now likely to be in receipt of grants predicated on helping the government achieve wide-ranging and regionally specific environmental objectives demanded by international and European Union agreements. Such agri-environmental grants have allowed successive governments to increase their control over the moorscape, which, as Natural England explained in 2011, aims to allow *'land managers* to survive in the harsh conditions of the South West uplands'.[94] Publicity material produced by the Dartmoor Commoners Council aligns with this policy agenda, explaining the environmental services provided by grazing and offering a more sophisticated expression of the DPA's 1972 argument. Grazing, the council boasts, maintains the 'rich mosaic' of plant and animal life on which Dartmoor's 'biodiversity value rests', keeps the moor accessible by maintaining vegetation below knee height, preserves the 'visual textures' of its 'singular beauty', and ensures the richest concentration of surface antiquities in Europe remains visible.

> Commoners are thus providing a huge social and scientific benefit while going about their own business, which is why government, on behalf of us all, makes agreements with groups of commoners to ensure that they can stay on the farm, stick to the task and continue to deliver these 'public goods'. Their 'own business' is of course the production of stock – sheep and cattle that form the first link in a food chain, which continues with the fattening of the same animals for meat in lower, more lush pastures.[95]

In the first decade of the twenty-first century environmental management schemes were agreed between individual commoners associations, the Countryside Commission and Natural England – successor organisation to Nature Conservancy and a part of DEFRA, the Department for Environment, Food and Rural Affairs. By 2010 approximately 75 per cent of the common was subject to agri-environment schemes under the Environmentally Sensitive Areas rubric; schemes related to less precious but still highly valued land on Dartmoor were funded under Natural England's Environmental Stewardship programme.[96]

An important component of this new approach was the passage of the Environment Act 1995, which greatly strengthened the power of the national park authority. For some the old fear that the commons registration process would see control of the common fall into the hands of a small number of unscrupulous graziers has been superseded by fear of an over-mighty park authority. Whether or not those fears are justified, there is no question that since the 1970s the evolving statutory framework has enhanced the authority's capacity to oversee many aspects of life within the park. With regard to the common, two official documents published in 2001 established a management agenda that holds for the foreseeable future. Both documents reflected the United Nations Convention on Biological Diversity agreed at the Earth Summit at Rio in 1992 and together formed one of the numerous biodiversity action plans commissioned by the British Government. *The Nature of Dartmoor. A Biodiversity Profile* was a joint production of English Nature and Dartmoor National Park Authority and constituted a clear statement of the ecological status quo on Dartmoor at the turn of the twenty-first century. It divided the Dartmoor Natural Area, which is slightly larger than the national park, into a series of distinct wildlife habitats, identifying key nature conservation objectives for each. The second document, *Action for Wildlife. The Dartmoor Biodiversity Action Plan*, was published by the Dartmoor Biodiversity Steering Group. Reflecting New Labour's target culture, it explained how the objectives outlined in the *Profile* would be met. Much emphasis was placed on the complexity of the interests involved, the need to work closely with local stakeholders and the importance of agri-environment schemes to the adoption of 'environmentally beneficial livestock farming systems and other land management practices'.[97]

The detail of these and successor documents do not invite simple summary, but a brief look at their assessment of the upland common gives some idea of how the biodiversity principle has come to shape the management of the common. As a mix of blanket bog and upland heathland, with smaller stretches of raised bog, upland grass moor, valley mire and upland oak wood, the prescription for the common is complicated. The devil is in the detail, and conservation professionals intent on maintaining bio-diversity must invest a great deal of time and energy in monitoring the health of individual wildlife populations. For instance, Dartmoor habitats support most of the world's population of Vigur's eyebright (a flower), Haliford's pygmy moth and a rare cave-dwelling shrimp. Damage caused to Dartmoor's distinctive habitats quite literally threatens the extinction of these species. Most of the UK's popula-tion of bog hoverfly, blue ground beetle, Deptford pink and flax-leaved St John's wort have also found a niche on Dartmoor. Whether species like these thrive is one way of measuring the NPA's perfor-mance.

Certain characteristic Dartmoor activities are now considered more damaging than officials previously acknowledged. Swaling, long thought essential to the maintenance of the moorscape, is judged to have 'badly degraded' blanket bog, removing the layer of bog moss that forms its living skin; artillery and mortar fire has damaged peat bog, causing erosion, although the renewal of the training licences in 2012–3 makes it unlikely the park authority's wish to see the military off the moor will be fulfilled in the fore-seeable future. More significant are the mixed effects of grazing. Whereas uneven grazing has led to the growth of long grass in central areas of the bog, reducing the habitat available to birds like the golden plover and dunlin, both conservation priorities, over-grazing has converted much upland heathland into grassland, encouraging heavy bracken and fern growth, which then requires swaling. Greater vegetation density also increases tick numbers, leading to a higher incidence in livestock of the tick-borne disease Louping ill. Restoring heather helps revive red grouse and skylark numbers and might even tempt merlin onto the moor, but all of this needs to be carefully calibrated for some Dartmoor grassland (particularly Rhôs pasture) supports rare chamomile and heath violet, and vulnerable populations of butterfly, notably the high brown fritillary and pearl-bordered fritillary, as well as the hornet robber-fly and whinchat. Valley mire, extensive thanks to pre-modern tin streaming, boasts bog orchid, Irish lady's-tresses and the keeled skimmer dragonfly, as well as less rare flora and fauna. It too does not look after itself.

Much of this biodiversity, including breeding pairs of curlew and lapwing, is threatened by water extraction, predatory corvids (crows), nutrient enrichment caused by heavy stocking and fertiliser run-off, as well as disturbance by people and their dogs. Finally, upland oak woods, including Wistman's Wood and Black Tor Copse, as Dartmoor's most species-rich ecosystems, need careful management. Mosses, lichens, insects and birds, including 'special species' like wild daffodil, string-of-sausage lichen, *Graphina pauciloculata* (a rare lichen), blue ground beetles, dormice and buzzards inhabit these extraordinary ancient oaklands and semi-natural woods. Thinning heavily coppiced woodland and controlling the influx of non-native tree and shrub cover is needed, but as important is continuing to replant conifer stands with native broadleaf trees and ensuring that grazing does not suppress seedlings and saplings. If woodland cover should be extended, particularly by connecting semi-natural woodland, it must not be at the expense of important existing habitats.

Commoners with stock on the common can no longer expect to be awarded the grants that keep them going if their activities conflict with these biodiversity aims. A sense of how the commoners have adapted to the new regime can be gleaned from a short document posted on the website run by the Belstone village community explaining 'the ownership, administration and management of land in and around the village'. It refers to the overlapping institutions responsible for Belstone Common, naming the Belstone Commoners Association, the Manor of Belstone, the Duchy of Cornwall,

Dartmoor National Park Authority, West Devon Borough District Council and Belstone Parish Council, and then notes that the agri-environmental schemes that have replaced subsidies 'strike the currently favoured balance between wildlife ("biodiversity") and amenity values and those of agriculture'. If a weary cynicism can be discerned in the reference to 'currently favoured' and the apos-trophising of biodiversity, this simply reflects how the survival of an early twenty-first century Dartmoor grazier depends on a great deal of form-filling and the effective application of the latest bureau-cratic language.

Although the Belstone document asserts the principle that Belstone and the other Commons of Devon 'are completely distinct commons with their own Lords of the Manor and distinct commoners' associations' and that many commoners own grazing rights, only six 'active' commoners were currently entitled to exer-cise those rights. The distinction between the passive ownership of rights of common, as established by the commons registration process and subsequent arbitration, and the active use of those rights holds throughout much of the moorscape. Of 850 registered Dartmoor commoners, about a quarter are active, while in recent decades stock levels have fallen by at least a quarter. This decline is borne out by official figures, which show hill farming's contribu-tion to Dartmoor's overall economic output to have fallen from 10 to 4.2 per cent between 1998 and 2008. The downward trend, marking a real-terms rather than a proportional reduction in output, was accelerated by the foot-and-mouth outbreak of 2001 but the underlying cause has been the falling value of state support.[98] And though a precipitous decline by its own standard, the drop of 5.8 per cent did not prove devastating to the overall local economy. This has obscured how serious the situation for Dartmoor is. If carefully controlled grazing is to remain the lynchpin of the management of the common as an ecosystem, the grant system needs to be looked at again. Only then can acute concerns about farm succession be met. Entrepreneurialism might account for a disproportionately high proportion of the Dartmoor economy, creating pockets of considerable prosperity, but if the sons and daughters of graziers do not keep up the old ways the common and the enclosed land beyond will change very quickly. Scrubbing up, the long-heralded consequence of reduced grazing, has been happening for some time. Bracken, gorse and secondary woodland growth is decreasing the area of open heather moorland, particularly in the more sheltered upland river valleys, and unchecked grass growth in the northern and southern plateaus is likely to significantly diminish the hiker's pleasure.[99]

The official mind considers these developments a bad thing, but is it necessarily so? Ecologists have begun to question the assumptions underpinning nature conservation in the UK, questioning the DPA's insistence that no sensible person could want 'Nature left to run its course'. In *Future Nature: A Vision for Conservation* (2003) Bill Adams highlights the radical possibilities provided by the declining importance of agriculture to the rural economy, especially in uplands. Adams asks whether conservationists should 'fight to maintain farming in marginal areas of the UK, or accept structural change and seek a new conservation landscape'. Losing subsidised sheep flocks would create 'new ecological possibilities and new landscapes' that could allow the 'large-scale restoration of wildlife habitats, such as upland scrub and woodland'.[100]

Peter Taylor develops the argument. Drawing inspiration from Moor Trees, a local initiative cultivating small plantations of native oaks in the moorscape, he suggests Dartmoor, along with Glen Affric in Caledon and Coed Eryri in Snowdonia, are among British sites most suited to 'rewilding'. Taylor argues that 'a hands-off approach to management, relying upon natural processes of regeneration, could significantly alter the balance of species and community structure: heathland would be colonised by birch and rowan, pastures by bracken and eventually both would become woodland'. Much of this is at odds with the conventional conservationist thinking that holds sway on Dartmoor, and though conceding that

allowing nature to 'do its thing' would be to the detriment of flora
and fauna communities identified as conservation priorities, Taylor
asks whether habitats marginal to the survival of certain species
should be maintained simply to sustain or slightly increase small
local populations. Taylor suggests that confining the rewilding
experiment to the southern quarter of Dartmoor would not undermine current priorities and could provide an inspiring example of
what might be possible. Once denuded of domestic stock, 'managed
regeneration' could see new tree growth established and mammal
species reintroduced, including herbivores and predators similar to
species that once naturally populated the moorscape. In parts of
continental Europe the successful reintroduction of wolves has
become something of a symbol for the rewilding movement, but
Taylor judges Dartmoor too intensively used, too small and too
isolated from other wild areas to support this charismatic predator.
The same goes for bears. Lynx, however, are another matter.[101]

Taylor's prescription for Dartmoor is part of a book-length
argument and was something of a standard-bearer for the rather
modest British component of the global rewilding movement that
has found more influence elsewhere in the developed world. His
unorthodox position on climate change does not endear him to
mainstream environmentalists. George Monbiot's *Feral* (2013), a
recent attempt to popularise the agenda, emphasises the psychological benefits of a 're-involvement' with nature for humans living
in the developed world, while his newspaper polemics complaining
of how Britain's most prized natural landscapes have been 'sheep-
wrecked' take the pragmatic line that uplands absorb less water
when denuded of trees, particularly if soils are compacted by sheep
hooves, leading to greater run-off and more flooding in lowlands.[102]
Whether this applies to Dartmoor is debatable, but the rewilding
agenda is a reminder that although amenity, preservation and
conservation might have been superseded by biodiversity as the
dominant environmental concept underpinning the park's management, landscape restoration projects aim to reverse the negative

effects of subsidy-driven intensive modern agriculture rather than to recover a past wild Dartmoor.

Concerns about climate change mean great emphasis is now placed on the 'ecosystem services' Dartmoor provides. Not only are the Dartmoor uplands an important 'sponge', but its soils, woodlands and grasslands are significant carbon stores, the deep peat holding more than ten megatonnes of carbon, the equivalent of a year's carbon dioxide output from UK industry.[103] That restoration has emphatically not been disconnected from 'delivering people's social and economic well-being while working within environmental limits', raises some interesting possibilities.[104] Grazing might be in decline but the anxieties of the twenty-first century have made the park more important than ever. Is there any reason why the commoners shouldn't be kept by the state as rewilders rather than stock raisers or landscape restorers? That would take a shift in sensibility as profound as any that has occurred with regard to the moor over the past 200 years.

Surveys suggest the word most visitors associate with Dartmoor is 'wild', but to re-imagine sheep and cattle as an alien presence on the moor and to recognise that the very look and experience of being there is contingent on a host of human activities brings much into question.[105] In an isolated spot in the north quarter when there is barely a sheep or cow to be seen and only the lark, a stream and the wind to be heard rewilding polemics can seem a little overwrought, perhaps more applicable elsewhere than here. But once dubious but resilient notions of Dartmoor as unspoilt nature are set aside and the significance of the 'huge time depth of the human signature' on the British landscape is gracefully accepted,[106] there is no need to follow Taylor and Monbiot all the way for the rewilding thought experiment to shape how we encounter the moor. As Peter Fowler, world heritage adviser to UNESCO, reminded readers of the *Guardian* in 2005, what we see is both anthropic and ancient:

The ubiquitous post-glacial forest began to be burnt down 10,000–5,000 BC by hunting, fishing and collecting communities – and, during the next 4,000 years, much British upland became treeless under the impact of early farmers. Their stone axes felled timber, and their crop-growing and grazing animals inhibited arboreal regeneration. Dartmoor . . . was divided up by territorial boundaries and walled field systems during an intensive phase of agriculture either side of 1300 BC which, abandoned, left those granite uplands looking pretty much as they do today.[107]

Fowler found the desire to recover a natural Britain faintly absurd, but to accept the implications of his observations is to leave behind the old preservationist agenda and to recognise that what we encounter on the moor is historically contingent, a thought at once liberating and troubling. Liberating because it restores choice to human agents worried that they must somehow divine and implement nature's blueprint, and troubling because in practice 'nature-at-will' requires a novel ecological managerialism. Lynx prowling Dartmoor's southern quarter on the hunt for deer grazing in new woodland is a thrilling thought, but the possibility should not obscure the degree to which rewilding raises not only the spectre of a new 'improvement' but also the likelihood of significant unintended consequences.

Epilogue: The Dog That Hasn't Barked

In all England there is no district more dismal than that vast expanse
of primitive wasteland, the moors of Dartmoor in Devonshire.
Opening titles to *The Hound of the Baskervilles* (1939),
Twentieth Century-Fox

In April 1901 Arthur Conan Doyle stayed at Rowe's Duchy Hotel
at Princetown. He wrote the following words to his mother, Mary
Doyle. 'We did 14 miles over the moor today and we are now
pleasantly weary. It is a great place, very sad & wild, dotted with
the dwellings of prehistoric man, strange monoliths and huts and
graves. In those old days there was evidently a population of many
thousands here & now you may walk all day and never see one
human being.'[1]

Conan Doyle had spent a few days on Dartmoor with Bertram
Fletcher Robinson, whom he met on board a passenger ship from
Cape Town to Southampton. He was gathering materials and
impressions for what would arguably become the most enduringly
popular crime novel in the English language and certainly the
most influential literary evocation of Dartmoor ever written.
Although Conan Doyle later played down Fletcher Robinson's
role in the conception of *The Hound of the Baskervilles* (1901), Conan
Doyle's biographers agree that he was inspired by stories Fletcher
Robinson told, in his own words, of 'the ghost hounds, of the
headless riders, and of the devils that lurk in the hollows – legends

upon which I had been reared, for my home lay on the borders of the moor'.[2]

Despite the exhaustive efforts of investigators it has proved difficult to precisely relate Conan Doyle's fictional locations to real Dartmoor places, just as it has proved difficult to identify a particular source story for the Baskerville legend – the hellhound has a rich place in myth and legend.[3] If Conan Doyle's Dartmoor is best treated as imagined, its construction owed much to the Holmesian method. 'It is the scientific use of the imagination,' Holmes explained to Watson; 'we have always some material basis on which to start our speculations.'[4] Conan Doyle had family connections in Devon and had known Dartmoor for some decades when he sat down to write *Hound* in the early summer of 1901. He had used Dartmoor at least twice before as settings for stories. In 1882, when he briefly practised medicine in Plymouth, his visit to Tavistock and Dartmoor helped inspire 'The Winning Shot', a short story published in *Bow Bells* in July 1883. Set in and around Roborough, a bleak encounter on the moor with the sinister Dr Octavius Glaster precipitates events that lead to the death of the hero. By contrast, the Dartmoor setting of 'Silver Blaze', first published in the *Strand Magazine* in 1892, is entirely incidental to the plot. It is quite the opposite with *Hound*, whose effectiveness relies heavily on Conan Doyle's ability to conjure up an evocative Dartmoor setting. A recent biographer has criticised *Hound* as an 'anachronism, full of Victorian novelistic devices, such as a hereditary curse and a rash of mistaken identities, as well as stock Victorian figures, including a doctor interested in craniology and a naturalist with his butterfly collection'.[5] It is open to question if a work as phenomenally successful as *Hound* could be described as an anachronism, though it is perhaps true that the 1889 setting licensed Conan Doyle to pastiche the sensational fiction of the late Victorian period. Either way, revisiting the story alert to some of the themes examined in *Quartz and Feldspar* reveals how effectively Conan Doyle synthesised much Dartmoor knowledge in his tale. By weaving much local

colour into his story, he magnified a set of long-established perceptions of 'the Moor', ensuring *Hound* became the reticule that carried
them into the Edwardian period and, thanks to numerous cinematic
and television adaptations, long beyond.

Certain characters are indeed stereotypical, though Conan Doyle
is always sensitive to the specificity of the Dartmoor setting.
Stapleton, the naturalist, lives near a disused granite quarry in a
house that once belonged to a grazier and has now been modernised for the incomer. Dr Mortimer's interest in craniology is indeed
evident from the off. When he first visits 221B Baker Street, he
mistakes Watson for Holmes, indelicately commenting that he 'had
hardly expected so dolichocephalic a skull or such well-marked
supra-orbital development'. That goes beyond the implicit craniology that often features in Conan Doyle's descriptions of criminals
and echoes the racial theories that had currency among Dartmoor
enthusiasts like Sabine Baring-Gould. Indeed, Mortimer even
observes of Sir Henry Baskerville that he has the 'rounded head
of the Celt, which carries inside it the Celtic enthusiasm and power
of attachment', whereas Sir Charles Baskerville's head 'was of a
very rare type, half Gaelic, half Ivernian in its characteristics'.[6] The
possibility that the ancestral fate of the Baskervilles was biologically
determined adds a scientific dimension to the Baskerville legend,
complicating the case by suggesting science is not just on Holmes
and Watson's side. Such thinking had a currency that went far
beyond Dartmoor but, as has been shown, it was a distinct interest
among its late-Victorian enthusiasts.

Also among the pantheon of characters that could be imagined
pontificating to the membership of the Dartmoor Preservation
Association or the Devonshire Association is Mr Frankland. He
derives his greatest pleasure from pursuing legal cases intended to
uphold manorial or common rights, and Mortimer's work as an
amateur archaeologist has brought him into conflict with this iras-
cible figure. 'Sometimes he will shut up a right of way and defy
the parish to make him open it,' explains Watson. 'At others he

will with his own hands tear down some other man's gate and declare that a path has existed there from time immemorial, defying the owner to prosecute him for trespass.'[7] Frankland is Conan Doyle's most obviously satirical creation in *Hound*, and his expertise and the ambivalence he is shown by local people make him an immediately recognisable Dartmoor type.

Primacy, however, must go to Conan Doyle's representation of the physical character of the moor, much of which is conveyed to the reader through Dr Watson's impressionable letters to Holmes. Dartmoor's capricious character gets an outing, and 'the fresh beauty' of its mornings cannot override the signifiers of Gothic menace that crowd Conan Doyle's descriptive writing. The exotic otherness of the location is established during discussions in London and confirmed by Watson's impressions of the journey from Waterloo to Baskerville Hall with Dr Mortimer and Sir Henry. As the metropolis is left behind, the country changes: brown earth becomes 'ruddy', brick is replaced with granite, and a 'more luxuriant vegetation spoke of a richer, if a damper climate'. Nearing their destination, Watson becomes aware that the bucolic countryside is overlooked by 'a grey, melancholy hill, with a strange jagged summit, dim and vague in the distance'. Melancholia is quickly established as the settled state of the moor, though as the narrative develops it shows other faces, generally of a more active severity. From the small station halt the three men proceed to Baskerville Hall in a wagonette, travelling along deep Devonshire lanes. Baskerville's delight in all he sees is offset by Watson's growing sense of alienation. Abundant foliage, like the 'dripping moss and fleshy hart's-tongue fern' that crowd the sunken roads, evokes an excessive fecundity disturbing to the metropolitan eye schooled in the tidy landscapes of improved agriculture. Above this 'fertile country . . . rose' the contrastingly 'huge expanse of the moor', a 'desolate plain', 'bleaker and wilder' the further they progressed. They pass a mounted soldier, the first of several figures in the story to appear dramatically silhouetted on the skyline. The driver explains that the soldier is watching for an escaped convict from Dartmoor Prison, introducing a key subplot that also reminds the reader that the moorscape hosts an institution of chilling repute. Contemplating the essential 'malignancy' of 'this fiendish man' at large in the upland, Watson implicitly eschews the reformist penal regimes that once held sway in Dartmoor Prison. He also recognises that the escaped prisoner was needed 'to complete the grim suggestiveness of the barren waste, the chilling wind, and the darkling sky'.[8] Clad in a heavy coat, inured against the threatening landscape by the promise of a warm bed and buoyed by his self-conscious rationality, Watson takes melodramatic pleasure in this Gothic *mise en scène*.

Watson's stolid equanimity is shaken by real fear at later points in the story, sometimes as a consequence of a direct threat, sometimes because of a spooky occurrence. These irrational fears destabilise Watson's sense of self, problematising the assumption that a gulf of perception separates Dartmoor's 'educated' residents and visitors from its credulous and superstitious 'peasants'; his fears suggest the locals can billet the moor's sinister quiddities in ways that the incomer, determined to maintain his haughty detachment, cannot. This reinforces the old notion that the moorscape's natural characteristics are peculiarly affecting, a theme that gathers particular significance during Watson's first encounter with Stapleton. The lapidopterist explains to Watson the dangers of Grimpen Mire and advises him never to venture there alone. The two men then witness a pony being sucked into the morass; its distressed cries deeply disturb Watson but leave Stapleton unaffected. Stapleton then boasts he can find his way to the 'heart' of the mire and 'return alive', and a moment later he does just that, chasing after a moth. 'Bounding from tuft to tuft,' his 'jerky, zigzag, irregular progress made him not unlike some huge moth himself.'[9]

With the essential exception of Holmes, Stapleton is the only character to dismiss as entirely risible supernatural explanations of the Baskerville mystery. And in the peculiar economy of the story his complacent nonchalance about the threats others imagine the moor poses combine with his callous disregard for the pony swallowed by the mire to provide not just evidence of his guilt but also of his diabolical personality, whereas the susceptibility of all other characters but Holmes to the affectivity of the moorscape is humanising. Stapleton is not only an archetypal Holmesian alter ego, the Moriarty of the moor, but by his possession of exclusive knowledge, he is also Druidical, the arch manipulator of the moorscape's affectivity. 'I feel a foil as quick and supple as my own,' boasts Holmes, pleased to have met his match.[10]

Above all, *Hound* is a parable of rationalism, and the revived hellhound of the Baskerville legend is of course eventually exposed as a half-starved bloodhound–mastiff cross that Stapleton keeps in a disused mine out in the mire, its mouth smeared with phosphorous to create a devilish effect. Holmes' rational approach to the mystery is immediately established on the first day of his investigation, which he spends at home in London drinking a large quantity of coffee, smoking a great deal of tobacco and scrutinising a copy of the Ordnance Survey map of Dartmoor. Having 'been to Devonshire' through the agency of the map, Holmes informs Watson he could now find his 'way about', an arrogance permitted by the cultural and political authority of the map as an accurate representation of the moorscape.[11] Trig points rather than myth are a superior way of knowing the moor, modern mapping radically simplifying a space that easily intimidates Watson but fails to alarm the better informed Holmes. But his cartographic preparation is fallible. As a recent critic observes of the sudden fog that disrupts Holmes' plans and endangers Baskerville's life, the weather on Dartmoor is 'an elementary piece of local knowledge neglected by the London specialist'.[12]

Improvement is another way to domesticate the moorscape, and its prospects give the investigation a greater purpose. In Baker Street Mortimer explains his fear that Sir Charles Baskerville's 'schemes of reconstruction and improvement' will come to a halt if fear of a foul end prevents his heir from taking up residence at

Baskerville Hall. Solving the mystery and banishing the legend of the hound is thus the essential prerequisite to the 'whole country-side' profiting from Sir Henry's 'good fortune'. The point is re-inforced by Sir Henry's first response on seeing his inheritance. He pledges to spend some of his South African gold having Baskerville Hall lit with 'electric lamps . . . inside of six months', and taming its threatening atmosphere with 'a thousand-candle-power Swan and Edison . . . in front of the hall door'.[13] Jarringly symbolic, this bespeaks as much Sir Henry's faith in technology as it does a source of enlightenment for the moor.

The hound is shot dead, Stapleton flees into the mire, and Holmes and Watson abandon the pursuit confident of Stapleton's end 'in the foul slime of the huge morass'. Watson describes how the mire's 'tenacious grip plucked at our heels as we walked, and when we sank into it it was as if some malignant hand was tugging us down into those obscene depths, so grim and purposeful was the clutch in which it held us.'[14] Watson's 'as if' exonerates him of the pathetic fallacy: Dartmoor is a natural phenomenon, not a thing of sinister intent, but what matters is that its repute gave Stapleton's scheme its plausibility.

There were few other locations that could have proved as affec-tive. Dartmoor's unimproved state allowed the Baskerville legend, told from a manuscript of 1742 but dating back to the Civil War of the 1640s, to be translated to the 1880s. This 'old-world narrative',[15] easily dismissed in London, proved peculiarly affecting once the human imagination, endlessly suggestible, contemplated its rami-fications on a dark night on the moor. This, of course, is internal to a story whose affects rely on an over-dramatised version of the moorscape and over-susceptible antagonists. But the reader's disbe-lief is only suspended when, as Holmes might have put it, there is a 'material basis' on which the imagination can work. *Hound*'s success relied on Conan Doyle's ability to manipulate the accumu-lated meanings associated with the moorscape, repackaging them for a readership hungry for fresh sensation.

Hound contains repeated references to nature: Mortimer refers to 'the settled order of Nature', Holmes to 'the ordinary laws of Nature', and Stapleton tells of how he and his sister are 'devoted to Nature'; even Baskerville's determination to claim his inheritance is thought 'natural'.[16] In these usages, nature is the obverse of supernatural, a benign agent whose determining force ensures stability. With the hellhound exposed as an anthropic illusion, Dartmoor's unimproved state exemplified a reassuring natural equilibrium. As ecologists now insist, that equilibrium is also an illusion. Nature-at-will, with or without the actions of human beings, is dynamic and evolving. The notion that on the moor – or anywhere else – there was ever a 'settled order of Nature' is a myth that needs to be extruded as vigorously as Holmes exorcised the Baskerville legend.

Notes

Note on sources.

All references that begin HC or HL refer to House of Commons or House of Lords debates and the referencing conventions used by the online edition of Hansard have been used (www.hansard. millbanksystems.com). References that begin with a date followed by a bracketed number, e.g. 1810–11 (241), refer to British Parliamentary Papers and the referencing conventions of www. parlipapers.chadwyck.co.uk have been followed. References that begin NA, ADM, BT, F, DEFE, MAF, FT, WO, COU and HLG refer to files held by the National Archives at Kew in London. Frequently used abbreviations include *DPA* (the newsletter of the *Dartmoor Preservation Association*) and *TDA* (*Transactions of the Devonshire Association*). The meaning of all other abbreviations, which mainly take the form of acronyms, should be evident from the text.

Introduction

1   William B. Bridges, *Some Account of the Barony and Town of Okehampton; its Antiquities and Institutions* . . . (Plymouth, 1839), 14.
2   Ibid., 76.
3   Tristram Risdon, *The Chorographical Description or Survey of the County of Devon* (London, 1811), 258.
4   Bridges, *Barony and Town of Okehampton*, 76.
5   Ibid.

6　J. W. Besley, *The 'Borough' Guide to Okehampton* (Cheltenham, 1919), 14–16, 30–33, 38.

7　W. G. Hoskins, *Devon* (Newton Abbott, 1972) 447.

8　Ibid.

9　Quoted many times during various parliamentary debates, including by Nicholas Ridley, HC Deb. 19 Nov. 1985, Vol. 87, 143.

10　Quoted in the excellent 'Okehampton Bypass Factsheet' produced by the Dartmoor National Park Authority: http://www.dartmoor-npa.gov.uk/__data/assets/pdf_file/0015/41262/lab-okebypass.pdf [accessed 30 Sep. 2013].

11　HC Deb., 19 Nov. 1985, Vol. 87, 87.

12　*Guardian*, 3 Apr. 1985.

13　*Guardian*, 11 Nov. 1985.

14　*The Times*, 13 Nov. 1985.

15　HC Deb., 19 Nov. 1985, Vol. 87, 221.

16　*The Times*, 30 Nov. 1985.

17　HL Deb., 3 Dec. 1984, Vol. 457, 1175.

18　See the *Dartmoor Preservation Association Newsletter*, Dec. 1986 & Aug. 1988.

Part I

1　Quoted in P. A. S. Pool, *William Borlase* (Truro, 1986), 149.

2　Eric Hemery, *High Dartmoor* (London, 1983), 813–17.

3　Samuel Rowe, *A Perambulation of the Antient & Royal Forest of Dartmoor* (Plymouth, 1848), 23, 110.

4　Ibid., 72, 78.

5　Ibid., i.

6　Ibid., ii–iii.

7　Ann Bermingham, *Landscape and Ideology. The English Rustic Tradition 1740–1860* (Berkeley, 1986), 12–13.

8　Tim Fulford, *Landscape, Liberty and Authority. Poetry, Criticism and Politics from Thomson to Wordsworth* (Cambridge, 1996), 117–20.

9　Edmund Burke, *A Philosophical Enquiry into the Sublime and Beautiful* (Abingdon, 1858, 2008), 51.

10　Rowe, *Perambulation*, iii; William Howitt, *The Rural Life of England*,

Vol. II (London, 1838), 378–9. The passage continues by describing Howitt's 'feelings of delicious entrancement'.

11 Rowe, *Perambulation*, iii.

12 Ibid., iv.

13 This paragraph summarises Rowe, *Perambulation*, 20–58.

14 Samuel Rowe, 'Antiquarian Investigations in the Forest of Dartmoor, Devon', in *Transactions of the Plymouth Institute* (1830), 179–212.

15 Roughly two thirds of the 400 attributed articles published in *Archaeologia* between 1770 and 1796 were written by clergymen. See Rosemary Sweet, *Antiquaries* (London, 2004), 49.

16 Sam Smiles, *The Image of Antiquity. Ancient Britains and the Romantic Imagination* (New Haven and London, 1994), 10–11.

17 Sweet, *Antiquaries*, 34–6.

18 Smiles, *Image of Antiquity*, 16.

19 Ibid., 17.

20 Ibid., 45; Stuart Piggott, *The Druids* (Harmondsworth, 1974), 119–20.

21 On Thomas Pownell's important stadial antiquarian writing, see Sweet, *Antiquarians*, 23–6.

22 Pool, *Borlase*, 2–6, 12, 15–18, 89–92, 99–100, 103, 116–17, 125; Sweet, *Antiquarians*, 53.

23 Piggott, *The Druids*, 118, 123; Sweet, *Antiquarians*, 125; Ronald Hutton, *Blood and Mistletoe. The History of the Druids in Britain* (New Haven and London, 2009), 89.

24 Alexandra Walsham, *The Reformation of the Landscape* (Cambridge, 2011), 566.

25 Stukeley quoted in Hutton, *Blood and Mistletoe*, 91.

26 Ibid., 91.

27 Sweet, *Antiquaries*, 124; David Boyd Hancock, *William Stukeley: Science, Religion and Archaeology in Eighteenth-Century England* (Woodbridge, 2002), 164.

28 Hutton, *Blood and Mistletoe*, 98.

29 Borlase, *Observations*, 31.

30 Ibid., 53.

31 Ibid., 55; Boyd Hancock, *William Stukeley*, 150–1.

32 Borlase, *Observations*, 56.

33 Borlase, *Observations*, 56, 59, 60.

34 Ibid., 61.

35 Walsham, *Reformation of the Landscape*, 531–47.

36 Borlase, *Observations*, 61.

37 They are summarised in Hutton, *Blood and Mistletoe*, 1–48.

38 Borlase, *Observations*, 63–7.

39 Ibid., 76–8.

40 Piggott, *The Druids*, 141.

41 Borlase, *Observations*, 87–91.

42 Ibid., 94–100.

43 Ibid., 107.

44 Burke, *Sublime*, 58–9.

45 Borlase, *Observations*, 121–3.

46 Ibid., 152.

47 Smiles, *Image of Antiquity*, 3.

48 Ibid., 4.

49 Borlase, *Observations*, 153.

50 Walsham, *Reformation of the Landscape*, 476–7.

51 Borlase, *Observations*, 172.

52 Ibid., 214–15; Sweet, *Antiquaries*, 133.

53 Borlase, *Observations*, 227–41.

54 Hutton, *Blood and Mistletoe*, 108–9.

55 Ibid., 111.

56 Cf. Walsham, *Reformation of the Landscape*, 308.

57 Borlase, *Observations*, 188.

58 Anon., 'Rev. Richard Polwhele', in *Public Characters of 1802–1803* (London, 1803), 254–67.

59 W. P. Courtenay (revised by Grant P. Cerny), 'Richard Polwhele (1760–1838)', in *ODNB*.

60 Richard Polwhele, *The History of Devonshire*, Vol. I (Dorking, 1977), 3–6.

61 Richard Polwhele, *Reminiscences in Prose and Verse; consisting of the Epistolary Correspondence of many Distinguished Characters. With Notes and Illustrations* (London, 1836), 43–5, 46, 47–9, 50–76. Yonge was governor of the Cape from 1799 but was recalled in 1801 following a disastrous period in office and accusations of corruption.

62 Richard Polwhele, *Historical Views of Devonshire in Five Volumes*, Vol. I (Exeter, 1793), 3–11.

63 Ibid., 11–12.

64   Ibid., 19.

65   Polwhele, *History of Devonshire*, 141.

66   Polwhele, *Historical Views*, 20; Polwhele, *History of Devonshire*, 151.

67   Polwhele, *Historical Views*, 20.

68   Risdon, *Survey of Devon*, 405.

69   Polwhele, *History of Devonshire*, 146–8.

70   Ibid., 146–8, 153; Polwhele, *Devonshire Views*, 48.

71   Polwhele, *History of Devonshire*, 64.

72   Ibid., 6, 7, 10, 44.

73   Risdon, *Survey of Devon*, 198–9: 'It is left us by tradition that one Childe, of Plimstoke, a man of fair possessions, have no issue, ordained, by his will, that wheresoever he should happen to be buried, to that church his lands should belong. It so fortuned, that he riding to hunt in the forest of Dartmore, being in pursuit of his game, casually lost his company, and his way likewise. The season then being so cold, and he so benumbed therewith, as he was enforced to kill his horse, and emboweled him, to creep into his belly to get heat; which not able to preserve him, was there frozen to death; and so found, was carried by Tavistoke men to be buried in the church of that abbey; which was not so secretly done but the inhabitants of Plimstoke had knowledge therefor; which to prevent, they resorted to defend the carriage of the corpse over the bridge, where, they conceived, necessity compelled them to pass. But they were deceived by a guile; for the Tavistock men forthwith built a slight bridge, and passed over at another place without resistance, buried the body, and enjoyed the lands; in memory of whereof the bridge beareth the name of *Guilebridge* to this day.'

74   Oberon, king of the fairies in Shakespeare's *A Midsummer Night's Dream*, is part of an older tradition.

75   *The Critical Review; or, Annals of Literature*, Vol. 27 (Sept. 1799), 47–57.

76   Anon., 'Rev. Richard Polwhele', in *Public Characters of 1802–1803* (London, 1803), 260.

77   Boyd Hancock, *William Stukeley*, 204.

78   Richard Warner, *A Walk Through Some of the Western Counties of England* (Bath, 1800), 172. Emphasis added.

79   John Britton and Edward Wedlake Brayley, *The Beauties of England*

and Wales; or, Delineations, Topographical, Historical, and Descriptive, of Each County. Embellished with Engraving, Vol. IV (London, 1803), 10, 231–3, 237–8.

80	N. T. Carrington, Dartmoor: a Descriptive Poem (London, 1826), vii.

81	Literary Gazette, No. 229 (9 June 1821).

82	The Times, 30 Oct. 1834.

83	Henry F. Chorley, Memorials of Mrs Hemans (London, 1836), 47.

84	Felicia Hemans, The Poetical Works of Mrs Felicia Hemans (1836), 174.

85	Carrington, Dartmoor, x.

86	Ibid., viii.

87	Ibid., xi.

88	Ibid., ix–xii.

89	Ibid., xiii, xvii, xix.

90	Ibid., 23, 33, 38, 50, 53, 55, 64.

91	Ibid., 87–8.

92	Ibid., 6.

93	Ibid., 7.

94	Ibid., 9.

95	Ibid., 7, 8, 10, 11, 18, 19, 22, 23, 26, 33, 37, 51, 53, 82, 91.

96	Ibid., 68.

97	Ibid., 82–3.

98	Simon Schama, Landscape and Memory (London, 1995), 135ff.

99	Fulford, Landscape, Liberty and Authority, 18–37.

100	Ibid., 217.

101	Carrington, Dartmoor, 89.

102	Raymond Williams, The Country and the City (London, 1973).

103	Robert Southey, The Life and Correspondence, Volume 6 (1850) 322–3.

104	John A. Kempe (ed.), Autobiography of Anna Eliza Bray (London, 1884), 336.

105	Ibid., 202–5.

106	Southey to Bray, 12 June 1832, 17 Feb. 1833 (Robert Southey Papers, British Library (MS. FACS. *615)).

107	Kempe (ed.), Autobiography of Anna Eliza Bray, 134.

108	Southey to Bray, 12 June 1832 (Southey Papers).

109	Southey to Bray, 14 July 1832 (Southey Papers).

110	Mrs Bray [Anna Eliza], Traditions, legends, superstitions and sketches of Devonshire (1838), 92–3.

111 For an extended discussion of these natural characteristics, see Ian Mercer, *Dartmoor* (London, 2009), 160–2.

112 Polwhele, *History of Devonshire*, 94.

113 The complexities of this process are summarised in Mercer, *Dartmoor*, 84–93.

114 Candida Lycett Green, *England. Travels Through An Unwrecked Landscape* (London, 1996).

115 Bray, *Traditions*, 98–9.

116 Ibid., 17.

117 Ibid., 18.

118 Ibid., 44–5.

119 Ibid., 43.

120 Ibid., 19.

121 Ibid., 46.

122 Southey had encouraged this collecting: see Mary I. Shamburger & Vera R. Lachman, 'Southey and "The Three Bears", in *The Journal of American Folklore*, Vol. 59, No. 234 (Oct.–Dec., 1946), 402.

123 See Williams, *Country and the City*.

124 Bray, *Traditions*, 61.

125 Ibid., 62–3.

126 The digitisation of books – rather than masochism – allows these figures to be easily calculated.

127 *Trewman's Exeter Flying Post or Plymouth and Cornish Advertiser*, 16 Aug. 1849 & 12 Sep. 1850; *The Literary Examiner*, 15 Sep. 1849.

128 See George W. Stocking, Jr, *Victorian Anthropology* (New York, 1987).

129 John Bowring, 'Presidential Address', TDA, 1862–1866, Vol. I, 9–10.

130 John Kelly, 'Celtic Remains on Dartmoor', TDA, Vol. I, Pt v (1866), 45–6.

131 Ibid., 47.

132 Ibid., 48.

133 Stocking, *Victorian Anthropology*, 72–4.

134 Kelly, 'Celtic Remains', 48.

135 C. Spence Bate, 'On the Prehistoric Antiquities of Dartmoor', in TDA, Vol. IV (1871), 496.

136 Ibid., 495.

137 Ibid., 498.

138 Ibid., 504, 509.

139   Ibid., 512.

140   Ibid., 503.

141   C. Spence Bate, 'Researches into some Antient Tumuli on Dartmoor', *TDA*, Vol. V (1872), 556.

142   Ibid., 555.

143   Ibid..

144   Ibid., 556.

145   C. Spence Bate, 'A Contribution towards determining the etymology of Dartmoor Names', in *TDA*, Vol. IV, Pt ii (1870–1), 527.

146   Ibid., 529.

147   Ibid., 530.

148   Ibid., 533.

149   Ibid.

150   Ibid., 534–5.

151   G. Wareing Ormerod, 'Rude Stone Remains Situate on the Easterly Side of Dartmoor', in *Archaeological Memoirs Relating to the East of Dartmoor* (Exeter, ?1876), 4–5.

152   Ibid., 7–8.

153   Ibid., 9–10.

154   Ibid., 10.

155   Ibid., 13–16.

156   C. Spence Bate, 'Grimspound and its Associated Relics', in *Annual Reports and Transactions of the Plymouth Institution and Devon and Cornwall Natural History Society*, Vol. V (1873–6), 37–8, 50.

157   Ibid., 52–3.

158   P. Pattison and M. Fletcher, 'Grimspound, One Hundred Years On', in Deborah Griffiths (ed.), *The Archaeology of Dartmoor. Perspectives from the 1990s* (Stroud, 1996), 24.

159   R. N. Worth, 'Were There Druids in Devon?' in *TDA*, Vol. XII (1880), 228.

160   Ibid.

161   Ibid., 233.

162   Ibid.

163   Ibid., 234.

164   Ibid., 238–42.

165   Ibid., 236.

166   Ibid.

167 Ibid., 237.

168 Ibid., 235.

169 Ibid., 237.

170 Birkenhead, Bolton, Cardiff, Cheltenham, Clitheroe, Darwen (Lancs.), Leeds, Oldham, Plymouth, Portsmouth, and Tavistock, plus the Torquay Natural History Library and the Washington Library of the Supreme Council of the Thirty-third Degree of A. and A. S. Rite.

171 *Western Antinquary [WA]*, Vol. I, Mar. 1881–Mar. 1882.

172 Ibid, Vol. VII, No. 3 (Aug. 1887).

173 Ibid, Vol. VII, No. 5 (Oct. 1887).

174 Ibid, Vol. VII, No. 6 (Nov. 1887).

175 John Lloyd Warden Page, *An Exploration of Dartmoor and Its Antiquities* (London, 3rd edn, 1892), 20.

176 Ibid., 80–1, 159, 185–9, 208–11, 220. Emphasis added.

177 Ibid., 285.

178 Ibid., 285.

179 Ibid., 286. 'And everything of the sort.'

180 *WA*, Vol. VIII, No. 12 (June 1889).

181 Samuel Rowe, *A Perambulation of Dartmoor* (Exeter, 1985), ix.

182 Page, *Exploration*, 284.

183 Ibid., xiv.

184 K. Baedeker, *Great Britain. England, Wales and Scotland as far as Loch Maree and the Cromarty Firth. Handbook for Travellers* (London, 1887), 94–5.

185 Ibid., 129.

186 Ibid.

187 S. Baring-Gould, *A Book of Dartmoor* (London, 1900), 75.

188 S. Baring-Gould, *A Book of the West Being An Introduction to Devon and Cornwall. Vol. I. Devon* (London, 1899), 175.

189 Baring-Gould, *Dartmoor*, 75.

190 Ibid., 35.

191 Ibid., 39.

192 Ibid., 39–40.

193 Ibid., 80–1.

194 Baedeker, *Great Britain*, 386.

195 Eden Phillpots, *Dartmoor Omnibus* (London, ?1933), 94–7.

196   T. D. Kendrick, *The Druids. A Study of Keltic Prehistory* (London, 1927), 1–11.

197   Alexandra Walsham, 'The Reformation of the Landscape: Religion, Identity and Memory in Early Modern Britain and Ireland', lecture, University College London, May 2010.

198   Ibid.

Part II

1   Hemery, *High Dartmoor*, 1029.

2   Gavin Daly, 'Napoleon's Lost Legions: French Prisoners of War in Britain, 1803–1914' in *History*, 89, 295 (July 2004), 366.

3   Ibid., 364.

4   1809 (124) 'The ninth report of the Commissioners for Revising and Digesting the Civil Affairs of His Majesty's Navy', 23–4.

5   Trevor James, *Dartmoor Prisoner of War Depot and Convict Jail* (Crediton, 2002), 11.

6   *Aberdeen Journal*, 30 Oct. 1805; also announced in *The Times*, 29 Oct. 1805 &, for example, in the *Derby Mercury*, 31 Oct. 1805.

7   *The Bury and Norwich Post*, 6 Nov. 1805 & *The Ipswich Journal*, 9 Nov. 1805.

8   *Trewman's*, 7 Nov. 1805.

9   Ibid., 25 May 1809.

10   *Bristol Times and Mirror*, 15 July 1805, quoted in Elisabeth Stanbrook, *Dartmoor's War Prison & Church 1805–1817* (Tavistock, 2002), 9.

11   Stanbrook, *Dartmoor's War Prison & Church*, 6–7; James, *Dartmoor Prisoner of War Depot and Convict Jail*, 8.

12   William Crossing, *Princetown: Its Rise and Progress* (Brixham, 1989), 14.

13   Ibid., 17, 19.

14   Ibid., 16. For good measure, Crossing repeated memories of local gossip that Tyrwhitt's visits to Tor Royal saw 'glorious times'. 'Those were the days of two- or even three-bottle men, when guests sat long at tables, and had often to be helped to bed.' 'If old gossips spoke truly,' he added, 'the young master of Tor Royal and his guests found as much delight in gazing upon the cherry

checks and bright eyes of the rustic beauties of the neighbourhood as they did in acquainting themselves with the agricultural capabilities of the great waste.' Perhaps so, though Eric Hemery later took Baring-Gould to task in a po-faced footnote for similarly repeating the 'calumnious rumour' that the Prince Regent enjoyed '"high revelry and debauches"' at Tor Royal. Hemery, *High Dartmoor*, 382.

15 Quoted in Sarah Wilmot, '"The Scientific Gaze": Agricultural Improvers and the Topography of South-West England', in Mark Brayshay (ed.), *Topographical Writers in South-West England* (Exeter, 1996), 115.

16 Jerome Blum, 'English Parliamentary Enclosure' in *Journal of Modern History*, Vol. 53, No. 3 (Sep. 1981), 479–80.

17 Wilmot, 'The Scientific Gaze', 114.

18 Sarah Wilmot, '"The Business of Improvement": Agriculture and Scientific Culture in Britain *c.1700–c.1870*', *Historical Geography Research Series*, No. 24 (1990), 41.

19 Charles Vancouver, *General View of the Agriculture of the County of Devon; with observations on the means of its improvement. Drawn up for the consideration of the Board of Agriculture, and Internal Improvement* (London, 1808), 275.

20 Ibid., 272–8.

21 Ibid., 275.

22 On Moretonhampstead see Francis Pryor, *The Making of the British Landscape* (London, 2010), 304, 341.

23 Vancouver, *General Survey*, 271.

24 In an indispensable review article Jerome Blum delineates the findings of post-war revisionist historians, exploring the strengths and weakness of their arguments. See his 'English Parliamentary Enclosure', in *Journal of Modern History*, Vol. 53, No. 3 (Sep. 1981), 477–504.

25 E. P. Thompson, *The Making of the English Working Class* (London, 1963), 221.

26 W. G. Hoskins, 'The Reclamation of the Waste in Devon, 1550–1800' in the *Economic History Review*, Vol. 13, No. 1/2 (1943), 80–92.

27 Wilmot, 'Scientific Gaze', 119–20.

28 Pryor, *Making*, 305.

29   Hoskins, 'Reclamation', 84–5.

30   Vancouver, *General Survey*, 278.

31   Crossing, *Princetown*, 19; James, *Dartmoor Prisoner of War Depot*, 11; Hemery, *High Dartmoor*, 382.

32   *Trewman's*, 7 Jan. 1802: the mail coach had to take on extra horses, the roads were 'so frozen' & 3 Oct. 1805.

33   *Hampshire Telegraph and Sussex Chronicle*, 2 Oct. 1809; *The Aberdeen Journal*, 11 Oct. 1809.

34   *Morning Post*, 23 Mar. 1810.

35   *Trewman's*, 25 Oct. 1809.

36   *Hull Packet*, 21 Apr. 1812.

37   *Morning Post*, 12 Dec. 1808.

38   ADM 98/225, 25 Feb., 7 Apr., 13 Apr. 1809.

39   ADM 98/225, 14 Apr., 16 Apr., 25 Apr., 17 May 1809.

40   *Lancaster Gazette*, 30 Apr. & 3 June 1809; *York Herald*, 7 May 1809.

41   *Jackson's Oxford Journal*, 26 Aug. 1809; *Morning Post*, 28 Aug. 1809; *Trewman's*, 31 Aug. 1809.

42   *Lancaster Gazette*, 21 Oct. 1809.

43   *Morning Post*, 28 Nov. 1810; *Jackson's Oxford Journal*, 1 Dec. 1810; *Royal Cornwall Gazette*, 30 May 1811 & 12 Nov. 1814; *Trewman's*, 10 Oct. & 17 Oct. 1811, 6 May, 1813; *The Times*, 18 Sep. 1812; *Lancaster Gazette*, 29 May 1813; *Morning Post*, 3 Mar. 1814.

44   ADM 98/227, 26 Apr. 1813, 5 May 1813.

45   ADM 1/5122/16.

46   ADM 98/225, 17 Feb. 1810, 13 Mar. 1810, 17 Aug. 1810, 3 Sep. 1810, 6 Dec. 1810, 12 Aug. 1811.

47   *Morning Chronicle*, 17 Sept. 1812; *Caledonian Mercury*, 19 Sep. 1812; *Jackson's Oxford Journal*, 19 Sep. 1812; *Liverpool Mercury*, 25 Sep. 1812.

48   *The Times*, 18 Sep. 1812.

49   ADM 98/226, 4 & 6 Jan. 1812.

50   ADM 98/225, 27 Dec. 1809, 3 May 1811, 28 May 1811.

51   ADM 98/225, 26 Sep., 3 Oct., 15 Oct., 17 Oct., 29 Oct., 7 Nov., 24 Nov. 1808.

52   Reports can be found in ADM 98/225, ADM 98/226, & ADM 98/227.

53   [Charles Andrews], *The Prisoners' Memoirs, or Dartmoor Prison* (New York, 1852), 53–4.

54 Josiah Cobb, *A Green Hand's First Cruise, roughed out from The Log-Book of Memory, for twenty-five years standing: Together with a Residence of Five Months in Dartmoor*, Vol. 2 (Baltimore, 1841), 184–5.

55 Daly, 'Napoleon's Lost Legions', 370–1, 376.

56 Ibid., 369.

57 ADM 98/225, 12 Feb. 1810 & 30 Jan. 1811.

58 ADM 98/227, 14 Jan. 1812.

59 ADM 98/227, 31 Mar. 1812.

60 Robin F. A. Fabel, '"Self-Help on Dartmoor": Black and White Prisoners in the War of 1812', in *Journal of the Early Republic*, Vol. 9, No. 2 (summer, 1989), 178–9, 189–90.

61 Joseph Valpey, *Journal of Joseph Valpey, Jr of Salem: November 1813-April 1815* (Michigan, 1822), 15–18.

62 ADM 98/228, 8 June 1814, 6 & 11 Dec. 1815.

63 See Michel Foucault, *Discipline and Punish. The Birth of the Prison* (London, 1977).

64 Andrews, *Prisoners' Memoirs*, 60.

65 Andrews, quoted in Justin Jones, 'The prison on the moor: A study of the American prisoner-of-war experience within Dartmoor prison, 1813–1815', M.A. dissertation Univ. of Texas at Arlington, 2011, 71; ADM 98/227, 11 July 1813.

66 ADM 98/227, 4 & 15 Apr. 1812.

67 ADM 98/225, 3 & 18 July 1809, 14 Dec. 1809.

68 ADM 98/227, 21 & 25 Aug. 1813, 3 Sep. 1813.

69 Andrews, *Prisoners' Memoirs*, 13.

70 ADM 98/225, 12 Aug. 1809.

71 ADM 98/226, 6, 12, 16 & 19 Feb. 1811, 25 Mar. 1812.

72 ADM 98/225, 10 Sep. 1810.

73 ADM 98/228, 14 & 17 Dec. 1814, 29 Dec. 1814.

74 ADM 98/225, 11 Aug. 1809, 22 Sep. 1809, 7 Dec. 1809, 28 Dec. 1810.

75 ADM 98/225, 21 Nov. 1809, 9 Jan. 1810; ADM 98/226 5 Feb. 1811, 6 Apr. 1811.

76 ADM 98/225, 24 Feb. 1810, 15 Mar. 1810, 22 & 28 Jan. 1812, 6 & 11 Feb. 1812, 28 May 1812; ADM 98/227, 4 & 16 Nov. 1813, 7 Jan. 1814, 2 Mar. 1814.

77 Andrews, *Prisoners' Memoir*, 60.

78 Fabel, 'Self-Help on Dartmoor', 179–80; Valpey, *Journal*, 18. There

were also cases of prisoners who had informed to the authorities having to be taken into protection.

79   ADM 98/225, 23 Jan. 1810, 2, 8, 10, 14 & 17 Feb. 1810, 25 June 1810.

80   *Morning Chronicle*, 19 May 1810.

81   *Hull Packet*, 9 Oct. 1810.

82   Ibid., 9 Oct. 1810.

83   Vancouver, *General Survey*, 280–3.

84   Much the same was reported in the *Agricultural Magazine, or Farmers' Monthly Journal of Husbandry and Rural Affairs*, Vol. VII, from July to December 1810.

85   *Aberdeen Journal*, 5 Dec. 1810.

86   *Cobbett's Weekly Political Register*, 29 Dec. 1810.

87   *Morning Post*, 17 June 1811; ADM 99/225, 31 Jan. 1811.

88   *Cobbett's Parliamentary Debates* (13 May 1811–24 July 1811), Vol. XX (London, 1812), Col. 634–8.

89   *Morning Post*, 19 June 1811; *Cobbett's Parliamentary Debates* (13 May–24 July 1811), Vol. XX (London, 1812), Col. 639.

90   1810–11 (241) (Prisoners of War) Transport Board, 18 June 1811, 'An account of the number of prisoners of war, in the prison of Dartmoor; from the first time when any were confined there, in every month, to the latest returns that have been received; shewing the number of deaths in each month'.

91   1810–11 (236) (Prisoners of War) Transport Board, 14 June 1811. 'An Account of the number of French Prisoners of War in England; distinguishing the prisons in which they were confined in the month of April 1810; and, according to the latest returns, distinguishing those in health, from the sick and the convalescents'.

92   Daly, 'Napoleon's Lost Legions', 375.

93   *Cobbett's Parliamentary Debates*, Vol. XX (London, 1812), Col. 638.

94   Quoted in Basil Thomson, *The Story of Dartmoor Prison* (London, 1907), 81.

95   Ibid.

96   Ibid., 82–3.

97   *Examiner*, 28 July 1811.

98   Ibid., 11 Aug. 1811.

99   Ibid., 25 Aug. 1811.

100   Ibid., 1 Sept. 1811.

101 Ibid., 29 Sept. 1811.

102 *Jackson's Oxford Examiner*, 30 Nov. 1811.

103 *Examiner*, 6 Oct. 1811.

104 Andrews, *Prisoners' Memoirs*, 17.

105 Ibid., 18.

106 Ibid., 18–19, 22.

107 Ibid., 22.

108 Ibid., 23–4. See also the 'Journal of Nathaniel Pierce', quoted in Justin Jones, 'The Prison on the Moor: A Study of the American Prisoner-of-War Experience within Dartmoor Prison 1813–1815', MA dissertation, University of Texas at Arlington, 2011, 51–2

109 Cobb, *A Green Hand*, Vol. 2, 6.

110 Ibid., 182–3.

111 James, *Dartmoor Prisoner of War Depot*, 74.

112 Andrews, *Prisoners' Memoirs*, 156–61; Cobb, *A Green Hand*, Vol. 2, 191.

113 Cobb, *A Green Hand*, Vol. 2, 190–1.

114 Andrews, *Prisoners' Memoirs*, 203–4; Cobb, *A Green Hand*, Vol. 2, 198.

115 Henry Tanner, *The Cultivation of Dartmoor. A Prize Essay* (London, 1854), 3.

116 Ibid., 8–9.

117 Ibid., 30–6.

118 Hadrian Cook & Tom Williamson (eds), *Water Management in the English Landscape. Field, Marsh and Meadow* (Edinburgh, 1999), 1.

119 Hadrian Cook, 'Soil and water management: principles and purposes', in Cook and Williamson, *Water Management*, 15.

120 A. D. M. Phillips, 'Arable land drainage in the nineteenth century', in Cook and Williamson, *Water Management*, 53, 65–8.

121 Joseph Bettey, 'The development of water meadows in the southern counties', in Cook and Williamson, *Water Management*, 192.

122 Tanner, *Cultivation*, 15.

123 G. A. Cooke, *Topography of Great Britain: or, British traveller's directory* (London, 1817) 27.

124 Ibid., 46.

125 Ibid., 46, 48–9, 68.

126 Thomas Tyrwhitt, *Substance of a Statement made to the Chamber of*

Commerce, Plymouth, on Tuesday, the 3rd Day of November, 1818, concerning the Formation of a Rail Road, from the Forest of Dartmoor to the Plymouth Lime-Quarries, and A Plan of the Intended Line. By Sir Thomas Tyrwhitt, 3. (NA ZLIB 29/67).

127   Ibid., 6–7.

128   Ibid., 8–13.

129   Ibid., 14–19.

130   Ibid., 19–20.

131   Ibid., 12–13, 26–8.

132   'Prospectus of the Plymouth & Dartmoor Rail Road.' (NA RAIL 1075/182)

133   NA RAIL 566/5.

134   *Morning Post*, 28 Jan. 1820.

135   Joseph Priestly, *Historical Account of the Navigable Rivers, Canals and Railways throughout Great Britain* (London, 1831), 555–7.

136   NA RAIL 1017/1/30.

137   Tyrwhitt, *Substance*, 24.

138   'Report from the Committee on the Prisons within the City of London and Borough of Southwark. 1. Newgate, &c.' (1818), 172–87, 194–5, 201–2.

139   *Gentleman's Magazine and Historical Chronicle* (July–December 1819), 632–3.

140   *The Times*, 10 & 24 Jan., 21 Feb. 1820.

141   With varying degrees of sycophancy and hysteria, some concerned personal debt, some sought an audience for their ideas, others warned His Lordship of dangers to his life.

142   Henry Wilson to Lord Sidmouth, 15 May 1820 (NA HO 44/145).

143   Henry Wilson, *Appeal to the Nobility, Clergy, Gentry, and Merchants of the United Kingdom, on the Proposed Plan for sending Pauper Children to Dartmoor Prison* (London, 1820), 20–22.

144   Ibid., 14–15.

145   Ibid., 19.

146   Ibid., 17.

147   Seán McConville, *A History of English Prison Administration. Vol. 1 1750–1877* (London, 1981), 187–9.

148   'Memorandum in reference to a Suggestion for the Establishment of a Depot for Convicts and SENTENCE OF TRANSPORTATION,

in the Prisons of War at Dartmoor, forwarded to the Secretary of State for the Home Department in Year 1826', quoted in 1831–32 (547) 'Report from Select Committee on Secondary Punishments; together with the minutes of evidence, an appendix, and index'.

149 *The Times*, 14 Aug. 1829.

150 Ibid.

151 1831–32 (547) 'Report from Select Committee on Secondary Punishments; together with the minutes of evidence, an appendix, and index', 16, 20, 135; McConville, *English Prison Administration*, 188.

152 NA TS 21/287, 25 July 1834.

153 1835 (438) (439) (440) (441) 'First Report of Select Committee of the House of Lords appointed to inquire into the present state of the several gaols and houses of correction in England and Wales; with the minutes of evidence and appendix', 325.

154 1836 (51) 'Convicts. Two Reports of John Henry Capper, Esq. superintendent of ships and vessels employed for the confinement of offenders under sentence of transportation; relating to the convict establishments at Portsmouth, Chatham, Woolwich; and at Bermuda', 1, 5.

155 *The Times*, 17 Aug. 1836.

156 1837 (80) 'Reports to the Secretary of State for the Home Department, relating to plans for a prison for juvenile offenders', 3–7.

157 McConville, *English Prison Administration*, 204–9.

158 NA HO 45/1075; 1851 (1419) 'Report of the directors of convict prisons on the discipline and management of the hulk establishment. For the year 1850', 20.

159 McConville, *English Prison Administration*, 191–5.

160 Ibid., 400.

161 NA PCOM 7/223; McConville, *English Prison Administration*, 400.

162 Ibid., 10–11.

163 1843 (325) 'Waste land. Inclosure Acts. Return of the estimated quantity of common or waste lands, not being held severalty, in every parish or title commutation district in England and Wales, up to March 1843'.

164 1844 (583) 'Report from the Select Committee on Commons' Inclosure; together with the minutes of evidence, and index', 195–7.

165   Ibid., 427–8.

166   BT 41/183/1046.

167   Michel Foucault, *Discipline and Punish: the Birth of the Prison* (London, repr. 1991).

168   McConville, *English Prison Administration*.

169   *The Times*, 7 Feb., 2 & 6 Sep. 1854, 29 Aug. 1855, 19 Jan. 1856, 23 April 1859.

170   NA PCOM 7/109.

171   1851 (1419)'Reports of the directors of convict prisons on the discipline management of the hulk establishment. For the year 1850', 20–22.

172   1852 (1524)'Reports of the directors of convict prisons on the discipline and management of Pentonville, Parkhurst, and Millbank Prisons, and of Portland and Dartmoor Prisons, and the Hulks, for the year 1851', 197–8.

173   1854 (1825)'Reports of the directors of convict prisons . . . for the year 1853', 163.

174   *The Times*, 30 Sept. 1853.

175   1857 Session 2 (2263) 'Reports of the directors of convict prisons . . . for the year 1856', 224.

176   1857–8 (2423)'Reports of the directors of convict prisons . . . for the year 1857', 191.

177   1854 (1825), 160.

178   1852 (1524)'Reports of the directors of convict prisons . . . for the year 1851', 192.

179   1856 (2126)'Reports of the directors of convict prisons . . . for the year 1855', 226.

180   1857 Session 2 (2263)'Reports of the directors of convict prisons . . . for the year 1856', 189.

181   1856 (2126), 221.

182   1854 (1825), 195.

183   1854 (1825), 194.

184   1854 (1825), 173–183.

185   1859 Session 2 (2556) 'Reports of the directors of convict prisons . . . for the year 1858', 220.

186   *The Times*, 1 Mar. 1862.

187   Ibid. 1 Sep. 1864.

188   NA PCOM 7/109.

189   *The Times*, 29 Sep. 1881.

190   Alan S. Baxendale, *Winston Leonard Spencer-Churchill. Penal Reformer* (Bern, 2010), 111–13; Philip Priestly, *Victorian Prison Lives: English Prison Biography, 1830–1914* (London, 1985), 69.

191   Basil Thompson, *The Story of Dartmoor Prison* (London, 1907), 266–7.

192   NA HO 144/10086; Baxendale, *Penal Reformer*, 114.

193   NA HO 144/10086.

194   *Daily Debates*, 23 Nov. 1910.

195   NA HO 144/10086; Baxendale, *Penal Reformer*, 117.

196   A. Scriven, *The Dartmoor Shepherd. 50 Years in Prison* (Oswestry, 1932).

197   Baxendale, *Penal Reformer*, 117–28.

198   Samuel Smiles, *The Life of Thomas Telford, Civil Engineer* (London, 1867), 33.

199   Smiles, *Telford*, 43–5.

200   BT 31/19579/110871.

201   BT 31/27333/183059.

202   *The Times*, 23 July 1869.

203   *The Times*, 16, 23 & 24 Jan. & 15 Mar. 1873.

204   Ibid., 15 Sep. 1873.

205   Ibid., 7, 13, 15, 16, 18, 20, 21, 22 & 23 Aug. 1873.

206   For the report, see *The Times*, 15 Apr. 1874, for the letter, 11 June 1874.

207   6 Aug. 1875.

208   *The Times*, 25 Jan. 1876.

209   Ibid., 30 Oct. 1875.

210   Ibid., 12 Nov. 1877.

Part III

1   These details are taken from Brian Le Messurier (ed.) *Crossing's Guide to Dartmoor* (Dawlish, 1965), repr. of 1912 edition.

2   J. F. Berger, 'IV. Observations on the Physical Structure of Devonshire and Cornwall', in *Transactions of the Geological Society*, Vol. I (1811), 116–22; Henry De la Beche, *Report on the Geology of Cornwall, Devon and West Somerset* (London, 1839).

3   Brian Le Messurier, *Crossing's Dartmoor Worker* (Dawlish, 1966), 146 (emphasis added).

4   Crossing, *Guide*, 69–70.

5   Ibid., 362, 372; Crossing, *Worker*, 60; Elisabeth Stanbrook, *Dartmoor Forest Farms: A Social History from Enclosure to Abandonment* (Tiverton, 1994), 85.

6   Crossing, *Guide*, 365–7.

7   Ibid., 32, 371, 386.

8   DPA Minute Book, 11 April 1910 & 23 Dec. 1912 (in the possession of the DPA).

9   Crossing, *Guide*, 434–5.

10   Ibid., 105.

11   Ibid., 104–5, which reproduces the explanation given in Crossing, *The Ancient Stone Crosses of Dartmoor and Its Borderland* (Exeter, 1902), 73–8.

12   Crossing, *Guide*, 106.

13   *Crossing's Dartmoor Worker*, 129.

14   William Crossing, *Amid Devonia's Alps* (Newton Abbot, 1974), 15–17, 19, 26–7.

15   Northcote said of Dartmoor: 'That is the only good of such places – that you are glad to escape from them, and look back to them with a pleasing horror ever after.' See William Hazlitt, *The Round Table: Northcote Conversations. Characteristics* (London, 1817, 1871), 349.

16   *Crossing's Dartmoor Worker*, 119–20.

17   Ibid., 11, 129–30; also, Crossing, *Amid*, 17.

18   W. F. Collier, 'Some Sixty Years' Reminiscences of Plymouth' in *TDA*, Vol. XXIV (1892), 86–95.

19   W. F. Collier, 'Dartmoor' in *TDA*, Vol. VIII (1876), 370–2, 379.

20   Ibid., 372–8.

21   *Crossing's Dartmoor Worker*, 130.

22   Brian Le Messurier, *Crossing's Hundred Years on Dartmoor* (Newton Abbot, 1967), 15.

23   *Crossing's Dartmoor Worker*, 104, 129, 131.

24   Paul Readman, *Land and Nation in England. Patriotism, National Identity, and the Politics of Land, 1880–1914* (Woodbridge, 2008), 116.

25   Paul Readman, 'Preserving the English Landscape, *c.* 1870–1914' in *Cultural and Social History*, Vol. 5, No. 2, 198.

26   Ibid., 201.

27   Collier, 'Dartmoor', 375.

28   William Crossing, *The Ancient Crosses of Dartmoor; With a Description of the Surroundings* (Exeter, 1887), v–vi, 112–16.

29   First Report of Committee on Dartmoor, *TDA*, Vol. IX (1877), 121 (emphasis added).

30   *Crossing's Dartmoor Worker*, 118.

31   Second Report of Committee on Dartmoor, *TDA*, Vol. X (1878), 110.

32   For example, Stiofán Ó Cadhla, *Civilising Ireland. Ordnance Survey 1824–1842. Ethnography, Cartography, Translation* (Dublin, 2007).

33   Crossing, *Ancient Stones Crosses* (rev. edn, 1902), 99.

34   Crossing, *Amid Devonia's Alps*.

35   *Crossing's Dartmoor Worker*, 89.

36   Later published as Brian Le Messurier (ed.), *Crossing's Dartmoor Worker* (Newton Abbot, 1992).

37   *Crossing's Dartmoor Worker*, 28, 50.

38   Crossing, *Guide*, 106.

39   Crossing, *Amid Devonia's Alps*, 22–3; *Crossing's Hundred Years on Dartmoor*, 59.

40   *Crossing's Dartmoor Worker*, 85; Crossing, *Amid Devonia's Alps*, 40.

41   *Crossing's Dartmoor Worker*, 43.

42   DPA Minute Book, 12 July 1909.

43   *Crossing's Dartmoor Worker*, 116, 127–8, 133–41, 156.

44   *Pall Mall Gazette*, 18 May 1891; *Western Gazette*, 20 May 1891

45   W. F Collier, 'Dartmoor for Devonshire', in *TDA*, Vol. XXVI (1894), 199.

46   Ibid., 202–6.

47   Ibid., 220.

48   Ibid.

49   W. F. Collier, 'Sport on Dartmoor', *TDA*, Vol. XXVII (1895), 121.

50   W. F. Collier, 'Dartmoor and the County Council of Devonshire', *TDA*, Vol. XXVIII (1895) 213.

51   Ibid., 217.

52   W. F. Collier, 'The Purchase of Dartmoor', *TDA*, Vol. XXVIII (1896), 202–8; *Trewman's*, 1 Aug. 1896.

53   *Trewman's*, 24 Mar. 1894.

54   *Trewman's*, 29 Jan. 1897.

55   *Trewman's*, 14 Dec. 1897.

56   *Western Daily Mercury*, 7 Sep. 1897.

57   *WMN*, 17 Sept. 1897.

58   All the material in this and following paragraph is taken from WO 21/7201.

59   John Sheail, *An Environmental History of Twentieth-Century Britain* (Basingstoke, 2002), 114–22.

60   National Parks Committee, Minutes of Meetings, 3 Dec. 1929 (HLG 52/717).

61   Submitted evidence (HLG 52/717 & 719).

62   The National Trust's anonymous undated submission can be found in HLG 52/718.

63   HLG 52/718.

64   Fourth revision, 103–4 (HLG 52/722).

65   'Report of the National Park Committee,' Cmnd 3851, HMSO, London, April 1931, 55.

66   Harry Batsford, 'Country and Coast' in James Lees-Milne (ed.), *The National Trust. A Record of Fifty Years' Achievement* (London, 1945), 11.

67   COU 1/449.

68   Sayer to Duff, 11 May 1950 (COU 1/449).

69   COU 1/450, 13 March 1951.

70   Report on Exeter public meeting, 19 June 1951 (COU 1/451).

71   COU 1/450.

72   COU 1/452.

73   Clark to Duff, 12 Sep. 1951 (COU 1/978).

74   Duff to Sayer, 9 Aug. 1951; Sayer to Duff, 13 Aug. 1951 (COU 1/449).

75   *DPA*, Aug. 1951, Dec. 1951.

76   H. G. Griffith, CPRE, to Abrahams, secretary of the Countryside Commission, 19 Mar. 1951 (COU 1/978).

77   Abrahams to Duff, 14 Sep. 1951; Duff to Clark, 18 Sep. 1951; H. A. Davis (clerk to DCC) to commission, 7 Nov. 1951; Duff to Slesser, 21 Nov. 1951; Sayer to Abrahams, 14 Nov. 1951; Sayer to Abrahams, 20 Nov. 1951 (COU 1/978); *DPA*, No. 3 (Dec. 1951) & No. 4 (Mar. 1952).

78   Clark to Duff, 20 Aug. 1951 (COU 1/454).

79 Confidential memo by Edward Bonong, superintendent engineer transmitters, to National Parks Commission, 22 Aug. 1951 (COU 1/454).

80 Duff's evidence before the public inquiry (COU 1/456).

81 *DPA*, Dec. 1951.

82 Sayer's note on a visit to the site, 11 Sep. 1952 (COU 1/454).

83 See Locke's note on 'Amenity Aspects' (COU 1/456).

84 Edward Bonong, comments to National Parks Commission, 6 Feb. 1952 (COU 1/454).

85 Notes of meeting with Smith-Rose & Saxton, 10 Mar. 1952; Smith-Rose to Abraham, 1 Apr. 1952; Smith-Rose to Sayer (copy), 27 June 1952 (COU 1/454).

86 Sayer to Abrahams, 26 & 30 June 1952; DCC Dartmoor Standing Committee Minutes, 15 July 1952; Sayer to Ferguson, 16 July 1952; Godsall to Abrahams, 18 July 1952 (COU 1/454).

87 H. G. Griffith (CPRE&W) to Abrahams, 30 June 1952; W. H. Williams (COSFPS) to Abrahams, 30 June 1952 (COU 1/454).

88 Duff to Haley, 24 July 1952; Duff to Macmillan, 24 July 1952; Macmillan to Duff, 31 July 1952; Haley to Duff, 6 Aug. 1952; Duff to Macmillan, 6 Aug. 1952; Macmillan to Duff, 12 Aug. 1952; Sayer to Macmillan, 14 Aug. 1952; Macmillan to Sayer, 30 Aug. 1952 (COU 1/454).

89 See letters from Exmouth Urban District Council (1 Aug. 1952), Exeter City Council (5 Sep. 1952) and Brixham Urban District Council (18 Sept. 1952 (COU 1/454)).

90 Sayer to Abrahams, 15 Oct. 1952, 6, 25 & 26 Nov. 1952 (COU 1/454).

91 Bishop to Duff, 26 June 1953; Sayer to Abrahams, 27 June 1953; CPRE to Abrahams, 20 July 1953; Duff to Macmillan, 17 July 1953; Sharp to Duff, 27 July 1953; Sayer to Abrahams, 20 July 1953; Macmillan to Abrahams, 6 Aug. 1953 (COU 1/455). H. G. Warren oversaw the local public inquiries into the designation of Hemel Hempstead as a new town, the use by the Air Ministry of the sky over Foreland Point, Devon in 1950 and the use of the Manhood Range, Selsey, West Sussex in 1955. See the Hemel Hempstead New Town Designation Order, 31 January 1947 and the *London Gazette*, 5 May 1950 & 7 Oct. 1955. On Sharp, see K. Theakston, 'Evelyn Sharp (1903–85)', in *Contemporary Record*, 7, 1 (1993), 132–48.

92 Smith-Rose to J. Locke (NPC), 5 Sep. 1953 (COU 1/456).

93 J. Locke to Sayer, 24 Aug. 1953; Locke to Smith-Rose, 27 Aug. 1953; Sayer to Locke, 3 Sep. 1953; Smith-Rose to Locke, 5 Sep. 1953; note on meeting between Abrahams, Locke & Smith-Rose, 9 July 1953 (COU 1/456).

94 Ritchie to Duff, 18 Sep. 1953; Osmond to Locke, 24 Sep. 1953; Locke on 'Amenity Aspects', undated (COU 1/456).

95 Sayer to Abrahams, 5 Feb. 1952 (COU 1/454).

96 Duff's evidence from the transcript of the inquiry (COU 1/456).

97 Sayer to Duff, 4 Oct. 1953; Duff to Sayer, 5 Oct. 1953; Duff to Watson-Watt, 1 Oct. 1953; Duff to Willis, 1 Oct. 1963 (COU 1/456).

98 WMN, 27 Jan. 1954.

99 DPA, Mar. 1954.

100 Sayer to Duff, 30 Jan. 1954 (COU 1/457).

101 COU 1/449.

102 HLG 71/844.

103 Mid Devon Advertiser, 1 Sep. 1951; Sayer to Abrahams, 1 Sep. 1951 (COU 1/449).

104 John Somers Cocks, 'Exploitation' in Crispin Gill (ed.), Dartmoor. A New Study (Newton Abbot, 1970), 268.

105 T. W. Birch, 'The Afforestation of Britain' in Economic Geography, Vol. 12, No. 1 (Jan. 1936), 1.

106 John, Environmental History, 82–90.

107 Birch, 'Afforestation', 11.

108 Ibid. 1.

109 I. G. Simmons, The Moorlands of England and Wales (Edinburgh, 2003), 156.

110 Simmons, Moorlands of England and Wales, 155–6, 159. Simmons also explains the effect on sediment yields and water chemistry. James C. Scott, Seeing Like a State. How Certain Schemes to Improve the Human Condition Have Failed (Yale, 1998), 19–21; Oliver Rackham, The History of the Countryside (London, 2000), 307.

111 Birch, 'Afforestation', 12, 14, 21; Simmons, Moorlands of England and Wales, 153.

112 Birch, 'Afforestation', 14.

113 Report and memo by O. J. Sanger, June & 21 July 1933 (F 43/125).

114 R. Hansford Worth to Captain McCormick, 21 Aug & Sept. 1930; Duchy to DPA, 10 Sep. 1930 (F18/612).

115 Lecture reprinted by the DPA and forwarded by Worth to the FC, 9 Oct. 1930 (F 18/612).

116 *WMN*, 14, 16 & 22 Oct. 1930.

117 Letters written by representatives of these organisations can be found in F18/612.

118 Simmons, *Moorlands of England and Wales*, 153.

119 F18/612.

120 DPA memo, Jan. 1953; FC internal memo by Popert, 21 Feb. 1953; O. J. Sangar to H. A. Turner, 2 Mar. 1953; E. M. Nicholson (HLG) to FC, 16 Mar. 1953 (F18/613).

121 Turner to Sangar, 8 Feb. 1954; Sanger to Turner, 10 Feb. 1955; extracts from minutes of National Committee for England [FC], 6 Apr. 1955; FC meeting minutes, 20 Apr. 1955 (F18/613); *DPA*, Mar–April 1955, June–July 1955.

122 Sangar to Turner, 10 Feb. 1954 (F18/613); *DPA*, Mar. 1956.

123 *DPA*, Oct. 1957.

124 Abraham to Turner, 29 Apr. 1958; Turner to Abraham, 15 May 1958; Report (NCE), 15 May 1958; Connell to Turner, 2 June 1958; Turnball to Abrahams, 27 May 1958; Turner to FC secretary, 22 Oct. 1958 (F18/612).

125 HL Deb. 25 Feb. 1959 Vol. 214 cc. 488–9.

126 DPA, Feb./Mar. 1959.

127 Minutes of the FC National Committee for England, 8 Apr. 1959 (F18/612).

128 *DPA*, Sep. 1959.

129 *DPA*, Dec. 1959 & Mar. 1960.

130 Sheail, *Environmental History*, 99–100.

131 National Parks Commission document on private forestry in Dartmoor National Park, 4 May 1960 (F18/612).

132 Marshall to Turner, 30 Jan. 1961; Turner to Ryle, 2 Feb. 1961; Ryle to Turner, 6 Feb. 1961 (F18/612).

133 Turner to Marshall, 26 Apr. 1961 (F18/613).

134 Sheail, *Environmental History*, 100.

135 A. N. Balfour to J. Adderley, 13 Feb. 1961 (F18/612); Abrahams to Malyon, 13 Apr. 1961; Turner to Connell, 27 Apr. 1961; Turner to Bell, 3 May 1961; Survey on Pudsham Down, 20 Mar. 1961 (F18/635).

136 Guardian, 2 Dec. 1960; *DPA*, Nov–Dec, 1962, Dec. 1964.

137   The map and accompanying notes can be found in F18/635.

138   Jim Thom to deputy DG, 16 Apr. 1964; Ryle to Connell, 27 & 28 Apr. 1964 (F18/635).

139   *Mid-Devon Advertiser*, 23 May 1964.

140   Lord Norton to Thom, 28 July 1964; Thom to Norton, 6 Oct. 1964; Report of the Joint CLA/TGO sub-committee on Dartmoor National Park Afforestation Survey, 18 Feb. 1965 (F18/635).

141   S. W. Rogers (district officer, DCC) to Connell, 8 Apr. 1965; Connell to Thom, 12 Apr. 1965; Thom to Connell, 14 Apr. 1965; Connell to Thom, 5 May 1965; Connell to Thom, 9 June 1965; Connell to Turner (Ministry of Land and Natural Resources), 14 July 1965; note on meeting between National Parks Commission & Forestry Commission, 8 July 1965 (F18/635).

142   *The Times*, 18 Sep. 1965. This was but the latest in a succession of preservationist objections to the survey. A year earlier the DPA had made a broadly twofold case against the survey. First, it observed that, by focusing on overcoming objections to planting in the national park, the Forestry Commission had not examined alternative areas outside the park, like the Culm Measures in the east of the county. Second, echoing its case against the television mast, the DPA argued that establishing the plantations had a much wider effect than the planting alone. Tracks would need to be expanded into roads, losing 'their character as Devon lanes', while 'any motorist who has been caught behind a Forestry Commission lorry crawling up the hilly lanes in the Chagford area will readily be able to visualise the potential traffic problem'. DPA comments on Voluntary Agreement Draft Forestry Map, DNP, July 1964 (F18/636).

143   Sayer & Somers Cocks to Willey, 17 Sept. 1965 (F18/636).

144   Ruck Keene to Sharp, 11 Nov. 1965 (F18/636).

145   Connell to Summers, 21 Apr. 1967; note of 20 June meeting FC and NPC; note of DCC Woodlands Sub-Committee meeting, 14 Nov. 1967; Connell to Damerel, 1 Feb. 1968 (F18/636).

146   Thurley to Chipperfield, 28 Aug. 1968; Chipperfield to Thurley, 13 Sep. 1968 (F18/1183).

147   Connell to Damerell, 17 Sep. 1968; Connell to FC Secretariat, 19 Sep. 1968; Damerell to Pickering, 24 Sep. 1968; M.F.B. Bell

(Countryside Commission) to Godsall (DCC), 7 Jan. 1960; Williams to Ruffell, 19 June 1969 [F18/1183]; G. D. Rouse, *The New Forests of Dartmoor* (London, 1972), 24–5.

148   Mercer, *Dartmoor*, 144, 146.

149   Ibid., 199–205; www.woodlandtrust.org.uk.

150   The full story can be followed in Harriet Ritvo, *The Dawn of Green. Manchester, Thirlmere, and Modern Environmentalism* (Chicago, 2009).

151   Ibid., 59, 71.

152   'Letter to Viscount Ebrington, MP, &c. &c. on Improving the Water Supply to the Borough of Plymouth' (1848). Bound with other papers as 'Miscellaneous pamphlets and leaflets concerning Plymouth Water Works (1888–93) (British Library).

153   Frank M. Law, 'Beardmore, Nathaniel (1816–1872),' *ODNB* (Oxford, 2004).

154   H. C. G. Matthew, 'Fortescue, Hugh, third Earl Fortescue (1818–1905)', *ODNB* (Oxford, 2004).

155   Jamie Benidickson, *The Culture of Flushing: A Social and Legal History of Sewage* (Vancouver, 2007).

156   Jamie Linton, *What is Water? The History of a Modern Abstraction* (Vancouver, 2010), 18–19.

157   'Letter to Viscount Ebrington', 265.

158   Ibid., 268.

159   The phrase comes from David Blackbourn, *The Conquest of Nature: Water, Landscape and the Making of Modern Germany* (London, 2006).

160   J. A. Hamilton, 'Lopes, Sir Manasseh Masseh, first baronet (1755–1831)', revised by Hallie Rubenhold, *ODNB*, Oxford University Press, 2004.

161   L. C. Sanders, 'Lopes, Sir Lopes Massey, third baronet (1818–1908)', revised by H. C. G. Matthew, *ODNB*, Oxford University Press, 2004.

162   The judgement was reprinted in R. N. Worth, 'Plain Facts on the Plymouth Water Question' (1886), 10–15.

163   R. N. Worth, *The History of Plymouth. From the Earliest Period to the Present Time* (Plymouth, 3rd edn, 1890), 451–5, a lively account of the conflict from which some of the detail is drawn.

164   *Exeter and Plymouth Gazette*, 7 July 1883; *Western Gazette*, 7 July 1883.

165   Lopes to J. Walter Wilson, town clerk of Plymouth, 23 Aug. 1886, reprinted in Worth, 'Plain Facts', 3–4.

166   *Western Daily Mercury*, 25 Aug. & 7 Sep. 1886, repr. in Worth, 'Plain Facts', 5–10.

167   'Report by Mr J. G. Inglis, CE, on the Harter and Head Weir Sites', 3, 5, 7–8. See note 152.

168   Figures from Worth, *Plymouth*, 455.

169   *Taunton Courier, and Western Advertiser*, 1 March 1891.

170   'Report on the Water Supply by the Water Engineer' (1891), 211–51. See note 152.

171   *Lincolnshire Chronicle*, 29 Sep. 1893.

172   *Western Gazette*, 25 Sep. 1898.

173   *British Medical Journal*, Vol. 2, No. 2129 (19 Oct. 1901), 1198.

174   Ibid. No. 1930 (25 Dec. 1897), 1868.

175   Ibid. No. 2221 (25 July 1903).

176   *WMN*, 11 Jan. 1901.

177   The issues are examined in Christopher P. Rodgers, Eleanor A. Straughton, Angus J. L. Winchester & Margherita Pieraccini, *Contested Common Land. Environmental Governance Past and Present* (London, 2011), 137–62.

178   *DPA*, Aug. 1951 (emphasis added).

179   Christine S. McCulloch, 'The Water Resources Board: England and Wales' Venture into National Water Resources Planning, 1964–1973', in *Water Alternative*, 2:3 (2009).

180   *DPA*, Dec. 1954, Mar. 1969.

181   *DPA*, Mar. 1958, summer, 1958.

182   *DPA*, autumn 1958, Feb./Mar. 1959, May 1959; Cocks, 'Exploitation', 266.

183   'Report on the Public Inquiry into the North Devon Water Board (Meldon Reservoir) Water Order and the Board's Planning Application called in for decision', March 1965 (HLG 127/906).

184   Cocks, 'Exploitation', 266.

185   *DPA*, summer 1958, autumn 1958, May 1959.

186   *DPA*, Nov–Dec. 1962.

187   *DPA*, Apr–May 1963.

188   http://www.legislation.gov.uk/ukpga/1963/38/pdfs/ukpga_1963 0038_en.pdf (accessed 24 Jan. 2013)

189   McCulloch, 'The Water Resources Board'.

190   *DPA*, Oct. 1966.

191 London and South Western Railway.

192 W. F. Beech to B. C. Wood & C. Johnson, 9 Mar. 1965 (HLG 127/906).

193 A. M. Walding-White to Crossman, 12 Mar. 1965 (HLG 127/906).

194 William Crossing, *Gems in a Granite Setting* (1905, reprinted Exeter, 1986), 38–44.

195 Report in NA HLG 127/906.

196 Mr Brain to J. E. MacColl, 2 May 1966 (HLG 127/906).

197 Note from B. D. Ponsford, 10 May 1966 (HLG 127/906).

198 Ritvo, *The Dawn of Green*, 168–70.

199 F. T. Willey to Crossman, 21 June & 7 July 1966 (HLG 127/906).

200 Note on meeting of 25 May 1966 (HLG 127/906).

201 Transcript of BBC Home Service, *Roundup*, 19 July 1966; *Sun*, 20 July 1966; *Daily Telegraph*, 20 July 1966; *The Times*, 20 July 1966; *Private Eye* quoted in *New Scientist*, 24 Aug. 1972; *Guardian*, 20 July 1966; R. H. Baker to Crossman, 25 July 1966; Brian Libby to Crossman, 6 Aug. 1966; postcard, 20 July 1966 (HLG 127/1138).

202 Letter to North Devon Water Board from T. & C. Hawksley, 1 Sept. 1967; Kennet to Greenwood, 23 Aug. 1968 (HLG 127/1138).

203 Beech to Wood, 30 April 1968 (HLG 127/1139).

204 Anon., *The Meldon Story* (Crapstone, 1972), 21–33; Rofe & Rafferty to chair of North Devon Water Board, 25 Oct. 1968; Kennet to Greenwood, 6 Nov. 1968 (HLG 127/1138); H. R. Slocombe to Woodward, 13 Nov. 1968 (HLG 127/1140); *WMN*, 7 Nov. 1968.

205 HLG 127/1138.

206 Godsall to Sir Matthew Stevenson, 28 & 29 Aug. 1968 (HLG 127/1138).

207 Unsigned overview on 'Water Supplies in Devon'; draft copy of 'Water for Plymouth' by Countryside Commission (HLG 127/1138); A. G. Rayner to Beddoe, 6 Nov. 1968; Beddoe to Chilver, 7 Nov. 1968; Kennet to Greenwood, 11 Nov. 1968 (HLG 1140/1140).

208 Rayner to Bell, 1 Oct. 1968 (HLG 127/1138); Rayner to Slocombe, 20 Nov. 1968; Slocombe to Rayner, 22 Nov. 1968; Report by Edward Compton, 12 Nov. 1969 (HLG 127/1140); HLG 127/1199.

209 HLG 120/1355.

210 Report of the Water Resources Board, Feb. 1970 (HLG 120/1351).

211 HLG 120/1352.

212 The petitions with comment are in HLG 120/1352.

213 The report is filed in HLG 120/1353.

214 'Regional Planning Division Report on Plymouth and South West Devon Water Bill (Swincombe Reservoir)', 6 Feb. 1970 (HLG 120/1353).

215 For the note on the meeting (16 Feb. 1970) see HLG 120/1354. For drafts of the speech and comment (14 Apr. 1970) see HLG 120/1351. For a brief account of committee proceedings, see DPA, May 1971.

216 Richard Crossman, *The Diaries of a Cabinet Minister. Volume One. Minister of Housing 1964–66* (London, 1975), 623.

217 Crossman, *Diaries*, 624.

218 HC Deb., 15 Jan. 1977, Vol. 809, Cols 439, 449.

219 http://www.environment–agency.gov.uk/business/topic s/water/135367.aspx; http://cdn.environment–agency.gov.uk/geho0212bwbl–e–e.pdf (both accessed 7 Jan. 1013)

220 John Somers Cocks, 'Exploitation', in Crispin Gill (ed.) *Dartmoor: A New Study* (Newton Abbot, 1970), 273–4.

221 Ibid.

222 See https://www.gov.uk/defence–infrastructure–organisation–and–the–defence–training–estate (accessed 24 Feb. 2013).

223 'Needs of the Armed Forces for Land for Training and Other Purposes', pamphlet presented to parliament by the Prime Minister, Dec. 1947, 4, 7–9, 11, 12. Archived copy in HLG 671/1622.

224 See Adrian Smith, 'Command and Control in Postwar Britain. Defence Decision-Making in the United Kingdom, 1945–1984', *Twentieth Century British History*, Vol. 2, No. 3 (1991), 302; a useful if dated overview is Franklyn A. Johnson, 'Politico-Military Organisation in the United Kingdom: Some Recent Developments', *Journal of Politics*, 27, 2 (May, 1965), 339–50.

225 Letters and enclosures written in October and November 1946, preserved in HLG 71/844; note on Dartmoor written after meeting between amenity societies and the Ministry of Agriculture and Fisheries, 26 Nov. 1946 (MAFE 143/35).

226 HLG 71/1622.

227 A copy of 'Use of land for the Admiralty and the War Department – Public Local Enquiry' is in HLG 71/1622.

228 Minto to Deans, 8 Feb. 1950 (HLG 71/1622).

229 HLG to DCC, 16 Mar. 1949 (draft); H. A. Davis (DCC) to Town and Country Planning, 4 Apr. 1949; Admiralty to HLG, 27 June 1949; L. E. Andrews (WO) to Minto (Admirality), 14 Sep. 1949; Admiralty to Andrews, 10 Oct. 1949; Andrews to Lang (HLG), 29 Oct. 1949; Minto to Deans (HLG), 8 Feb. 1950; Davis to HLG, 24 Mar. 1950; Lang to Andrews (28 Mar. 1950); Lang to Smith (Admiralty), 28 Mar. 1950; Smith to Lang, 29 July 1950; Andrews to Lang, 3 Aug. 1950 (NLG 71/1622).

230 *WMN*, 19 Oct. 1950.

231 *WMN*, 19 Oct. 1950.

232 Sayer to Abrahams, 23 Oct. 1950; Abrahams to Sayer, 15 Nov. 1950; J. D. W Jones (LGP) to Abrahams, 8 June 1951 (COU 1/464).

233 *DPA*, Mar. 1952.

234 These exchanges can be followed in COU 1/470.

235 Clark to Abrahams, 11 May 1955; J. D. W. Jones (HLG) to Godsall, 19 August 1955; note by T. G. Guillian [?] on meeting between HLG and Mr Best (WO), 9 Sep. 1955; Abrahams to Godsall, 15 Feb. 1955; Godsall to Abrahams, 23 Feb. 1956; Williams to Abrahams, 5 Mar. 1956; Sayer to Abrahams, 9 Mar. 1956; to Sayer, 15 Mar. 1956 (COU 1/980); *DPA*, Mar. 1956.

236 Note of meeting held 3 Chester Gate, on Wednesday, October 24, by the National Parks Commission with representatives of the Dartmoor National Park Committee and the War Office. COU 1/980.

237 Strang to Playfair, 31 Oct. 1956; Strang to Sharp, 31 Oct. 1956; Sharp to Strang, 3 Dec. 1956; Clark to Commission, 12 Dec. 1956; S. G. Jenkins, Tavistock Urban District Council, to Clark, 21 Dec. 1956; Percival D. Sugars, Tavistock Rural District Council, to Clark, 7 Feb. 1957; Playfair to Strang, 22 May 1957; Godsall to Abrahams, 14 Nov. 1957; Playfair to Strang, 9 Dec. 1957 (COU 1/980).

238 Martin S. Navias, 'Terminating Conscription? The British National Service Controversy 1955–56', *Journal of Contemporary History*, 24, 2 (Apr. 1989).

239 Matthew Grant, 'Home Defence and the Sandys Defence White Paper, 1957', *Journal of Strategic Studies*, 31, 6 (Dec. 2008), 925–7; for long term context, see Smith, 'Command and Control'.

240 *DPA*, Mar.–Apr. 1957, June–July 1957.

241 J. Hastings (WO) to Godsall, 21 Jan. 1958 (WO 32/20944).

242 Andrews to N. H. Calvert, 21 Apr. 1958; Herbert Griffin (CPRE) to Calvert, 29 Apr. 1958 (COU 1/982).

243 *DPA*, Sep. 1952, Dec. 1953, Mar., June, Sep. & Dec. 1954, Oct. 1955.

244 Godsall to Captain E. L. Carter, 5 Sep. 1957; Andrews to Godsall, 21 Mar. 1958; Godsall to Andrews, 2 Apr. 1958; Andrews to W. B. Harvey (Admiralty), 16 Feb. 1959, 11 Mar. 1959; Admiralty to Carter, 14 Apr. 1959; Somers Cocks to Godsall, 7 Aug. 1959; Godsall to Andrews, 12 Aug. 1959; Godsall to Andrews, 9 Dec. 1959; Andrews to Godsall, 7 Jan. 1960; Godsall to Andrews, 15 Jan., 21 July 1960; J. Scobey (WO) to Godsall, 11 Aug. 1960; Godsall to Andrews, 12 Aug. 1960; Gosdall to Andrews, 13 Dec, 1960 (WO 32/20944).

245 HC Deb., 30 Apr. 1958, Vol. 587 cc 364–5.

246 Abrahams to Andrews, 18 June 1958 (WO 32/20944).

247 Godsall to Andrews, 6 July 1960 (WO 32/20944).

248 Stokes to under secretary of state for war, 21 Sep. 1960; Andrews to Stokes, 12 Oct. 1960; Stokes to Andrews, 19 Oct. 1960; Andrews to Stokes, 18 Nov. 1960; Stokes to Andrews, 26 Nov. 1960; Moore to WO, 5 Jan. 1961; Andrews to Stokes, 17 Feb. 1961; Stokes to Andrews, 3 Mar. 1961; Miller to Stokes, 24 Mar. 1961; Stokes to Miller, 28 Mar. 1961; Burridge (NPC) to Andrews, 5 April 1963; Andrews to Burridge, 21 May 1963 (WO 32/20944).

249 Note of Trip to Dartmoor, 30 Aug. 1961; Jellicoe to Brigadier J. M. Thomas, 13 Nov. 1961; Acland to Abrahams, 29 Jan. 1961; Jellicoe to Abrahams, 31 Jan. 1962; Abrahams to Acland, 2 Feb. 1962; V. G. F. Bovenizer (WO) to Herbert Griffin, 17 July 1962; Sayer to Abrahams, 2 Aug. 1962 (COU 1/1156).

250 DPA objectives, summer 1961 (COU 1/1156); DPA, 'National Land Use and the Dartmoor National Park', Oct. 1964; minute by Walter Luttrell, 24 Nov. 1964 (COU 1/983).

251 DPA, 'Misuse of a National Park'; Sayer to Bell, 4 Mar. 1964; Sayer to Strang, 30 Sep. 1964; Sayer to Bell, 26 Oct. 1964; Sayer to General Tailyour, 30 Aug. 1964 (COU 1/983).

252 *WMN*, 21 Dec. 1959. HC Deb 27 Jan 1960 vol 616 cc40–1w.

253 'Notes of a meeting held in Miss McNicol's Room, Queen Anne's Mansions, on Friday, 29th July 1966' (COU 1/984).

254 Patrick Kingsley (duchy) to James Dunnett (MoD), 14 Aug. 1968; MoD minute, 23 Aug. 1968 (DEFE 51/91).

255 John Diamond, chief secretary to the Treasury, to Anthony Greenwood, minister of HLG, 22 Oct. 1968; secretary of state of economic affairs to MOD, 31 Oct. 1968; note of meeting between minister and ministers of HLG, 15 Nov. 1968; comment from HQ, Southern Command, 22 Nov. 1968; MoD minute, 26 Nov. 1968 (DEFE 51/91).

256 'Note of Meeting about Dartmoor Training Area Held in Minister (A)'s office on 26 Feb. 1969' (DEFE 51/92); 'Presentation to Lord Foot, Officials of the DCC, and Local MPs', 11 Mar. 1969 (DEFE 51/93).

257 Kingsley (Duchy) to Sir Arthur Drew (MoD), 21 Apr. 1969; WMN, 30 April 1969; confidential minute, 5 May 1969; MoD to Duchy, 27 May 1969 (DEFE 51/93).

258 Report on meeting of 15 May 1969; 'Dartmoor Training Area' by E. H. Palmer 27 May 1969; MoD comment, 27 May 1969 (DEFE 51/93).

259 Paper by Palmer, 22 Aug. 1969; draft of letter to Lady Wootton; press release, 11 Sep. 1969; WMN, 12 & 13 Sep. 1969 (DEFE 51/93); DPA, Nov. 1969.

260 J. D. Boles (NT) to L. H. Ford (Department of Environment), 14 Jan. 1975; Sir Frederick Bishop (NT) to Sir John Wilson (MoD), 15 Jan. 1975 (DEFE 70/390); Silkin to Sharp, 27 Nov. 1975 (DEFE 70/310).

261 'Dartmoor. Report by Lady Sharp G.B.E. to the Secretary of State for the Environment and the Secretary of State for Defence of a public local inquiry held in December, 1975 and May, 1976 into the continued use of Dartmoor by the Ministry of Defence for training purposes', 77. Copy filed in DEFE 70/215.

262 Ibid.

263 Ibid., 12, 83.

264 Ibid., 14–15.

265 Ibid., 40–2.

266 Ibid., 34–7.

267 Ibid., 82.

268 Foot presented on behalf of the DPA. See DPA, Apr. 1976 for his closing comments.

269 Sharp Report, 10–11.

270 DEFE 70/215.

271 Marianna Dudley, *An Environmental History of the UK Defence Estate. 1945 to the Present* (London, 2012), 76–89.

272 Dartmoor National Park Plan (Devon County Council, 1977), 54.

273 Ibid., 17.

274 Ibid., 13.

275 Ibid., 24.

276 Ibid., 24.

277 Conversation with Tom Greeves.

Part IV

1 Garrett Hardin, 'The Tragedy of the Commons', in *Science*, 162 (1968), 1243–8, reprinted in G. Hardin & J. Baden (eds), *Managing the Commons* (San Francisco, 1977).

2 Thomas Dietz, Nives Dolšale & Elinor Ostrom, 'The Drama of the Commons' in Dietz, Dolšale & Ostrom (eds), *The Drama of the Commons* (Washington, 2002), 18.

3 Quoted in Lara Phelan, 'Economy to Amenity: The Commons of the New Forest and Ashdown Forest, 1851–1939' DPhil, Sussex 2002, 20–5; Dietz, Dolšale & Ostrom, 'Drama of the Commons', 12–13, 18–19.

4 Dietz, Dolšale & Ostrom, 'Drama of the Commons', 12–13, 18–19.

5 See Arun Agrawal, *Environmentality. Technologies of Government and the Making of Subjects* (Durham & London, 2005).

6 W. G. Hoskins and L. Dudley Stamp, *The Common Lands of England and Wales* (London, 1963), 80–1; Eleanor A. Straughton, *Common Grazing in the Northern English Uplands, 1800–1965. A History of National Policy and Local Practice with Special Attention to the Case of Cumbria* (Lampeter, 2008), 58–60.

7 Percival Birkett, 'A Short History of the Rights of Common Upon the Forest of Dartmoor and the Commons of Devon', xvii–xiv, published as a pamphlet by the Dartmoor Preservation Association with the 'Report of Mr Stuart A. Moore, and Appendix of Documents' (1889).

8   Hoskins and Stamp, *Common Lands of England and Wales*, 186.

9   Detail from Ian Mercer, *Dartmoor* (London, 2009), 279–80, 293.

10   Harold Fox, *Dartmoor's Alluring Uplands. Transhumance and Pastoral Management in the Middle Ages* (Exeter, 2012), 48.

11   Ibid. 49–55.

12   Birkett, 'Short History', xi.

13   Ibid., viii, xxx–xxxi.

14   W. F. Collier, 'Venville Rights on Dartmoor', *TDA*, Vol. XIX (1887), 383.

15   Ibid., 385.

16   W. F. Collier, 'The Duchy of Cornwall on Dartmoor', *TDA*, Vol. XXI (1889), 298.

17   Mercer, *Dartmoor*, 295, 302–3.

18   Collier, 'Venville Rights', 380.

19   Mary Sylvia Calmady-Hamlyn, *The Dartmoor Pony* (Buckfast, 1952).

20   Sylvia Calmady-Hamlyn to R. A. Brown (British Horse Society), 12 Jan. 1949; Brown to H. F. Lovell (MAF), 13 Jan. 1949; Lovell to Brown, 18 Jan. 1949; Calmady-Hamlyn to Brown, 14 Jan. 1949; Maher (MAF) to Calmady-Hamlyn, 4 Feb. 1949; Wilcox (MAF) to Brown, 11 Mar. 1949; Brown to Wilcox, 11 Mar. 1949; MAF (Scottish office) to Maher, 18 July 1949; Calmady-Hamlyn to Maher, 24 Mar. 1949; Calmady-Hamlyn to Maher, 29 Apr. 1949; and Lovell to Calmady-Hamlyn, 13 May 1949 (MAF 121/47).

21   Lady Elizabeth Pease to Dugdale, 11 Jan. 1952; Gainford to Dugdale, 20 Feb. 1952; Gainford to Dugdale, 12 Mar. 1952; *WMN*, 17 Aug. 1950 (MAF 121/47).

22   E. A. Farey to Austin Jenkins, 15 June 1950; Farey to Jenkins, 21 Aug. 1950; Jenkins to Farey, 22 May 1951; H. J. Bevan (livestock officer) to Jenkins, 28 July 1951 (MAF 121/47).

23   J. H. Locke, minute, 10 Sep. 1951 (MAF 121/47).

24   Ferguson (NPC) to J. H. Locke, 26 Nov. 1951; Locke to Ferguson, 11 Dec. 1951; Nature Conservancy to G. S. Dunnett (MAF), 2 Jan. 1952; for NPC meeting see MAF 171/7.

25   Additional to letters already cited, Dugdale to Studholme & Pease, 25 Feb. 1952; Dugdale to Gainford, 10 Mar. 1952 (MAF 121/47). The Duchy was told the same in an exchange in July 1950 (MAF 235/81).

26   'Report of the First Meeting of the Panel set up to consider the

formation of a Dartmoor Commoners' Association . . .' (MAF 171/7).

27   Note by Jenkins; Macfarlane (Country Agriculture Officer) to J. A. Payne (MAF), 24 Mar. 1952 (MAF 235/81).

28   Dartmoor Commoners Association. Evidence before Royal Commission on Common Land, 1157–1165 (MAF 96/50).

29   'Evidence submitted to the Royal Commission on Common Land by the Dartmoor Preservation Association' (April 1956), 1228, 1238 (MAF 96/69).

30   R. V. Vernede to L. A. Freeman, MAF, 18 May 1954; Dorothy Paton to Dugdale, 9 July 1954; G. L. Wilde (MAF) to George Cockerill, 14 June 1954 & Wilde to Paton, 30 July 1954 (MAF 235/81).

31   *Tavistock Gazette*, 8 July 1955.

32   DCA evidence, 1165–7, 1170–3 (MAF 96/50).

33   R. Hansford Worth to Calmady-Hamlyn, 18 Aug. 1949 (MAF 121/47).

34   Brown to Christopher Soames, 22 Sep. 1961; G. L. Wilde (MAFF) to Brown, 2 Oct. 1961; 'Note of a meeting held on 17th November, 1961 . . .'; Wilde to Brown, 21 Dec. 1961; 'Note of meeting held on 22 January, 1962 . . .' (MAF 292/12).

35   R. E. Booth, 'The Severe Winter of 1963, compared with other cold winters, particularly that of 1947', *Weather*, Vol. 23, No. 11, 477–9.

36   *Daily Telegraph*, 15 Mar. 1963. Despite the lurid headline, this article by Guy Rais was a balanced account of the different points of view.

37   HC Deb., 30 Jan. 1963, Vol. 673, c. 919.

38   HC Deb., 26 Feb. 1963, Vol. 673, c. 1,200.

39   For detail in this and following paragraphs see brief by Mr Holmes, county advisory officer, 6 Feb. 1963 (MAF 109/289).

40   Note, 1 & 4 Feb. 1963 (MAF 109/289); HC Deb., 26 Feb. 1963, Vol. 673, cc 1205–6.

41   Draft speech for the adjournment debate (MAF 109/289).

42   HC Deb., 26 Feb. 1963, Vol. 673, cc 1223–4.

43   'Plans to minimize losses of livestock on Dartmoor', note by G. W. Ford, 8 Aug. 1963 (MAF 109/289).

44   'Report of a meeting . . . to discuss Winter Emergency Procedures, with particular regard to Dartmoor', 27 June 1963 (MAF 109/289).

45 *WMN*, 30 Sep. 1963; on the DCA and RSPCA, see Holmes to Ford, 30 Sept. 1963 & 1 Nov. 1963, Ford to John Barrah (MAFF), 20 Nov. 1963 (MAF 109/289).

46 'Minutes of the Inaugural Meeting of the Dartmoor Livestock Protection Society', 21 Oct. 1963; note by G. W. Ford, 1 Nov. 1963 (MAF 109/289).

47 DPA, April–May, 1963.

48 *Animal World*, Nov. 1967; MacDonald to Reid, 10 Apr. 1968; MacDonald to C. Hughes MP, 10 Apr. 1968; *WMN*, 10 Apr. & 1 May 1968 (MAF 369/122).

49 Reports by A. E. Ridgeway (3 May 1968), J. C. Cotter (7 May 1968), U. R. Perry (6 May 1968), P. D. Sullivan (6 May 1968); O'Sullivan to Abercrombie, 21 Apr. & 7 May 1968 (MAF 369/122).

50 H. D. Abercrombie to A. B. Hoare, 7 May 1968 (MAF 369/122).

51 H. L. Watkins to minister [?], 7 May 1968; draft replies & replies to Watkins and MacDonald, 27 & 30 May 1968 (MAF 369/122).

52 MacDonald to H. B. Fawcett, Principal Animal Health Division (MAFF), 16 June 1968; J. R. Potts (MAFF) to MacDonald, 2 July 1968; MacDonald to Potts, 2 July 1968; CPRE *et al* to Greenwood, 23 July 1968 (MAF 369/122).

53 *WMN*, 18 Sept. 1968; K. Harrison Jones (regional controller, MAFF) to George Lace (Animal Health, MAFF), 25 Sep. 1968; K. W. Evans (HLG) to A. Foreman (MAFF), 19 Nov. 1968; A. P. Stevens (MAFF) to Foreman, 13 Dec. 1968; John Mackie (MAFF) to Kenneth Robinson, 4 Feb. 1969 (MAF 368/122).

54 MacDonald to Reid Chambers, 12 Mar. 1969; MacDonald to Godsall, 14 April 1969; H. Biscoe (Ramblers Association) to Godsall, 21 April 1969; Brien le Messeurier to Godsall, 13 April 1969; Godsall to MAFF, 5 May 1969; H. Pryce (MAFF) to MacDonald, 30 May 1969; note on meeting by Harrison Jones, 30 June 1969 (MAF 369/123).

55 Report by G. W. Ford, 8 Aug. 1963 (MAF 109/289).

56 Holmes brief, 7 Feb. 1963 (MAF 109/289); Spencer to Stevens, 28 Aug. 1968 (MAF 369/122).

57 Evidence before Royal Commission on Common Land, 1241–2 (MAF 96/69).

58 It's Mercer's phrase and is kindly, *Dartmoor*, xiv.

59 Donald Denman, Robert Alun Roberts & H. J. F. Smith, *Commons*

and village greens: a study of land use, conservation and management based on a national survey of commons in England and Wales, 1961–1966, financed by the Nuffield Foundation (London, 1967), 413.

60   DPA, Aug. 1975.

61   *Dartmoor National Park Plan*, 34.

62   Sheail, *Environmental History*, 153.

63   Draft copy of bill preserved in FT 9/66.

64   C. Ford (Environment) to E. Bristow (Nature Conservancy), 21 July 1977; Bristow to P. B. Nicholson, 2 Aug. 1977; Bristow to Nicholson, 5 Aug. 1977; Nicholson to Bristow, 11 Aug. 1977 (FT 9/66).

65   Lamerton to Nicholson, 22 Sept. 1977; Nicholson to Brown (DCC), 23 Nov. 1977 (FT 9/66).

66   Nicholson to Bristow, 8 June 1978; Nicholson to Browne, 29 June 1978; Devereux to B. Forman, 31 Oct. 1978; Statement by Mercer and Bennett (undated); confidential note, 3 Nov. 1978; Devereux to John Goldsmith (Environment), undated; Devereux note, 24 Nov. 1978 (FT 9/66).

67   Copies of the petitions are preserved in FT 9/67.

68   House of Lords, Minutes of Evidence taken before the Committee of the Dartmoor Commons Bill, Friday, 20 July, 1977, 3–32.

69   House of Lords Minutes, 19 June 1979, 9–11, 15.

70   Ibid., 46ff (esp. 67–8).

71   Ibid., 20 June 1979, 39–49.

72   Ibid., 25.

73   Ibid., 41ff (esp. 55).

74   Ibid., 25 June 1979, 11.

75   Ibid. 20 June 1979, 21.

76   Ibid. 25 June 1979, 27ff, 35ff, 47–9; 25 June 1979, 2ff.

77   HC Deb. 24 April 1980 Vol 983 c. 973.

78   Ibid. cc 789, 794.

79   Ibid. cc 795, 797.

80   Ibid. cc 800–2.

81   Ibid. cc 803–5, 807–10.

82   Ibid. cc 815–16.

83   Ibid. cc 830–4.

84   HC Deb. 25 June 1984 Vol 62 c. 743.

85   Ibid. c. 744.

86   Ibid. cc 735, 737.

87   Ibid. c. 736.

88   *DPA*, Apr. 1983.

89   Ibid., Dec. 1983.

90   HC Deb. 25 June 1984 Vol 62 c. 750.

91   Ibid. c. 738, 741–2.

92   Ibid. c. 752.

93   *DPA*, May 1972.

94   'State of the natural environment in the South West', Natural England (2011), 24 (emphasis added).

95   http://www.dartmoorcommonerscouncil.org.uk/menu_page. php ?id=52 (accessed 15 Aug. 2013).

96   'Dartmoor State of the Park Report 2010', 37; Managing for Ecosystem services, 34 [http://publications.naturalengland.org. uk/publication/4700I?category=38019 accessed 15 Aug. 2013].

97   'Action for Wildlife', Dartmoor Biodiversity Steering Group (2001), 28.

98   'Dartmoor State of the Park Report 2010', 4.

99   Mercer, *Dartmoor*, 190.

100 W. M. Adams, *Future Nature: A Vision for Conservation*, revised edn (London, 2003), 183.

101 Peter Taylor, *Beyond Conservation. A Wildland Strategy* (London, 2005), 49–55.

102 George Monbiot, *Feral. Searching for enchantment on the frontiers of rewilding* (London, 2013), 11; *Guardian*, 30 Jan & 18 Feb. 2014. See also Peter Taylor (ed.), *Rewilding* (Walton Hill, 2011).

103 See 'ecosystem services', part of the Dartmoor National Park Management Plan (www.yourdartmoor.org/theplan/sustain/ natural-network (accessed 17 Nov. 2014).

104 'Dartmoor National Park Management Plan 2007–2012', 18, available at http://www.dartmoor–npa.gov.uk (accessed 17 Sept. 2013).

105 'Your Dartmoor', 10.

106 Adams, *Future Nature*, 196–8.

107 *Guardian*, 31 Oct. 2005.

Epilogue

1 J. Lellenberg, D. Stashower and C. Foley (eds) *Arthur Conan Doyle: A Life in Letters*, London 2007, 479–80.
2 Ibid.
3 Philip Weller, *The Hound of the Baskervilles. Hunting the Dartmoor Legend* (Tiverton, 2001); Kelvin I. Jones, *The Mythology of The Hound of the Baskervilles* (1986); Ruth E. St Leger-Gordon, *Witchcraft and Folklore of Dartmoor* (New York, 1965).
4 Arthur Conan Doyle, *The Hound of the Baskervilles* (London: Penguin English Library, 2012), 33.
5 Andrew Lycett, *Conan Doyle. The Man Who Created Sherlock Holmes* (London, 2007), 267–8.
6 Conan Doyle, *Hound*, 6, 54.
7 Ibid., 80.
8 Ibid., 54–5, 57.
9 Ibid., 68, 70.
10 Ibid., 49.
11 Ibid., 26.
12 Douglas Kerr, *Conan Doyle: Writing, Profession and Practice* (Oxford, 2013), 78.
13 Conan Doyle, *Hound*, 13, 58–7.
14 Ibid., 164.
15 Ibid., 9.
16 Ibid., 21, 27, 73, 23.

Index